G. Bernaschek · J. Deutinger · A. Kratochwil

Endosonography in Obstetrics and Gynecology

With 271 Figures in 392 Separate Illustrations

Springer-Verlag
Berlin Heidelberg New York
London Paris Tokyo
Hong Kong

Univ.-Doz. Dr. Gerhard Bernaschek
II. Universitäts-Frauenklinik Wien
Spitalgasse 23, A-1090 Wien

Dr. Josef Deutinger
II. Universitäts-Frauenklinik Wien
Spitalgasse 23, A-1090 Wien

Univ.-Prof. Dr. Alfred Kratochwil
Allgemeines öffentliches Krankenhaus
der Kurstadt Baden
Wimmergasse 19, A-2500 Baden

ISBN-13:978-3-642-74113-5 e-ISBN-13:978-3-642-74111-1
DOI: 10.1007/978-3-642-74111-1

Library of Congress Cataloging-in-Publication Data. Bernaschek, G. (Gerhard), 1950- Endosonography in obstetrics and gynecology / G. Bernaschek, J. Deutinger, A. Kratochwil. p. cm. Includes bibliographical references. ISBN-13:978-3-642-74113-5 1. Endoscopic ultrasonography. 2. Gynecology - Diagnosis. 3. Obstetrics - Diagnosis. I. Deutinger, J. (Josef), 1954- . II. Kratochwil, Alfred. III. Title. [DNLM: 1. Endoscopy. 2. Genital Diseases, Female - diagnosis. 3. Pregnancy Complications - diagnosis. 4. Ultrasonic Diagnosis. WQ 240 B523e] RG107.5.E48B47 1990 618'.047543 - dc20 DNLM/DLC for Library of Congress 89-26161 CIP

Softcover reprint of the hardcover 1st edition 1990

2121/3130-543210

Preface

More than 25 years ago, when ultrasound diagnostic methods were first introduced into gynecology and obstetrics, few of the pioneers of these techniques suspected what importance sonographic diagnosis was destined to assume.

It was soon recognized that the organs of the lesser pelvis could be visualized to much greater advantage by inserting probes into the natural bodily orifices than by abdominal sonography. Full exploitation of the physical properties of ultrasound had to wait, as so often in the history of sonography, for technological advances.

Endosonography in the form available to us today combines the advantages of endoscopy and sonography. The next light-reflecting surface, once the limit of endoscopy, represents no barrier to ultrasound. A whole range of both diagnostic and therapeutic procedures can be sonographically guided. Blood flow in vessels lying deep in the lesser pelvis can now be measured by means of vaginal duplex sonography.

So much material has now been published that it has become well-nigh impossible for any one individual to sift out what is directly relevant for their own work, and it is with this in mind that we set out to write this book. The book is intended not only to describe our own experience with the methods we helped to shape, but also to take stock of developments so far. We are convinced that endosonography will become established as an integrative gynecological/obstetrical investigation, and for this reason we have laid special emphasis on the practical utility of endosonography in this field.

We sincerely hope that this volume will assist practising obstetricians and gynecologists to familiarize themselves with the techniques of endosonography and to apply them to the benefit of their patients.

Gerhard Bernaschek
Josef Deutinger
Alfred Kratochwil

Contents

History of Endosonography

Twenty-five years ago, when ultrasound was introduced as a diagnostic tool in medicine, its noninvasive character was considered to be a great advantage (Kratochwil et al. 1966) (Fig. 1). Even then it was apparent, however, that the examination of intrapelvic organs could be significantly improved by passing ultrasound transducers into natural body orifices so that the region of interest could be scanned at closer range (Kratochwil et al. 1969) (Fig. 2). The rationale for this modality, called *endosonography,* is based on the relation of the penetration and resolution of ultrasound waves to their frequency. While lower frequencies show good penetration but relatively poor resolution, higher frequencies penetrate less deeply but show very good resolution. It was reasonable, then, to attempt to improve the imaging of intrapelvic organs by placing higher-frequency transducers in closer proximity to the region of interest.

The primary goal in obstetrics was to record fetal heart activity at the earliest possible stage. This was relatively easy to accomplish once the pregnant cor-

Fig. 1. First A-scan instrument, used originally for material testing and applied in 1964 to examinations in obstetrics and gynecology

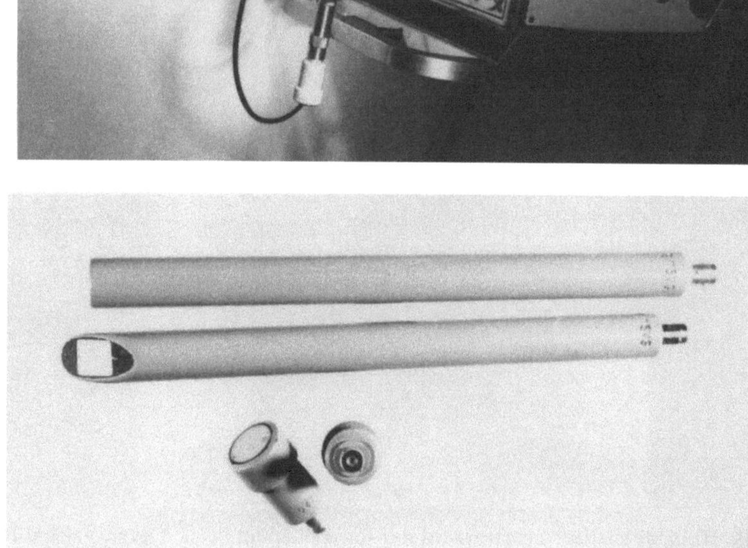

Fig. 2. Rod-shaped transducers for vaginal and rectal sonography

pus uteri had left the lesser pelvis (Kratochwil and Eisenhut 1967). But scanning the uterus in very early pregnancy, when it is still within the lesser pelvis, required the use of rod-shaped transducers with various scan angles that could be introduced into the vagina. Even a retroverted uterus in early pregnancy could be successfully examined from the vaginal fornices (Fig. 3). These early instruments used the A-mode technique, which displayed fetal heart activity as a synchronous, rhythmic sequence of echo pulses. It was now possible to detect fetal heart activity a mere 32 days after fertilization, providing a major advance in the diagnosis and management of early pregnancy disorders.

Soon this technique was applied to the detection of small ovarian masses (Kratochwil 1968) (Fig. 4). However, orientation was often very difficult, and it became necessary to combine the ultrasound scan with a bimanual pelvic examination to improve orientation and aid interpretation. For this purpose a

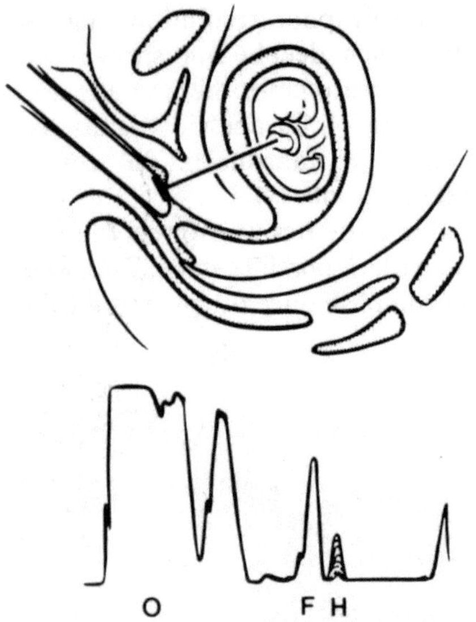

Fig. 3. Vaginal A-mode examination of early pregnancy from the anterior vaginal fornix. *O*, emitted signal; *F*, fetus; *H*, heart

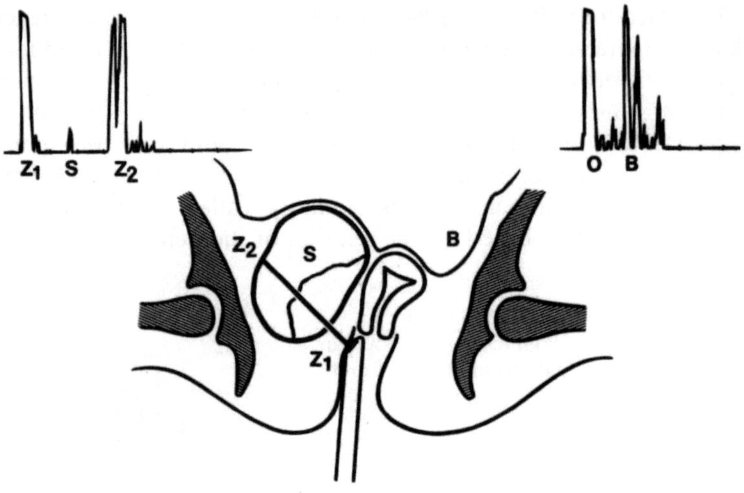

Fig. 4. A-mode examination of a septated ovarian cyst. Z_1, Z_2, cyst walls; *S*, septum; *B*, pelvic wall

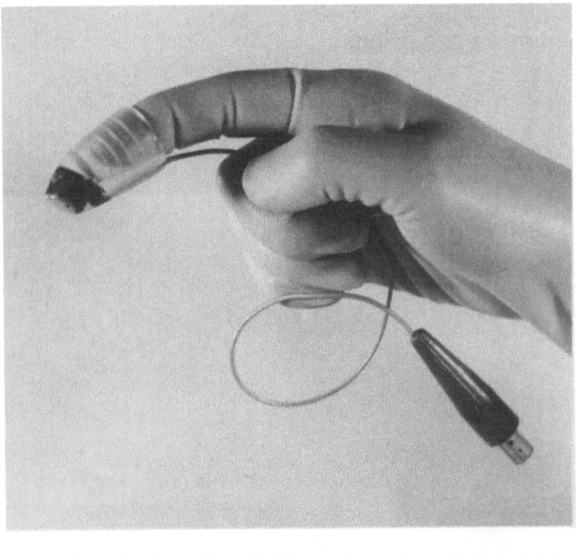

Fig. 5. Thimble-shaped vaginal transducer for concurrent bimanual examination and sonography

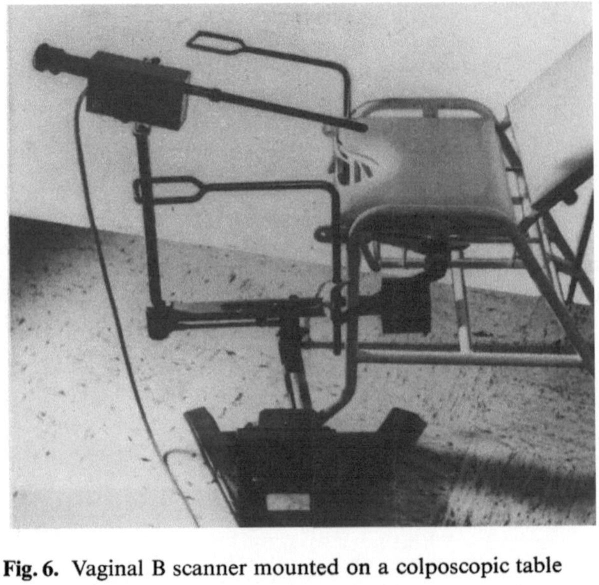

Fig. 6. Vaginal B scanner mounted on a colposcopic table

thimble-shaped transducer was designed that could slip over the palpating finger without interfering with the pelvic exam (Fig. 5). The A-mode technique was still employed.

A significant advance in endosonography came with the development of techniques for displaying echoes in two dimensions (Kratochwil 1969). For some time these techniques used A-mode-type transducers linked to a mechanical system that allowed longitudinal motion and 360° rotation of the transducer. The mechanical system was mounted on a colposcopic tripod. For proper geometric reproduction, the transducer was mounted at the end of the rod and emitted sound waves perpendicular to the rod axis (Fig. 6). Instead of x-y coordinates, polar coordinates were used like those in radar systems. The instrument could also be applied rectally by using a coupling path consisting of a condom or glove finger filled with degassed water. Probes for cystosonography examinations were constructed by reducing the diameter of the intracorporeal rod.

Fig. 7. Rectosonography: measurement of the interspinous distance. *O*, transducer; *SCH*, fetal head; *P*, sacral promontory

Fig. 8. Vaginasonography: examination of an ovarian cyst using simultaneous A- and B-mode scanning

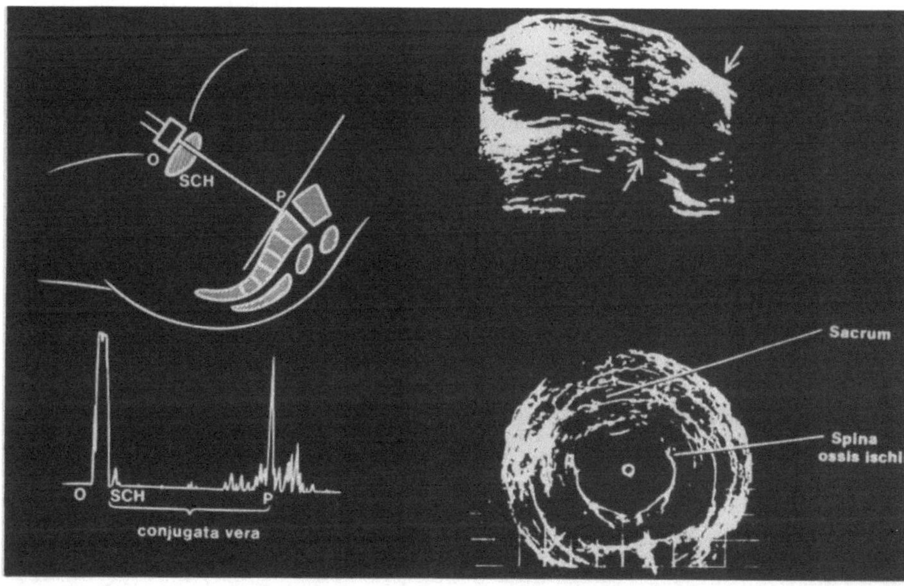

Fig. 7

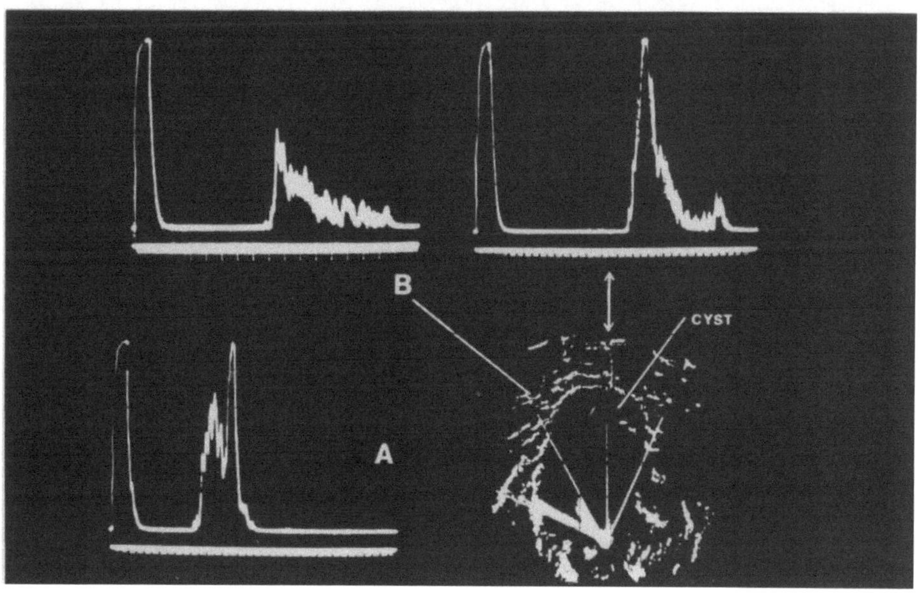

Fig. 8

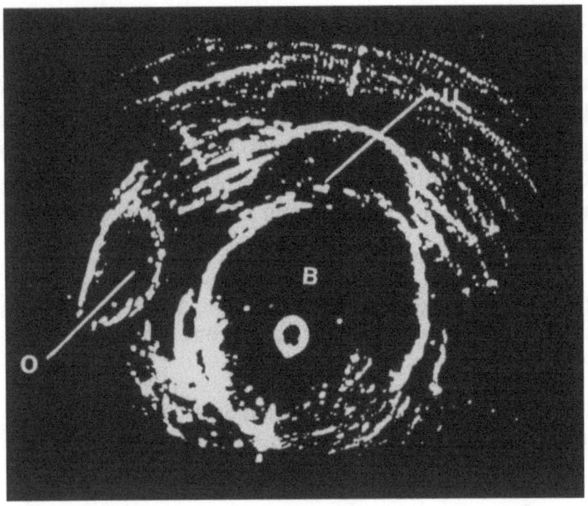

Fig. 9. Vesical scan demonstrating the uterus *(U)* and ovary *(O)*. *B*, bladder

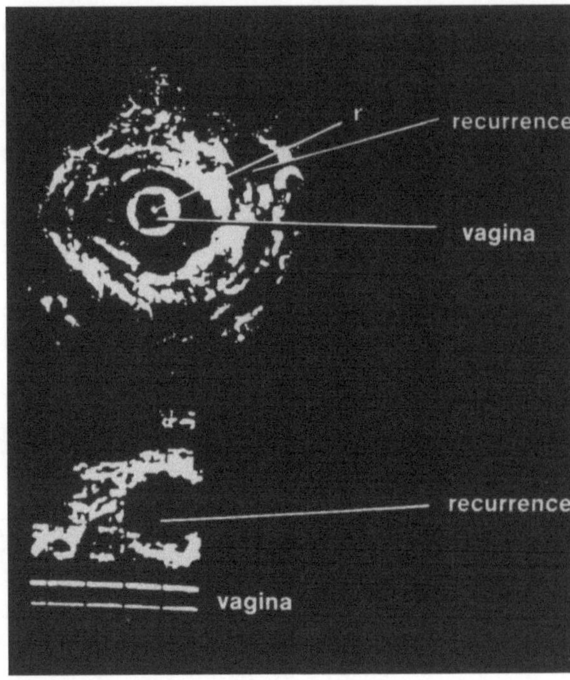

Fig. 10. Rectal scans showing postoperative recurrence of ▶ cervical carcinoma. The *upper* transverse scan localizes the tumor to the pelvic wall; the *lower* scan defines the longitudinal extent; *r*, rectum

Fig. 11. Tumor demonstrated by abdominal and rectal scans. *ABW*, abdominal wall; *Bl*, bladder; *F*, femur; *Tu*, tumor; *PW*, pelvic wall; *R*, rectum
▼

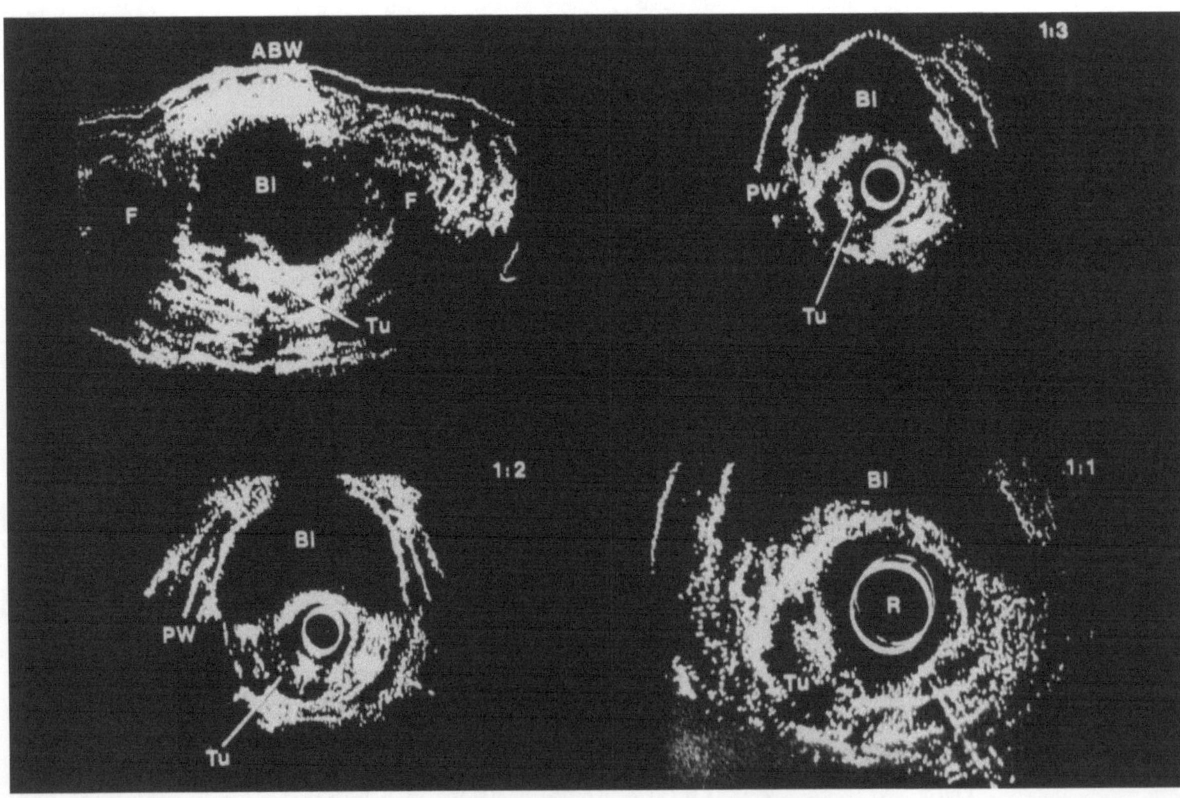

Rectal scanning was introduced primarily for examinations of the female pelvis and especially for measuring the interspinous distance to evaluate for early midpelvic cephalic disproportion (Fig. 7) (Kratochwil 1969). Both vaginal and rectal techniques were used for the evaluation of genital tumors (Fig. 8). The vesical route was used mainly to demonstrate the ovaries and their physiologic changes (Fig. 9). Although the two-dimensional images made interpretation difficult and even cumbersome because of the unusual scan levels, satisfactory results could be achieved when the scans were evaluated in the context of clinic findings.

Soon, rectal scanning proved superior to clinical examination for detecting recurrent disease high on the pelvic wall and inaccessible to finger palpation following the radical excision of cervical carcinoma (Kratochwil et al. 1974). The relation of the tumor to the pelvic wall and to neighboring organs could be accurately assessed. Even a tumor compressing the ureter at the ureterovesical junction could be recognized. The ultrasound findings not only confirmed the clinical diagnosis but also provided a basis for the more rational planning of radiation therapy (Kratochwil et al. 1970, 1971) (Figs. 10, 11).

As early as 1966, von Micsky developed an elaborate vesical scanning system for the ultrasound diagnosis of cervical carcinoma (von Micsky 1966).

Kratochwil et al. (1970) applied the rectal technique to the examination of the prostate gland (Fig. 12). The examination was sometimes painful due to pressure from the rotating probe on the rectal mucosa. This was later remedied by housing the instrument in an inflexible casing with an ultrasound window (von Micsky 1974; Watanabe et al. 1974; Holm and Northeved 1974). Also, the transducer was rotated by a small electric motor instead of manually. To facilitate insertion, a device was constructed that allowed the patient himself to insert the probe according to his own pain tolerance (Watanabe et al. 1974).

Following the introduction of rectal, vaginal, and vesical sonography, it was a short step to combining endosonography with laparoscopy (Kratochwil 1977). The primary goal of this technique was to make precise measurements of the ovarian follicles; it was not intended as an imaging method, nor was this feasible at that time. Therefore the probe was designed for ease of insertion through the second sleeve of the trocar of the laparoscope (Fig. 13). The scan was performed at a frequency of 4 MHz. The transducer was placed in direct contact with the organ under examination, and the dimension of interest was read directly from the A-mode trace (Fig. 14). This examination was easily extended to include upper abdominal organs such as the gallbladder, liver, and pancreas (Fig. 15).

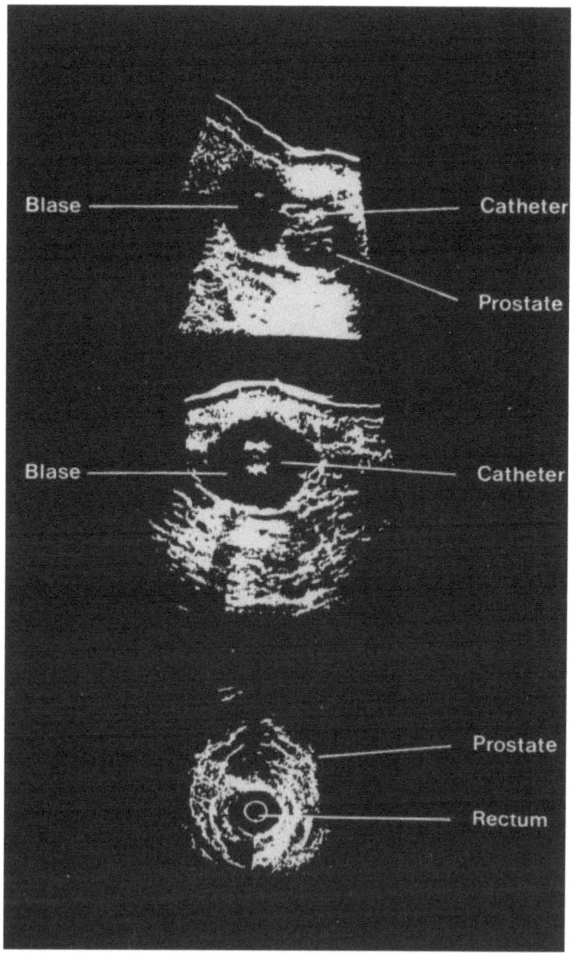

Fig. 12. First examination of the prostate gland. *Top:* longitudinal and transverse abdominal scans; *bottom;* rectal scan

All the foregoing techniques were significantly improved by the introduction of gray-scale imaging and real-time scanning. In urology, endosonographic methods have been used for some time to measure the depth of infiltration of bladder and prostatic tumors (Holm and Northeved 1974; Watanabe et al. 1975; Hanada et al. 1979; Gammelgaard and Holm 1981), and similar methods are currently employed for staging rectal tumors. In internal medicine, gastroscopes have been combined with sonography (Lutz et al. 1983) for the detection of gastric cancer and for examinations of the pancreas and liver.

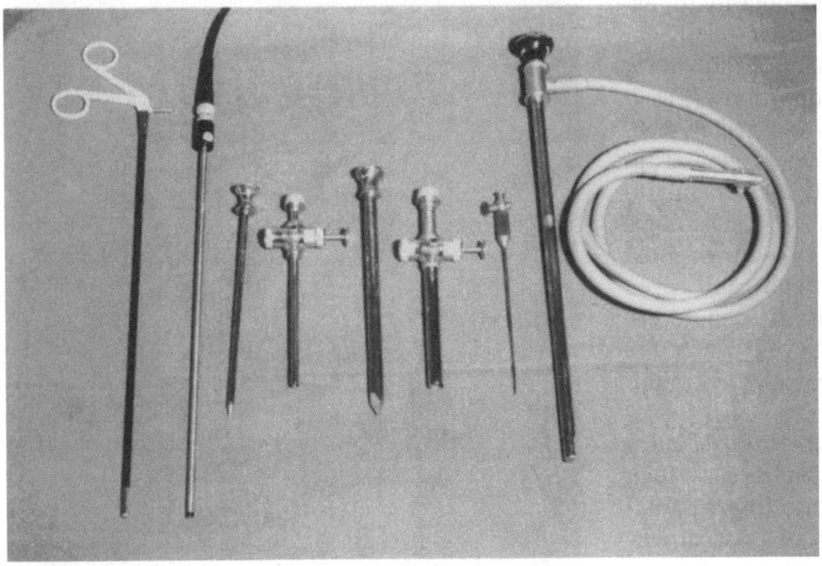

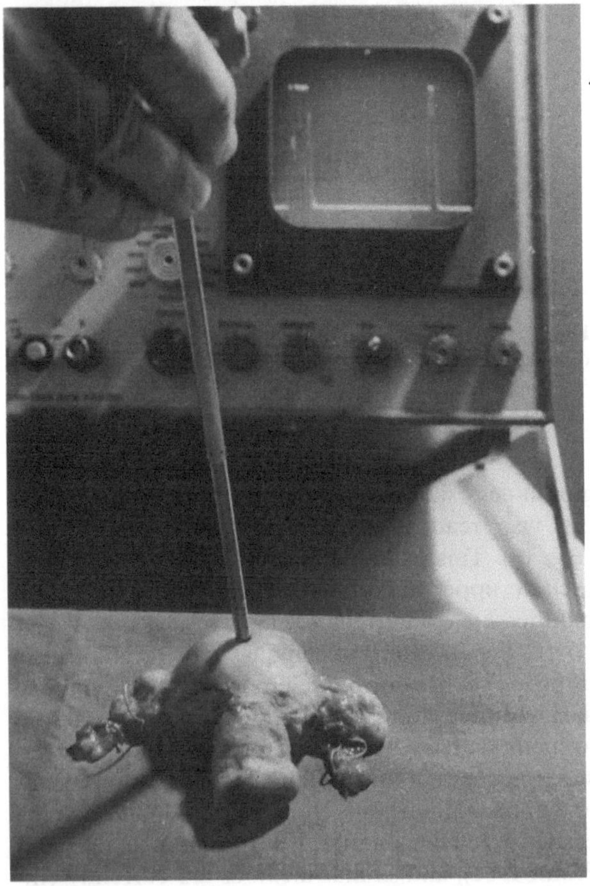

Fig. 13. Transducer for use through the second sleeve of a laparoscope

◀ **Fig. 14.** Reading the result in the A-mode technique

Fig. 15. *Top* Diagram of an A-mode examination of the gallbladder during laparoscopy. *Bottom* Ultrasound examination (A-mode) of the genital organs in conjunction with laparoscopy ▶

In recent years new sonographic techniques have been adapted for use in gynecology, starting with the development of uterine probes that, when passed into the uterine cavity, can measure the depth of infiltration of endometrial cancer.

Special transducers needed to be designed for vaginal sonography. While small, linear-array scanners give only a limited view of the internal genital organs, an excellent survey is provided by the panoramic scanners described by Popp et al. (1985).

At present much work is being done in all fields of endosonography to improve methodology and expand applications, and continued progress with this modality is anticipated in the future.

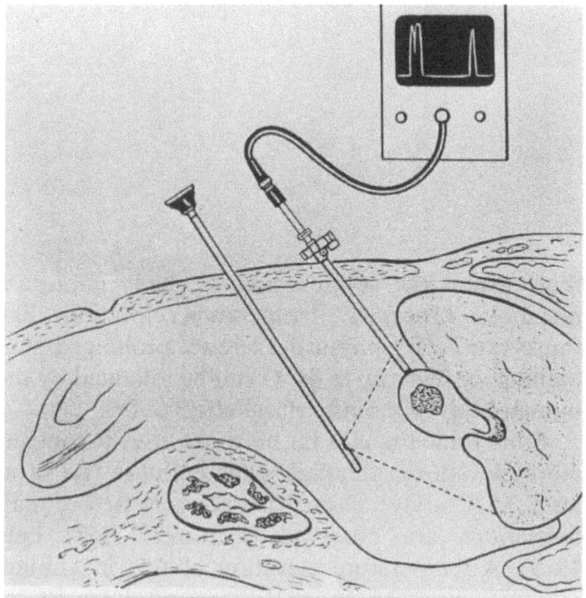

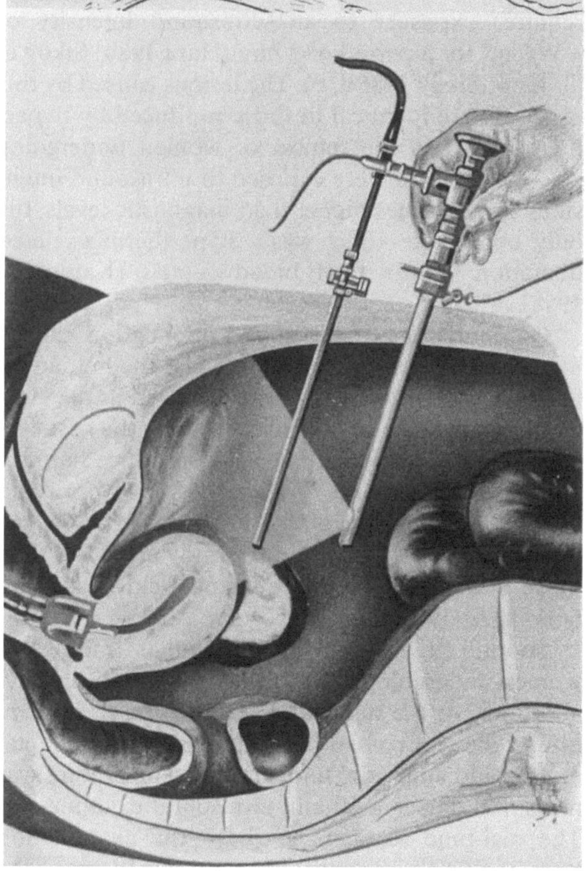

References

Gammelgaard J, Holm HH (1981) Transurethral scanning of the prostate. In: Watanabe H, Holm HH (eds) Diagnostic ultrasound in urology and nephrology. Igaku Shoin, Tokyo, p 153

Harada K, Igari B, Tanahashi Y (1979) Gray scale transrectal ultrasonography of the prostate. J Clin Ultrasound 7: 45-49

Holm HH, Northeved A (1974) A transurethral ultrasonic scanner. J Urol 111: 238-241

Kratochwil A (1966) Möglichkeiten der Ultraschalldiagnostik in der Geburtshilfe und Gynäkologie. Wien Klin Wochenschr 78: 190-191

Kratochwil A (1968) Ultraschalldiagnostik in Geburtshilfe und Gynäkologie. Thieme, Stuttgart

Kratochwil A (1969) Ein neues vaginales Schnittbildverfahren. Geburtshilfe Frauenheilkd 29: 379-385

Kratochwil A (1977) Vergleichende Laparoskopie und Ultraschalluntersuchungen bei hormonell stimulierten Frauen. Acta Endoscopica et Radiocinematographica VII: 21

Kratochwil A, Eisenhut L (1967) Der früheste Nachweis der fetalen Herzaktion durch Ultraschall. Geburtshilfe Frauenheilkd 27: 176

Kratochwil A, Beck A, Lim-Rachmat F (1969) Die Ultraschall-Beckenmessung. Proceedings of 1st World Congress on Ultrasound, Vienna

Kratochwil A, Gasser G, Mar HG (1970) Die Ultraschalldiagnostik in der Urologie. Wien Klin Wochenschr 82: 795-800

Kratochwil A, Alth G, Hofner W (1971) Zur Herdsuche in der gynäkologischen Strahlentherapie. Geburtshilfe Frauenheilkd 31: 833

Kratochwil A, Stöger H, Zeibekis N (1974) Die Darstellung von Beckenwandrezidiven im Ultraschalltomogramm. Geburtshilfe Frauenheilkd 34: 742

Lutz H, Lux H, Heyder N, Borburgh J (1983) Transgastric sonography in pancreatic diseases. In: Lerski RA, Morley P (eds) Ultrasound 82. Pergamon, Oxford, pp 465-467

Popp LW, Lemster S, Hinrichs S, Heesen te D, Müller-Holve W, Martin K (1985) Intravaginale Ultraschalldiagnostik (Vaginosonographie) - erste Erfahrungen mit dem Panoramasektor. In: Judmaier G, Frommhold H, Kratochwil A (eds) Ultraschalldiagnostik 84. Thieme, Stuttgart, pp 320-322

von Micsky LI (1966) Transvesical pelveosonography - a new theoretical and experimental approach to the investigation of gynecological cancer. Obstet Gynecol 27: 597-598

von Micsky LI (1974) Gynecological ultrasonography. In: King DI (ed) Diagnostic ultrasound. Mosby, St. Louis, pp 113-148

Watanabe H, Igari D, Tanahashi Y et al. (1974) Development and application of new equipment for transrectal ultrasonography. J Clin Ultrasound 2: 91-98

Watanabe H, Igari D, Tanahashi Y et al. (1975) Transrectal ultrasonography of the prostate. J Urol 114: 734-749

Safety Aspects of Endosonography

1 Biologic Effects of Ultrasound

In any type of sonographic examination, including endosonography, ultrasound waves are subject to certain physical laws. As the waves pass through the human body, interaction of the sound field with the tissues can produce various biologic effects (bioeffects), not all of which are harmful. The tissue responses to ultrasound are based upon thermal effects as well as other phenomena, most notably acoustic cavitation. The relevant quantity in terms of bioeffects is the ultrasound energy, which is measured in W/cm^2.

The intensity of continuous-wave ultrasound is stated in terms of its spatial average (SA) intensity for a precisely designated area. The highest intensity within this sound field is called the spatial peak (SP). When pulsed ultrasound is used, it is necessary to consider temporal intensity as well as spatial intensity. Thus, the intensity of pulsed ultrasound may be expressed using four different combinations of parameters:

1. SATA: spatial average - temporal average
2. SATP: spatial average - temporal peak
3. SPTA: spatial peak - temporal average
4. SPTP: spatial peak - temporal peak

In equipment for diagnostic imaging, the SATA intensity is very low and usually ranges from 1 to 10 W/cm^2. However, the SPTA values may be as high as 1000 W/cm^2 and therefore are significant in terms of bioeffects. High peak intensities are especially likely to occur with duplex scanners.

1.1 Thermal Effects

As ultrasound waves travel through the body, their kinetic energy is transformed by friction into heat. This heating effect varies in different tissues as a function of their absorption coefficient. Heat generated in tissue is eliminated by diffusion.

Natural thermoregulatory mechanisms are able to maintain the body temperature within a range that ensures the continuation of vitally important metabolic processes. Temperature elevations like those caused by physical exercise, prolonged sunbathing, or fever up to 40 °C can be tolerated by the human body without harmful effects.

Embryonic tissue is far more sensitive to temperature elevations, although a temperature rise of at least 2.5 °C above normal is necessary before malformations are observed (Edwards 1969; Lele 1979). A temperature elevation of this magnitude requires exposure to an ultrasound intensity of 3 W/cm^2 for a period of 3 min (Hara 1980; Sikov et al. 1984; Sikov 1986 a, b). The lesions caused by this exposure are identical to those produced by hyperthermia. When the fetuses of women undergoing elective abortion were exposed to ultrasound intensities several times higher than diagnostic levels, the only observable effect was a hyperthermia-related dilatation of the fetal blood vessels (Kamoscay 1958).

Computations of the sound attenuation caused by tissue between the transducer and the human fetus have shown that the intensities used in diagnostic ultrasound are not sufficient to heat the tissue by absorption to a degree that could produce harmful effects (Ter Haar and Williams 1981; Nyborg and Steel 1983). Because of the intervening tissue, the intensities measured within the body are lower than those measured in the free sound field in a water tank (Lizzi 1988).

Soothill et al. (1987) made similar findings in women undergoing elective first or second trimester abortions. At the start of the termination a temperature probe was passed percutaneously into the fetal muscle and amniotic fluid, and the temperature was measured during a 10-min ultrasound examination. The real-time scanners used for the examination had a SPTA intensity of 1.1-3.0 mW/cm^2, the Doppler scanners 112 mW/cm^2. A mean temperature of 36.0 °C was measured in the fetal skeletal muscle during the examination, and a mean temperature of 36.6 °C in the amniotic fluid. The temperature gradient from fetus to amniotic fluid indicated that the fetus may lose much of the muscle heat produced by its metabolism to the amniotic fluid. Thermal effects of sonography were ruled out as a potential cause of fetal damage.

Focused transducers lead to a temperature rise of 0.4 °C in the focal zone (Carsons 1988), while unfocused transducers produce greater heating. The heating effect of duplex scanners is frequency-dependent. The higher the frequency, the greater the temperature rise, although this increase is not linear (Lizzi 1988).

In summary, it may be said that diagnostically applied ultrasound is harmless as long as the anticipated temperature rise does not exceed 1 °C.

1.2 Cavitation

Increasing attention is being given to the phenomenon of acoustic cavitation as a precipitating cause of adverse bioeffects from ultrasound exposure. Acoustic cavitation refers to the formation of cavities in gasless liquids exposed to ultrasound. The sound wave may be characterized as a alternating sequence of compressive and tensile stresses acting through a medium. Molecular assemblies in the medium may become disrupted in the positive pressure phase, causing the formation of cavities which collapse again in the next pressure phase. This collapse generates a shock wave whose energy may be many times greater than that of the exciting wave and can generate transient pressures as high as 1000 atm (Matauschek 1961). This tremendous energy can produce a number of secondary effects such as temperature elevation, chemical reactions, free radical formation, and chemoluminescence. Shear forces, moreover, can induce microstreaming of the cytoplasm within cells. However, these effects are highly localized and are confined to a volume of several cubic micrometers (Lizzi 1988).

Pseudocavitation refers to the interaction of the sound field with small gas bubbles already present in the liquid medium. Bubbles smaller than the resonance size ("nuclei") are stimulated to undergo nonharmonic oscillations (Nyborg and Steel 1983). Since the internal strength of the bubbles is less than that of the medium, diffusion of the gas occurs during the positive pressure phase. In the subsequent expansion phase the bubble surface enlarges, and gas reenters. Just before the peak pressure is reached, the bubbles collapse violently, creating a shock wave (transient cavitation). Bubbles larger than the resonance size undergo a harmonic oscillation that does not generate shock waves (stable cavitation).

Thus, two conditions are necessary in order for cavitation to occur:

1. A high peak intensity of 2–50 W/cm^2, depending on the carrier frequency. The value of the negative acoustic alternating pressure is an even better predictor in this regard.
2. The presence of bubbles in the medium (Carstens and Gates 1984).

With ultrasound pulses of longer duration or with continuous-wave ultrasound, the threshold for cavitation is lower.

The likelihood of cavitation activity increases with the pulse length and with the sound frequency. Although modern, high-frequency ultrasound instruments utilize short pulse lengths, the possibility of cavitation cannot be entirely ruled out (Flynn 1952; Apfel 1986). While the temporal average intensities of these systems are low, temporal peak intensities may be as high as 1000 W/cm^2, and temporal peak pressures may reach 5 MPa.

The dilemma of cavitation research is that there is still some doubt whether the bubbles necessary for cavitation are actually present in the human body (Carstens and Gates 1984). So far, cavitation activity has been demonstrated only in plants (Miller 1977) and in *Drosophila* (Child et al. 1981), organisms that are permeated extensively by gas-filled cavities.

In summary, it is reasonable to assume that cavitation does not occur under the conditions of diagnostic ultrasound (Hill and ter Haar 1982).

1.3 Chromosome Damage

A number of in vitro studies have been published dealing with the question of chromosome abnormalities following ultrasound exposure. An advantage of in vitro studies is that the essential parameters of the experiment can be varied as desired; their disadvantage is their proneness to methodologic errors. Some groups of workers experimented with single-cell organisms whose mutability was unknown. When salmonella strains of known sensitivity were treated with pulsed ultrasound, no increase in mutagenicity was observed (Barnett et al. 1982).

In many in vitro studies, blood has been exposed to ultrasound and the lymphocytes subsequently examined for chromosome aberrations. But even doses well above diagnostic levels have failed to produce detectable changes (Fisher et al. 1969). In vitro studies of human fetal umbilical cord blood after sonography (Abdulla et al. 1971; Boyd et al.

1971) and after Doppler fetal cardiac monitoring (Lucas et al. 1972) likewise have shown no injurious effects.

Much concern was aroused by MacIntosh and Davey (1970), who reported chromosome abnormalities following exposure to ultrasound levels comparable to those used for fetal cardiac monitoring. In this case a methodologic error was made which resulted in exposure to standing waves. The very high energy levels occurring at the nodal points of these waves accounted for the changes that were observed. Fisher et al. (1969) described a similar experiment in which adverse effects were caused by excessive heat generation at the interface between the cell culture and the bottom of the petri dish. After other authors showed that 300 W of ultrasound produced no chromosome changes in exposed lymphocytes (Coakley et al. 1972) or in cell suspensions in plastic films (Rott and Soldner 1973), MacIntosh et al. (1975) found it necessary to revise their original findings.

1.4 Sister Chromatid Exchange Rate

The sister chromatid exchange (SCE) rate has been used by many as an indicator of mutagenic activity. Thus, Liebeskind et al. reported in 1979 on an increased SCE rate following exposure to diagnostic ultrasound. Although the authors were very cautious in interpreting their findings, the report was picked up by the lay press, where it was implied that ultrasound causes chromosome damage. Naturally this report prompted widespread concern.

In 1984 the Bioeffects Committee of the AIUM critically evaluated a total of 14 studies published on this subject (Goss 1984). Only three of those studies showed a positive correlation, with a slight but statistically significant increase in SCEs (Brulfert et al. 1984); the remaining 11 were unable to show any such connection.

Although most investigators believe that an increased SCE rate indicates chromosome injury, the biologic significance of the SCE rate remains unclear (Gebart 1981; Jacobson-Kram 1984). Thus, while a number of extremely potent mutagenic factors cause only a slight rise in SCEs, there are relatively weak carcinogens that produce a substantial increase. It is known, moreover, that a baseline SCE rate exists even in environments that contain no SCE-inducing substances. Given the potential for errors and disturbances in available test systems, it is currently believed that the increase in SCE rates must equal 100%–150% before mutagenic activity may be inferred (Haupt et al. 1981). What is more, the experiments that appear to confirm SCE induction used entirely different methodologies and therefore cannot be compared with one another.

In summary, then, many more carefully conducted and better coordinated studies are needed before it can be stated whether or not diagnostic ultrasound leads to an increase in SCEs.

1.5 Epidemiologic Studies

Despite lack of evidence of harmful side effects in experimental animals, epidemiologists have repeatedly stated that diagnostic ultrasound cannot be declared safe until long-term clinical studies have been done to investigate the direct effects of ultrasound exposure on the population.

In questions of epidemiology, a distinction must be drawn between a real and potential risk. While a real risk can be plainly recognized and generally can be associated with a specific injurious effect, a potential risk implies only that future research may uncover a previously unknown and unspecified risk, even though the probability of such a risk may be small.

We can better appreciate the caution of epidemiologists by noting that some adverse side effects of medical therapies and procedures have taken years to surface, such as the increased incidence of leukemia in the children of women who underwent X-ray pelvimetry during pregnancy and the increased rate of vaginal carcinoma in the daughters of women treated with diethylstilbestrol for threatened abortion.

Beginning with Hellman et al. in 1970, a number of authors have concluded that ultrasound has no adverse effects on embryonic development and that it does not induce fetal malformations (Scheidt et al. 1978; Stark et al. 1984; Baketeig et al. 1985; Mukubo et al. 1985). This negative result is not surprising, since most studies were performed after the 10th week of gestation and thus after the conclusion of organogenesis.

Large surveys on the safety of ultrasound during pregnancy like those conducted in the United States (Ziskin 1972, 1987) and Canada (Environmental Health Directorate 1981) and covering a total of 1.2 million examinations have so far failed to disclose any harmful effects.

The latest concerns about the use of ultrasound in pregnancy relate to a possible reduction in birthweight, the occurrence of mild neurologic deficits, and the development of carcinoma in early childhood. The potential link to birthweight was disclosed by an animal study using higher-than-diagnostic intensities. Reduced body weight was noted in the offspring of exposed pregnant mice (O'Brien 1983). The randomized studies of Wladimiroff and Laar (1980), Lyons et al. (1980), and Baketeig et al. (1985) showed no adverse effects of ultrasound on birthweight at intensities in the diagnostic range. This is consistent with the results of Scheidt et al. (1978), who used ultrasound for placental localization in early amniocentesis. However, Moore et al. (1982), using multivariant analysis in a retrospective study, found a marked difference in birthweight relative to an untreated control group. Stark et al. (1984) found no correlation between birthweight and ultrasound exposure. It is reasonable to assume that cases where differences were noted may have involved children who were at risk for intrauterine growth retardation.

Fears of neurologic impairment are based on the fact that ultrasound examinations are frequently performed between the 14th and 22nd weeks of gestation, and thus during a period of neuronal migration in the fetal brain. Indeed, this type of abnormality was an important focus of the epidemiologic studies of Stark and Scheidt. While Scheidt described a significantly increased rate of abnormal grasp reflexes and tonic neck reflexes, Stark found an increased incidence of dyslexia relative to controls. However, the fact that these changes were found only in children of lower birthweight makes the results difficult to interpret.

1.6 Carcinogenesis

Two large, carefully planned studies were unable to show any link between fetal ultrasound exposure and an increased incidence of cancer in early childhood (Cartwright et al. 1984; Kinnier Wilson and Waterhouse 1984).

The value of all the foregoing studies is called into question by the fact that only a few provide adequate information on the ultrasound intensity and the gestational age at the time of exposure.

1.7 Statistical Considerations

Adverse effects, if they exist, are not obvious and require statistical methods for their disclosure. The more subtle these effects are, the more difficult they are to recognize. A statistically valid conclusion would require an extremely large number of random samples, especially if the effects consist of mild chemical changes, minor behavioral alterations, or long-term effects. Even then, the changes could still remain undetected. Nevertheless, many experts, most notably epidemiologists, continue to argue that the inability to demonstrate an effect does not exclude the possibility of its occurrence.

This is a good example of the scientific principle that when a hypothesis is raised, it persists until it is disproved or replaced by a better hypothesis. If a hypothesis resists attempts at refutation, the likelihood increases that the hypothesis is correct. An essential feature of a scientific hypothesis, then, is its "falsifiability"; theses that can be falsified have no scientific value. We can illustrate this by comparing the hypothesis that dinosaurs are extinct with the hypothesis that they are not. These hypotheses may appear similar at first sight, but they differ greatly in their falsifiability. The first can be disproved by the discovery of a single living specimen; the second can be confirmed but never disproved (Rott 1981, 1988).

Today we find ourselves in a similar situation regarding the hypothesis that ultrasound causes long-term adverse effects. The long-term prospective studies called for by epidemiologists to refute this hypothesis would be very costly and labor-intensive; short-term evidence would be of no value. Technical innovations preclude complete certainty. The only valid question can be whether the potential risks inherent in such procedures are acceptable when weighed against the proven (real) benefits of the examination.

1.8 Summary

After 25 years of use, neither clinical nor epidemiologic studies have demonstrated adverse side effects from diagnostic ultrasound (Lyons 1981).

The 1984 Consensus Statement of the National Institute of Health caused much concern by denying that diagnostic ultrasound had real risks while conceding the possibility of potential risks and especially by claiming that routine sonography is of no proven benefit during pregnancy.

In response, various competent groups issued their own statements on the safety of diagnostic ultrasound. The British Medical Ultrasound Society (1984), the Royal College of Obstetricians (1984), the British Institute of Radiology (1987) and the European Federation of National Ultrasound Societies (1988) emphasized that after more than

20 years' experience with ultrasound diagnosis and bioeffects research, there was no known evidence of harmful side effects or risks that would justify a change in the current usage of the modality.

The American Institute of Ultrasound in Medicine and Biology also confirms in its safety statement (AIUM 1984) that there are no known bioeffects associated with diagnostic ultrasound, and that continuous-wave and pulsed Doppler ultrasound offer many advantages for the detection and supervision of high-risk pregnancies. Given the substantial number of studies, some randomized, which support the value of this method, the AIUM even recommends the use of Doppler examinations during pregnancy. Its recommendation for "wise use" of diagnostic ultrasound is based on the inability to quantify the unnecessary and excessive use of the modality. Because many medical procedures can have deleterious effects of one kind or another, the maxim of "first do no harm" has been replaced by the concept of causing the patient as little harm as possible while achieving the desired benefit for his health. Today we may assume in good conscience that there is no other procedure in clinical medicine that offers greater patient safety than diagnostic ultrasound (Ziskin 1987).

2 Sterilization of Vaginal Probes

The following recommendations for the sterilization of vaginal transducers were made by the Australian Society for Ultrasound in Medicine:

1. Some form of disposable cover should always be used on probes for vaginal sonography.
2. Gloves should be worn during the examination.
3. Ensure that the cover used is of adequate size, length, and strength to reduce the risk of rupture during use.
4. Remove the cover carefully to ensure mucus is contained.
5. If the cover is not damaged during the examination, there is NO need to sterilize the probe between examinations. It will need some cleaning to remove any coupling medium.
6. If the cover is damaged:
   a) Ensure all mucus and debris are carefully removed from the probe. Soap and water used with a brush are quite adequate.
   b) Soak in stabilized glutaraldehyde, as instructed by the manufacturer, for at least 5 min. It would be prudent to check with the manufacturer as to whether this is suitable for the particular transducer.

Comments: While ordinary glutaraldehyde and hypochloride solution are acceptable, stabilized glutaraldehyde is considered preferable as it is easier to prepare, remains effective for some days after opening, and has a more tolerable odor when used in a confined space.

Papillomaviruses (some groups of which are associated with cervical cancer) can remain infectious for up to 48 h, even outside the body. There is at present no known killing time for papillomavirus, but based on animal studies, 5 min seems reasonable for killing this virus and all other known organisms provided the transducer is adequately cleaned.

References

Abdulla U, Campbell S, Dewhurst CJ, Thalbot D, Lucas M, Mullorky M (1971) Effect of diagnostic ultrasound on maternal and fetal chromosomes. Lancet II: 829-831

American Institute of Ultrasound in Medicine (1984) Statement on clinical safety S 3. In: Safety considerations for diagnostic ultrasound. AIUM, Bethesda (also AIUM Lett October 1988)

Apfel RE (1986) Possibility of microcavitation from diagnostic ultrasound. IEEE Trans Ultrasound Ferroelectrics Frequency Control 33: 139-142

Baketeig LS, Eik-Nes S, Jacobson S (1985) A randomized controlled trial of ultrasonographic screening in pregnancy. Lancet II: 207-210

Barnett SB, Banin A, Mitchell G (1982) An investigation of mutagenic potentials of pulsed ultrasound. Br J Radiol 55: 501-504

Boyd E, Abdulla U, Donald J, Fleming IEE, Hall AJ, Ferguson MA (1971) Chromosome breakage and ultrasound. Br Med J 2: 501-502

British Institute of Radiology (1987) The safety of diagnostic ultrasound. Br J Radiol [Suppl] 20

British Medical Ultrasound Society (1984) Safety of diagnostic ultrasound. Press release, London

Brulfert A, Ciavarino V, Miller MW, Manlik D, Carstensen E (1984) Diagnostic insonation of extra utero human placentas: no effect on lymphocyte sister chromatid exchange. Hum Genet 66: 289-291

Carsons PL (1988) Medical ultrasound fields and exposure measurements. In: Proceedings of 22nd Annual Meeting of National Council on Radiation Protection and Measurements. NCRP, Bethesda MD

Carstens EL, Gates AH (1984) The effects of pulsed ultrasound on the fetus. J Ultrasound Med Biol 3: 145-147

Cartwright RA, McKinney PA, Hopton PA (1984) Ultrasound examination in pregnancy and childhood cancer. Lancet II: 999-1000

Child SZ, Carstens EL, Smacho K (1981) Effects of ultrasound on drosophila. III: Exposure of larvae to low tem-

poral average intensity pulsed irradiation. Ultrasound Med Biol 7: 167-173
Coakley WT, Slade JS, Braeman JM, Moore JL (1972) Examination of lymphocytes for chromosome aberrations after ultrasonic irradiation. Br J Radiol 45: 328-332
Edwards MJ (1969) Congenital defects in guinea pigs: fetal resorptions, abortions and malformations following hyperthermia during gestation. Teratology 2: 323-328
Environmental Health Directorate (1981) Safety Code 23: Guidelines for the safe use of ultrasound. Part I: Medical and paramedical application. Ottawa Environmental Health Department, Health Protection Branch, report 8
European Federation of Sections of Ultrasound in Medicine and Biology (1988) Sicherheits-Statement der Watchdog-group Ultraschall Med 9: 52-53
Fisher P, Golob E, Kratochwil A, Kunze-Mühl E (1967) Chromosomenuntersuchungen nach Ultraschalleinwirkung. Wien Klin Wochenschr 79: 436-437
Flynn HG (1952) Generation of transient cavities in liquids by microsecond pulses of ultrasound. J Acoust Soc Am 72: 1926-1932
Gebart E (1981) Sister chromatid exchange (SCE) and structural chromosome aberration in mutagenicity. Hum Genet 58: 235-254
Goss A (1984) Sister chromatid exchange and ultrasound. Report of the Bioeffects Committee of the AIUM. J Ultrasound Med Biol 3: 463-470
Hara K (1980) Effects of ultrasonic irradiation radiation on chromosomes, cell division and developing embryo. Acta Obstet Gynecol Jpn 32: 61-68
Haupt M, Martin AO, Simpson JL, Iqbal MA, Elias S, Dyer A, Sabbagha RE (1981) Ultrasonic induction of sister chromatid exchange in human lymphocytes. Hum Genet 59: 221-226
Hellmann LM, Duffus GM, Donald I, Sunden B (1970) Safety of diagnostic ultrasound in obstetrics. Lancet I: 1133-1135
Hill CR, ter Haar GR (1982) Nonionizing radiation protection. WHO Europe, Copenhagen, pp 199-228
Jacobsen-Kram D (1984) The effects of diagnostic ultrasound on sister chromatid exchange frequencies: a review of the recent literature. J Clin Ultrasound 12: 5-10
Kamoscay Gy (1958) Ultrasound in gynecology. Am J Phys Med 37: 196
Kinnier Wilson LM, Waterhouse JA (1984) Obstetric ultrasound and childhood malignancies. Lancet II: 997-999
Lele PP (1979) Safety and potential hazards in the current applications of ultrasound in obstetrics and gynecology. J Ultrasound Med Biol 5: 307-320
Liebeskind D, Bases R, Elequin F, Königsberg M (1979) Sister chromatid exchange in human lymphocytes after exposure to diagnostic ultrasound. Science 205: 1273-1275
Lizzi F (1988) Bioeffect considerations for the safety of diagnostic ultrasound. J Ultrasound Med Biol 14: 1-38
Lucas M, Mullorky M, Abdulla U (1972) Study of chromosomes in the newborn after ultrasonic fetal heart monitoring. Br Med J 3: 795-796
Lyons EA (1981) First symposium on safety and standardisation of ultrasound in medicine. Proceedings, World Federation of Ultrasound in Medicine, Sydney
Lyons EA, Coggrave M, Brown RE (1980) Follow up study in children exposed to ultrasound in utero: analysis of height and weight in the first six years of life. Proceedings of 25th Annual Meeting of AIUM, New Orleans, p 49
MacIntosh JC, Davey DA (1970) Chromosome aberrations induced by an ultrasonic fetal pulse detector. Br Med J 4: 92-93
MacIntosh JC, Brown RC, Coakley WT (1975) Ultrasound and "in vitro" chromosome abberations. Brit J Radiol 4: 230-232
Matauschek J (1961) Einführung in die Ultraschalltechnik, 2nd edn. Verlag Technik, Berlin
Miller DL (1977) The effects of ultrasonic cavitation of gas bodies in Elodea leaves during continuous and pulsed irradiation at 1 MHz. J Ultrasound Med Biol 3: 221-240
Moore RM, Barrick MK, Hamilton PM (1982) Ultrasound exposure during gestation and birthweight. J Ultrasound Med Biol 8: 34
Mukubo M, Okai T, Kozuma S (1985) Safety of diagnostic ultrasound during pregnancy on fetus and child development. Proceed 85 WFUMB Sidney. p 481
National Institutes of Health (1984) NIH Consensus Statement on diagnostic ultrasound in pregnancy. US Department of Health and Human Services, Bethesda
Nyborg WL, Steel RB (1983) Temperature elevation in a beam of ultrasound. J Ultrasound Med Biol 9: 611-620
O'Brien WD Jr (1983) Ultrasonic dose dependent effect of ultrasound on fetal weight in mice. J Ultrasound Med Biol 2: 1-8
Rott HD (1981) Zur Frage der Schädigungsmöglichkeit durch diagnostischen Ultraschall. Ultraschall 9: 56-64
Rott HD (1988) Diagnostischer Ultraschall: biologische Wirkungen, potentielle Risiken. Ultraschall 2-4
Rott HD, Soldner R (1973) The effect of ultrasound on human chromosomes. Humangenetik 20: 100-112
Royal College of obstetricians and Gynaecologists (1984) Report of the RCOG working party on routine ultrasound examination in pregnancy. Chameleon Press, London
Scheidt PD, Stanley F, Bryla DA (1978) One year follow up of infants exposed to ultrasound in utero. Am J Obstet Gynecol 121: 742-748
Sikov MR, Collins DH, Carr DR (1984) Measurement of temperature rise in prenatal rats during exposure of the exteriorized uterus to ultrasound. IEE Transonics Ultrasonics Suppl 31: 497-503
Sikov MR (1986a) Effects of ultrasound on development. Part I: introduction and studies in inframammalian species. J Ultrasound Med Biol 5: 577-583
Sikov MR (1986b) Effects of ultrasound on development. Part II: studies in mammalian species: on overview. J Ultrasound Med Biol 5: 651-661
Soothill PW, Nicolades KH, Rodeck CH, Campbell S (1987) Amniotic fluid and fetal tissues are not heated by obstetric ultrasound scanning. Br J Obstet Gynaecol 94: 675-677
Stark CR, Orleans M, Haverkamp AD, Murphy J (1984) Short- and long-term risks after exposure to diagnostic ultrasound. Obstet Gynecol 63: 194-200
ter Haar G, Williams AR (1981) Biophysical and physiological consequences of ultrasonic irradiation of tissue. In: Kurjak A, Kratochwil A (eds) Recent advances in ultrasonic diagnosis. Excerpta medica, Amsterdam pp 33-36
Wladimiroff JW, Laar J (1980) Ultrasound measurement of fetal body size: a randomized controlled trial. Acta Obstet Gynecol Scand 59: 177-179
Ziskin MC (1972) Survey of patient exposure to diagnostic ultrasound. In: Reid JM, Sikov MR (eds) Interaction of ultrasound and biological tissues. Department of Health, Education and Welfare (FDA) Publication 78-8008, Government Printing Office, Washington DC, p 203
Ziskin MC (1987) The prudent use of diagnostic ultrasound. J Ultrasound Med Biol 9: 415-416

Advantages and Disadvantages of Endosonography

Endosonography occupies something of an intermediate position between endoscopy and abdominal sonography, for it can visualize the surface of hollow organs as well as define structures behind the surface, doing so with an optimum level of detail. This is due chiefly to the physical properties of the ultrasound beam.

1 Advantages

The proximity of the transducer to the target organ allows the use of generally higher frequencies (5–10 MHz) in endosonography than in abdominal ultrasound. While this limits the penetration of the beam, it significantly enhances resolution and image quality, and the proximity of the probe to the region of interest allows for small-part scanning of the pelvic organs. The optimum frequency for the examination is determined mainly by the distance from the probe to the organ of interest and secondarily by the density of the intervening tissue, which influences sound attenuation as a result of absorption, divergence, and scatter.

Another advantage of endosonography is that the intracorporeal placement of the transducer bypasses the obstacles that ordinarily exist between an externally applied transducer and the organ of interest. This can eliminate "blind spots" like those caused by overlying bone or air-filled organs. Image improvement is particularly marked in women with extensive adhesions following previous surgery or radiotherapy or with inflammatory disease or endometriotic foci in the lesser pelvis. Endosonography also eliminates the problems of obesity or an inadequately filled urinary bladder in abdominal sonography.

A significant advantage of endosonography is that the distance from the probe to the organ of interest is approximately equal in every patient. Thus the examination can be conducted with largely standardized instrument settings, enabling the visual comparison of documented findings.

2 Disadvantages

The size of the transducer makes endosonography difficult in children and in patients with strictures.

The examination cannot be directly extended to the upper abdomen, if such scanning is required.

Because of the relatively poor penetration of the high-frequency beam, endosonograms can usually demonstrate only a portion of larger lesions. Structures more distant from the transducer may not be visualized at all.

Scanner Types

The sonographer who is not well versed in endosonography should become familiar not only with the advantages and disadvantages of the method but also with the advantages of the various scanner types before acquiring his own experience. Otherwise he may discover too late that the scanner he has purchased is not the best type for his own particular requirements.

The scanner types used in endosonography are basically the same as in abdominal sonography, with most examiners preferring sector scanners or curved-array scanners over linear-array devices.

1 Linear-Array Scanners

Linear-array scanners image a rectangular area that is perpendicular to the long axis of the probe. Thus, scans are made predominantly on the longitudinal axis of the body orifice into which the probe is inserted. A flat contact surface of sufficient length must be available for application of the transducer. The lateral scan direction (90° to the probe axis) makes it difficult to examine structures located higher in the pelvis. By contrast, short, front-viewing linear-array scanners can be used for endosonography, but their field of vision is so narrow that orientation can become extremely difficult when evaluating intrapelvic structures. Linear-array probes are most commonly used for rectosonography and less frequently for vaginosonography, where they are used mainly for evaluating the vagina itself and the urethra.

2 Curved-Array Scanners

These devices combine the excellent near-field resolution of linear scanners with the front-viewing capability of sector scanners, allowing for better evaluation of structures located higher in the pelvis. The curvature of the transducer array provides a larger scan angle. To maximize the field of view, most curved-array scanners have a varying radius of curvature, i. e., are curved slightly at the sides and strongly in front. The resulting asymmetric emission of sound waves increases the field of view, with potential 180° rotation of the probe, but precisely this 180° rotation may hinder spatial orientation, leading to confusion of left and right and of above and below.

3 Sector Scanners

A common type of sector scanner, the "wobbler," houses one or more transducers that move back and forth at high speed. Most sector scanners used in endosonography are symmetrically front-viewing with a 30°-100° angle of vision. Other types are transaxial-viewing; these are mainly used for rectosonography.

"Rotary" scanners contain several transducers that rotate rapidly about an axis that either coincides with the long axis of the probe, yielding transaxial images, or is perpendicular to the probe axis and thus allows forward scanning. The aperture angle is usually 360° for transaxial scanners but cannot exceed 270° in front-viewing scanners due to the presence of the scanner shaft. The panoramic view on both sides of the probe permits the use of high frequencies in rotary scanners and also provides an optimum survey of the lesser pelvis by defining numerous anatomic landmarks. In contrast to traditional sector scanners, the rotary scanner (panoramic scanner) effectively doubles the depth of sound penetration by its capacity for imaging on both sides of the probe. The slower frame rate of the rotary scanner, at approximately 8/s, does not degrade the morphologic definition that is desired in most examinations. Also, this frame rate is more than adequate for the purely visual monitoring of dynamic events such as cardiac activity. On the other hand, the "freeze frame" button captures fewer images of motion sequences than in ordinary sector scanners, resulting, for example, in the documentation of fewer anatomic structures in the individual phases of cardiac activity. The flicker rate of the human eye is approximately 15/s. Because the frame rate of the scanner is inversely proportional to the number of image lines displayed on the monitor, a slower frame rate is necessary in order to display the increased number of lines contained in the large image area of the panoramic sector scanner. However, this is a minor objection in most cases compared with the panoramic view and the resulting improvement in orientation, which frequently makes it easier to learn the endosonographic examination technique and sometimes shortens examination time considerably.

While wobblers are used chiefly in vaginosonography, rotary scanners have applications in vaginosonography, hysterosonography, rectosonography, and cystosonography.

Scan Planes

For those accustomed to the standard planes of section in abdominal ultrasound, we present diagrams that will aid in understanding the various transducer-dependent image planes that are employed in endosonography. An accurate description of the scan planes is crucial for spatial orientation in the lesser pelvis and for the interpretation of recorded images.

The linear-array scanners most commonly used for endosonography have a lateral scan direction. These devices can provide longitudinal scans in a 360° sweep around the probe axis, including an image on the sagittal plane (Figs. 1, 2). Vertical and lateral movements of the probe, though limited by the dimensions of the body orifice, can shift the scan plane sufficiently to give additional information on pelvic structures (Fig. 3).

Front-viewing rotary scanners (rotational axis perpendicular to probe axis), wobblers and curved arrays can provide coronal scans that include images on the sagittal and horizontal planes (Fig. 4). All intermediate planes can be visualized by rotation of the probe. Lateral or vertical movement of the probe within the body orifice, though limited, can demonstrate pelvic structures on additional planes (Fig. 5). Front-viewing rotary scanners with a large field of vision (panoramic sector scanners) basically combine the scanning capabilities of a linear-array scanner and a sector scanner with a small field of vision.

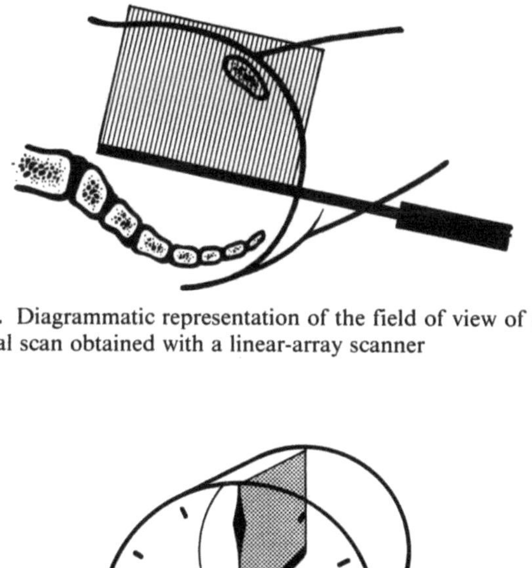

Fig. 1. Diagrammatic representation of the field of view of a sagittal scan obtained with a linear-array scanner

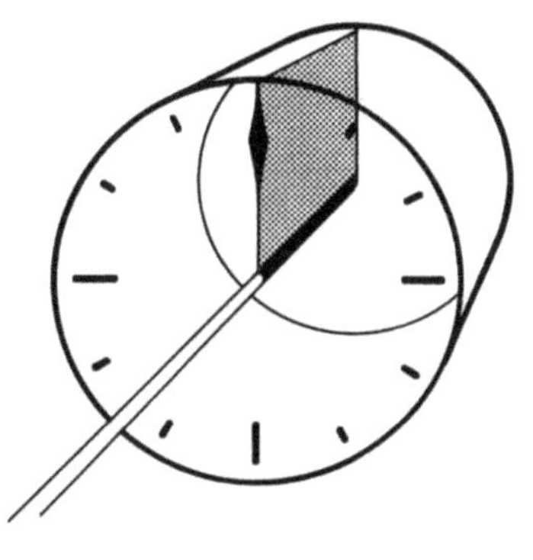

Fig. 2. Just as the hand on a clock can point in any direction, a linear-array scanner can be set to image any plane in a 360° sweep

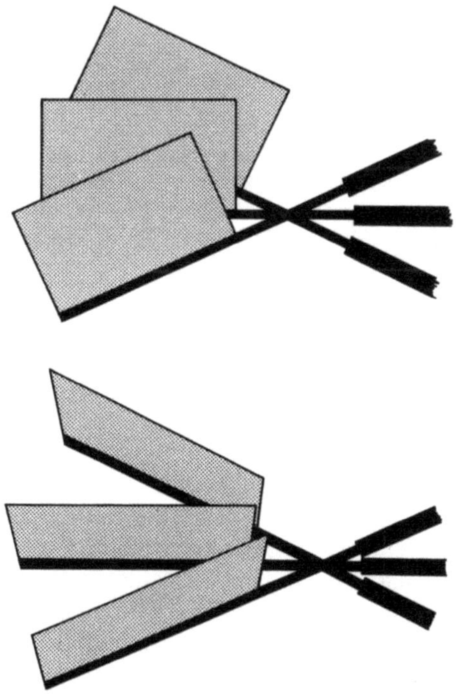

Fig. 3. The field of view of a linear-array scanner can be extended by raising and lowering the scanner shaft *(top)* or by moving the shaft laterally *(bottom)*

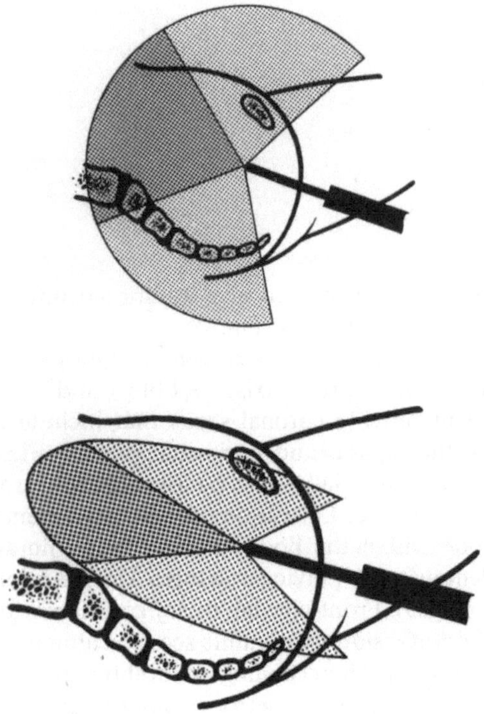

Fig. 4. Sector scanner with a small *(heavily shaded)* or large *(lightly shaded)* field of view. *Top:* sagittal scan; *bottom:* horizontal scan

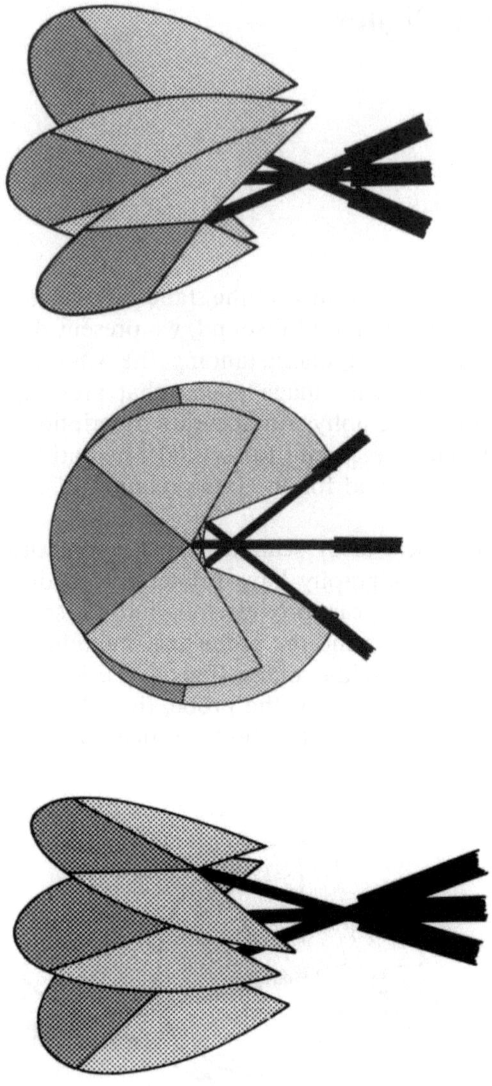

Fig. 5. The field of view of a sector scanner can be altered by ▶ moving the scanner shaft laterally starting from the sagittal plane *(top)*, by raising or lowering the shaft starting from the sagittal plane *(center)*, or by raising or lowering the shaft starting from the horizontal plane *(bottom)*

Rotary scanners with a transaxial scan direction (coaxial rotor and shaft) yield cross-sectional images on the transverse plane (Fig. 6). Vertical or lateral movements of the probe produce a corresponding angulation of the image plane (Fig. 7).

Endosonographic probes with variable scan angles allow all planes to be scanned in one examination without the need to exchange one probe for another. Probes whose scan direction can be varied from forward to transaxial are the most versatile, although preference in the individual case will be given to the scan that provides the most complete field of view.

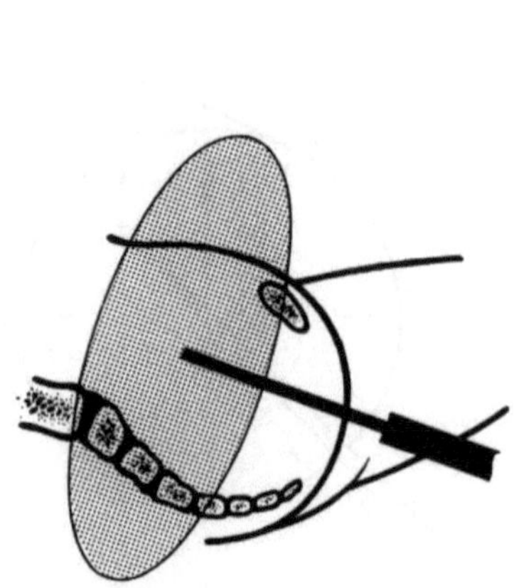

Fig. 6. Transaxial rotary scanner: the scan direction is perpendicular to the probe axis

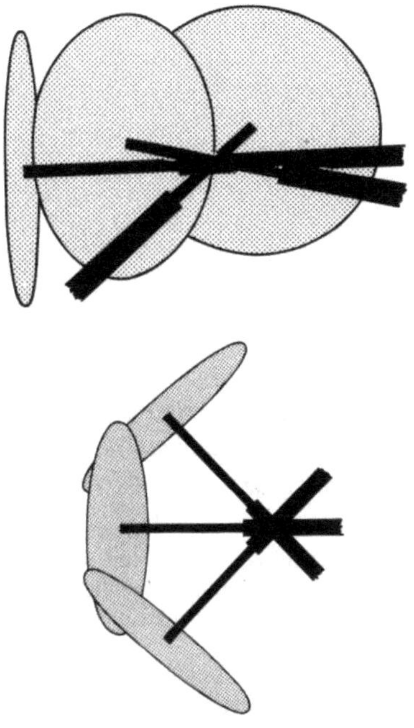

Fig. 7. The field of view of a transaxial rotary scanner can be altered by moving the scanner shaft laterally *(top)* or vertically *(bottom)*

1 Definition of Scan Directions

Lateral: Sound waves emitted at a 90° angle to the probe shaft (Fig. 1).

Frontal: Sound waves emitted forward from the probe tip, in line with the probe shaft (Fig. 4).

Transaxial: Sound waves emitted from the side of the probe tip, perpendicular to the probe axis (Fig. 6).

2 Definition of Scan Planes

Coronal: Planes displayed by a frontal scan direction; they include the horizontal scan (parallel to the table, not possible with abdominal ultrasound) and the sagittal scan (longitudinal scan), which is comparable to the longitudinal scan in abdominal ultrasound (see Fig. 4).

Transverse: Planes displayed by a transaxial scan direction, comparable to the transverse scan in abdominal ultrasound (see Fig. 6).

Orientation of Scan Planes

Endosonography has assumed much importance in recent years, especially in obstetrics and gynecology. This is evidenced by the many innovations in the design of vaginal probes and especially by the growing number of scientific papers on the subject. However, the documentation of endosonograms is so poorly standardized that even examiners familiar with the method may experience difficulties in interpretation. For example, the transducer is displayed at the top of some scans and at the bottom of others (Fig. 1). Similar discrepancies exist with regard to the sides of the image (Fig. 2). Since endosonography is still in its early stages, this orientation problem could have a considerable limiting effect on the wider implementation of the method.

Any attempt to standardize the documentation of endosonograms must satisfy two criteria:

Fig. 1. Longitudinal vaginal scan of the uterus with a panoramic sector scanner. *Left:* transducer displayed at the bottom of the image; *right:* transducer displayed at the top

Fig. 2. Longitudinal vaginal scan of the anteverted uterus. *Left:* left side of image is posterior (see filled bladder); *right:* left side of image is anterior

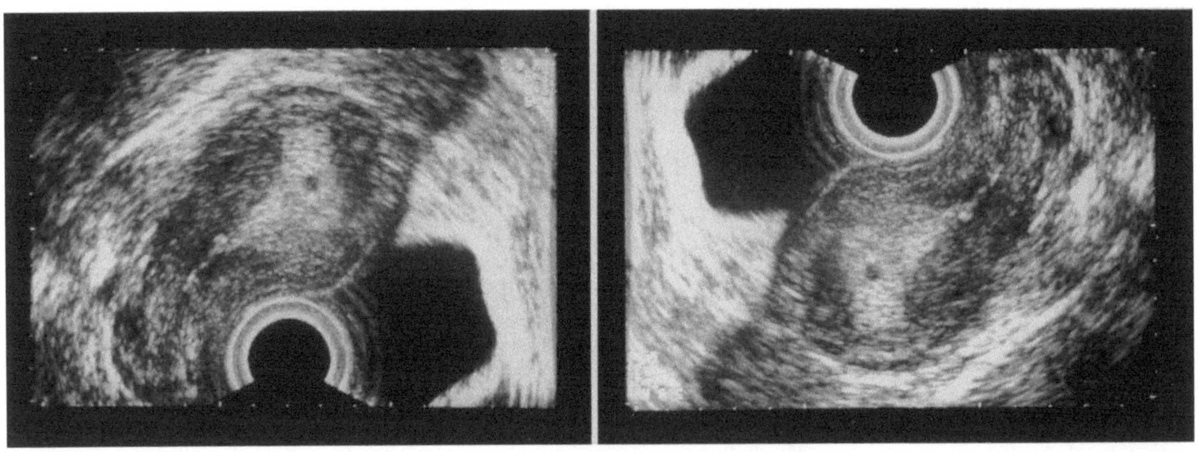

F

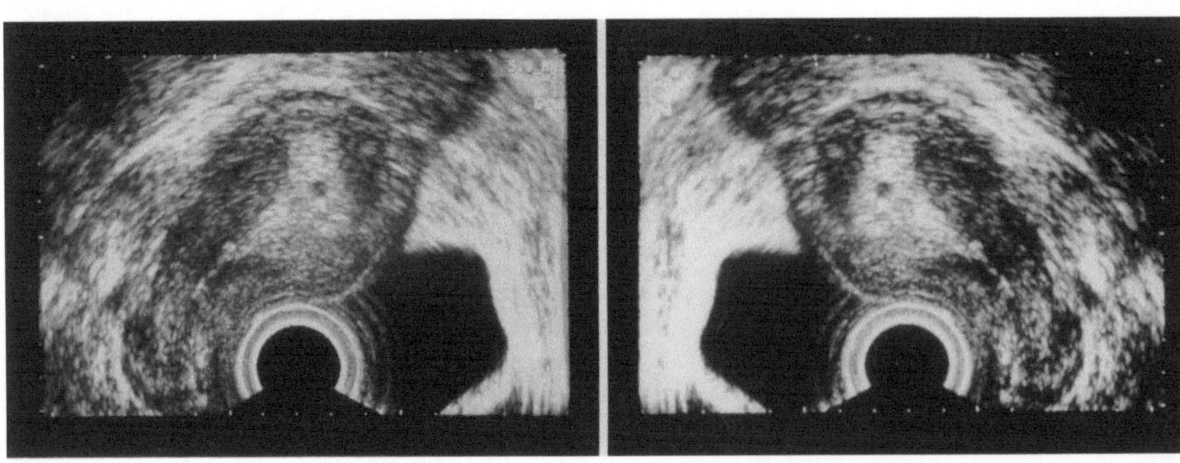

F

1. The endosonogram must be clearly distinguishable from a transabdominal scan.
2. It must be possible to correlate endosonograms with transabdominal findings.

We therefore propose the following guidelines with regard to the standardization of endosonographic documentation (Bernaschek and Deutinger 1989):

1. The transducer should always be displayed at the *bottom* of the scan, regardless of whether a linear array, curved array, or sector scanner is used. Thus, the monitor image will always be generated from below upward and is easily distinguished from the top downward image generation of abdominal scanning (Fig. 3). With the more commonly used front-viewing sector scanners and curved arrays, caudal would then be at the lower edge of the scan and cranial at the top (Fig. 4). With the less commonly used linear arrays, caudal would be on the right side of the image and cranial on the left (Fig. 5).
2. To preserve the topography familiar from abdominal ultrasound, which depicts the urinary bladder on the right side of the image, sagittal scanning with a front-viewing probe should be performed so that the left side of the scan is posterior and the right side is anterior (Fig. 6). With transaxial rotary scanners for vesical and rectal examinations, the upper edge of the scan is anterior and the lower edge posterior (Fig. 7).
3. On all other planes the right and left sides of the scan should be designated (Fig. 8) following the convention used in abdominal ultrasound, where the left side of the monitor corresponds to the right side of the patient and vice versa (Fig. 9).

The first priority in documentation is to allow immediate visual discrimination between an endosonogram and an abdominal scan, for otherwise every endosonogram would have to be individually labeled to specify the method used (e.g., transabdominal vs vaginal). This is the main justification for displaying scans with the transducer at the bottom. This convention also helps standardize documentation and interpretation; without it, the transducer would appear on the right side of a sagittal scan but at the bottom of a horizontal scan (see Fig. 3). While this would be correct in a topographic sense, it would compromise standardization and would also cause difficulties due to the short vertical dimension of the ultrasound monitor screen.

The documentation format proposed here for endosonography is analogous to that of abdominal sonography, in which the image is always displayed

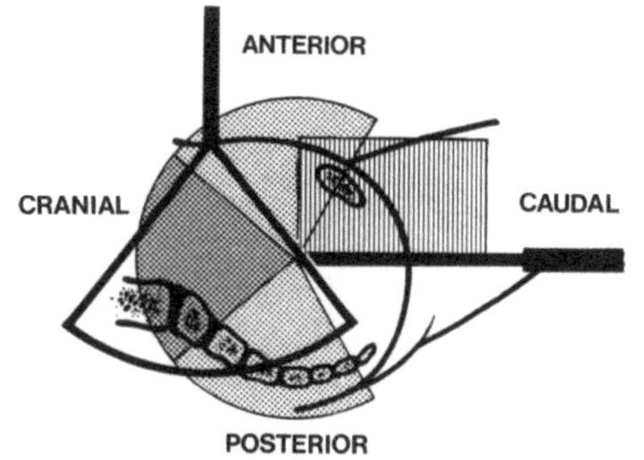

Fig. 3. Orientation of anterior-posterior and cranial-caudal on the sagittal plane for an abdominal sector scanner *(heavy border)*, a front-viewing endosonographic probe (sector scanner with small and panoramic fields of view, *shaded*), and a linear-array endosonographic probe *(hatched)*

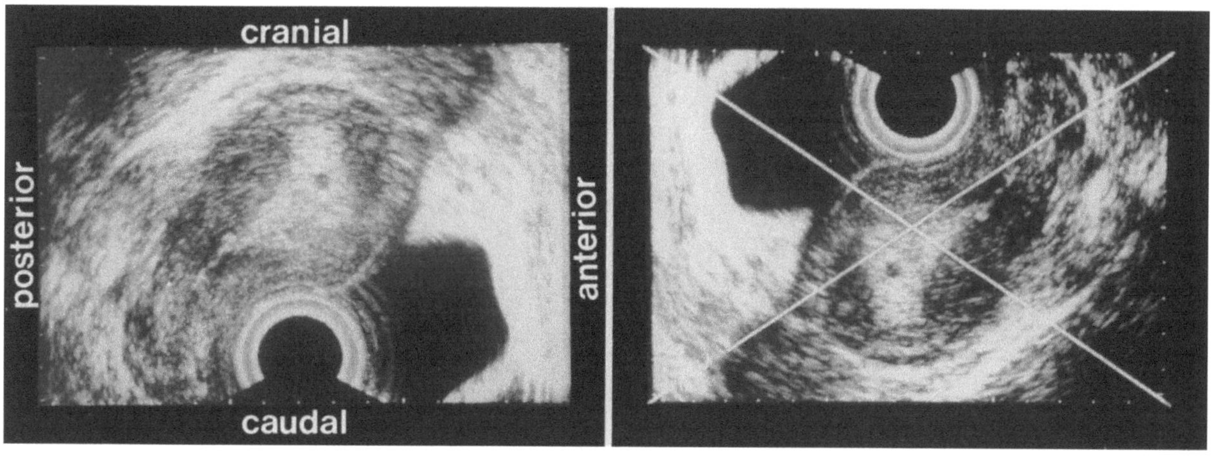

Fig. 4. Longitudinal vaginal scan of the anteverted uterus with a front-viewing panoramic sector scanner. *Left:* correct orientation; *right:* incorrect orientation

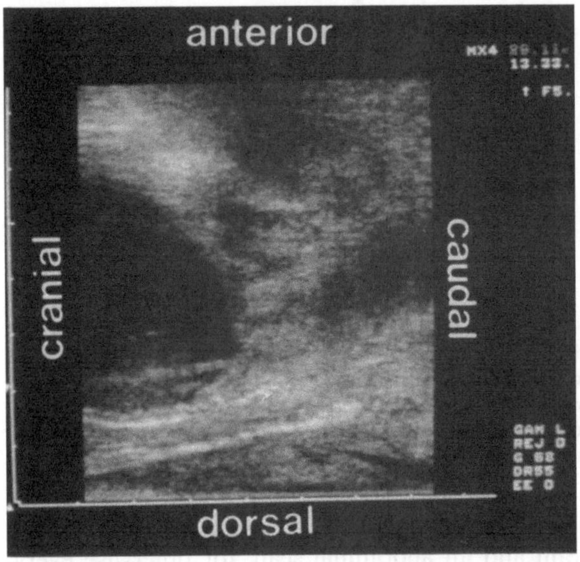

with the transducer at the top, regardless of whether the kidneys, for example, are scanned from the posterior, anterior, or lateral aspect.

Standardization should pose no difficulties, since all conventional ultrasound equipment can be adjusted so that the transducer appears at the bottom of the image, the left side of the monitor is posterior, and the right side is anterior.

◀ **Fig. 5.** Sagittal rectal scan with a linear-array probe and corresponding orientation

Fig. 6. Longitudinal vaginal scan of the anteverted uterus with a front-viewing panoramic sector scanner. *Left:* correct orientation; *right:* incorrect orientation
▼

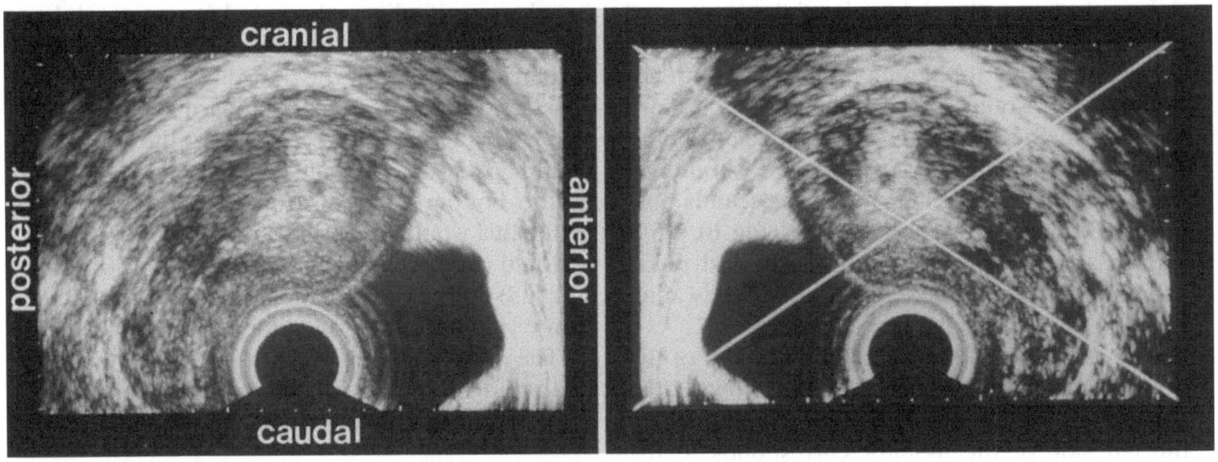

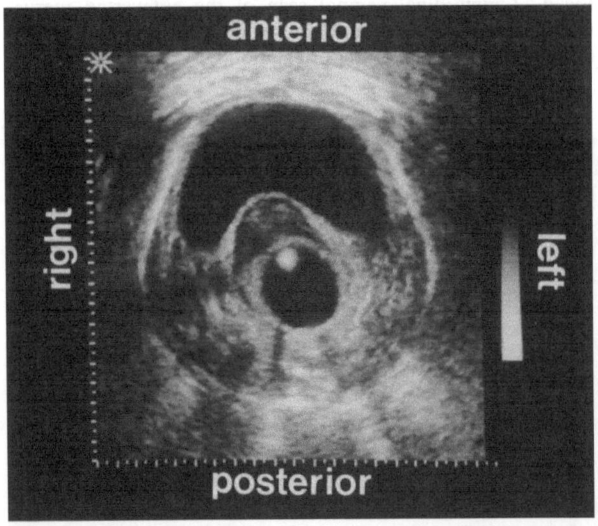

Fig. 7. Transverse rectal scan through the pelvis with a transaxial rotary scanner

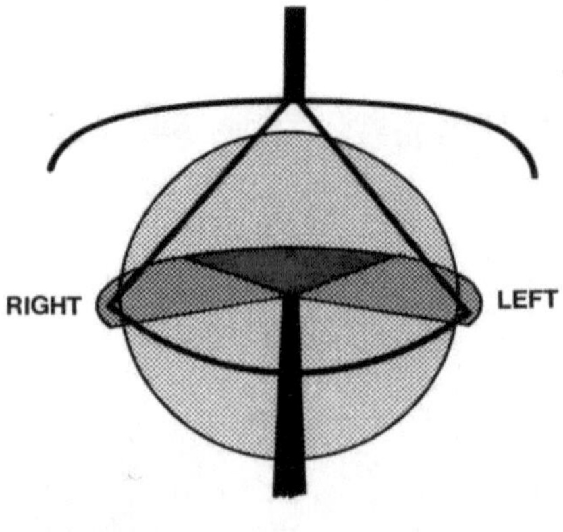

Fig. 8. Right-left assignment on the transverse and horizontal planes for an abdominal sector scanner *(heavy border),* a front-viewing endosonographic probe (sector scanner with small and panoramic fields of view, *heavy shading*), and a transaxial probe *(light shading)*

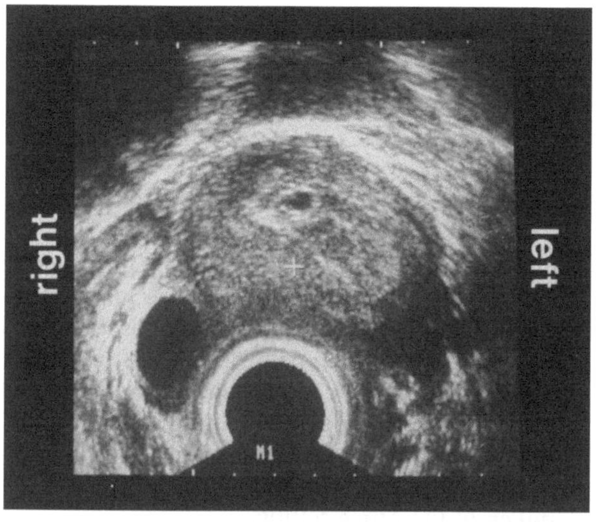

Fig. 9. Horizontal vaginal scan of the uterus, indicating right/left orientation

Reference

Bernaschek G, Deutinger J (1989) Endosonography in obstetrics and gynecology: the importance of standardized image display. Obstet Gynecol 74: 817-820

Endosonographic Procedures

The obvious advantages of endosonography give it a large range of potential applications. Vaginosonography and hysterosonography are of particular interest in obstetrics and gynecology, although rectosonography and cystosonography, long practiced routinely in urology (Watanabe et al. 1974, 1977; Holm and Northeved 1974; Gammelgard and Holm 1980), also offer significant advantages in examinations of the female pelvis. Other endosonographic procedures, such as transesophageal sonography of the heart (Strohm et al. 1980; Reifart and Strohm 1982) and transgastric or transduodenal scanning of the pancreas, porta hepatis, gastric wall, and other upper abdominal organs (Lutz and Rosch 1976; Yamanaka et al. 1982), with the possible exception of laparoscopic applications, do not expand the spectrum of ultrasound diagnosis in obstetrics and gynecology.

In the interest of clarity and uniformity, the terms for the various endosonographic examinations should follow the standard nomenclature used for endoscopic procedures (Deutinger and Bernaschek 1989):

Vaginosonography	(vaginal sonography)	vaginoscopy
Hysterosonography	(uterine sonography)	hysteroscopy
Rectosonography	(rectal sonography)	rectoscopy
Cystosonography	(vesical sonography)	cystoscopy

1 Vaginosonography

While hystero-, recto-, and cystosonography tend to have specialized applications and are reserved essentially for the hospital setting, vaginosonography opens up a wealth of new options to the office practitioner as well. A major reason for this is that vaginosonography does not require a full urinary bladder and thus avoids the organizational problems that can arise when abdominal scanning is undertaken in the office setting.

Linear-array probes are basically as effective as nonpanoramic sector scanners and curved-array systems for imaging the anteverted uterus in gynecologic patients and in early pregnancy (Fig. 1). A linear array usually cannot make a complete longitudinal scan of the uterus, however, because the cervix and body of the uterus cannot be imaged concurrently when significant anteflexion is present. The linear-array scanner is of almost no value for examining the remaining structures in the lesser pelvis, for it is blind to organs situated above the end of the vagina. The advantage of front-viewing sector scanners and curved arrays lies in their ability to image a retroverted uterus and examine the cervical region, especially in advanced pregnancy. In addition, these scanners are markedly superior to linear arrays for demonstrating extrauterine changes, especially those involving the adnexa. Forward-viewing sector and curved-array scanners with

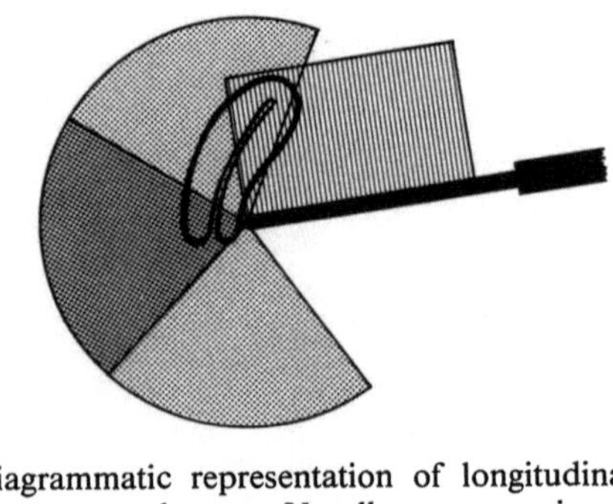

Fig. 1. Diagrammatic representation of longitudinal scans through an anteverted uterus. Usually a panoramic scanner is needed to image the whole organ. *Hatching,* linear-array scanner; *light and heavy shading,* sector scanner with small or panoramic field of view respectively

an aperture angle of 100° are the most commonly used, but their relatively narrow field of vision appears inadequate for many applications of vaginosonography, such as pelvimetry. Another drawback of these wobblers is the point origin of the field in the near zone - a significant disadvantage considering the proximity of the organs to the transducer. These scanners are decidedly inferior to "panoramic" scanners, whose 240° field of view gives better visualization of structures in the near zone and enables better orientation while providing the same amount of image detail. Panoramic scanners, then, are considered optimum for gynecologic examinations. Nonpanoramic sector scanners (wobblers) are advantageous only in that their higher frame rates permit observation of fast-moving structures. Curved array scanners have a wider origin of the field in the near zone and provide excellent resolution in this area.

Examinations of the vagina itself can be performed with linear-array scanners or with transaxial rotary scanners (rectal probes). Following the intravaginal placement of a fluid-filled balloon, these devices afford excellent detail of the vaginal wall.

Technique

The vagina provides the best access for endosonography by virtue of its size and mobility. In addition, vaginosonography is the least painful of all the forms of gynecologic endosonography. The centralized location of the upper end of the vagina within the lesser pelvis allows optimum visualization of that region using the highest possible transducer frequency. Another advantage is the natural moisture of the vaginal milieu, which eliminates the need for coupling gel.

Vaginosonography is performed with an empty or almost empty urinary bladder. This eliminates the problems of an inadequately filled bladder in abdominal sonography and avoids the well-known difficulties of having a full bladder at the right time. This also makes it possible to perform vaginosonography immediately after the bimanual pelvic examination, or even concurrent with it if desired. The examiner's free hand can be used to displace the organs of interest closer to the transducer. Another advantage of an empty bladder is the lack of distortion of intrapelvic anatomy, which can thus be evaluated in correlation with bimanual findings. This contrasts markedly with traditional abdominal ultrasound, in which the full bladder, necessary as an acoustic window, can cause significant anatomic distortion and alter the relationships of the pelvic organs. The close proximity of the vaginal probe to the organs of the lesser pelvis enables a kind of "small part scanning" of the pelvic organs with all its advantages.

The lithotomy position is best for vaginosonography, as it allows the greatest freedom of scanner manipulation. Before each examination a condom is slid over the probe as a hygienic measure and to allow immediate reuse of the probe in the next patient. A small amount of ultrasound gel is placed in the tip of the condom before it is applied (condoms without a reservoir have proven best for front-viewing scanners). Intravaginal coupling gel is not essential, though it may facilitate insertion in some patients. The scanner tip is positioned in the anterior vaginal fornix for examination of the uterus, or in the posterior fornix if the uterus is retroverted. The adnexa are scanned from the lateral vaginal fornices.

Vaginosonography is well tolerated by almost all patients after brief preliminary counseling. Most patients compare the procedure to ordinary bimanual palpation. The great majority of women who have had numerous vaginal and transabdominal ultrasound examinations, especially for daily supervision of follicular growth, express a preference for vaginosonography. The examiner, too, comes quickly to appreciate the short examination time and the absence of the delay between the bimanual examination (with an empty bladder) and the traditional ultrasound examination (with a full bladder). When the bimanual examination is completed, the patient can simply remain on the table for vaginosonography. Any extra time required for the endosonographic procedure is usually offset by the ability to make a more rapid and accurate diagnosis and a more rapid causal assessment of abnormal palpable findings. One might say that vaginosonography first gives the "gynecologist's eye" as the palpating finger is sometimes called, its ability to see clearly.

On the negative side, the current size of vaginal probes still precludes vaginosonography in patients with an intact hymen or a very narrow vagina (vaginal stenosis).

Another factor to be considered is the limited depth of sound penetration caused by the higher transducer frequency. Large lesions, such as tumors, can be detected by vaginosonography only if they project low enough into the pelvis or can be displaced into that region with the free hand.

Examination Procedure

The scanner is inserted into the vagina until the structures of interest are visualized in the focal zone of the transducer.

The examination begins with a sagittal view of the lesser pelvis, whereupon the scanner shaft is angled laterally (left and right) to scan through the uterus on its longitudinal axis. Demonstration of the endometrial echo is usually helpful at this stage. The back-and-forth, fanlike scanning movement provides all essential information that is available on the sagittal plane.

When a sector or curved-array device is used, the scanner shaft is next rotated 90°. In the anteverted uterus, the cervical region is visualized on a horizontal coronal scan, and then the probe handle is gradually lowered to move the scan plane up through the uterus, ending with an almost transverse scan of the fundus (Fig. 2).

A significant advantage of vaginosonography in patients with pain in the lesser pelvis is that the site of maximum pain can be localized by applying gentle pressure with the probe tip and then evaluated under vision. Moreover, a sudden, gentle thrusting movement of the probe, comparable to percussion, can assist the differential diagnosis of adhesions in the lesser pelvis. When adhesions are present, the individual organs will not glide on the pelvic wall or one another, as is normally the case, when pressure is applied to the uterus or adnexal region.

Whereas the anteverted uterus is scanned from the anterior fornix, the retroverted uterus is examined from the posterior fornix. Again the organ is surveyed in transverse sections by starting with a coronal scan through the cervical region, but this time the probe handle is gradually raised until the fundus is reached.

Following evaluation of the uterus, the adnexa are scanned from the lateral vaginal fornices. Usually the examination begins with a horizontal coronal scan, which is then systematically moved through the adnexa by angling the probe upward and downward. Structures of particular interest in the adnexal region are then evaluated on vertical coronal scans while the scanner shaft is angled to the left and right.

2 Hysterosonography

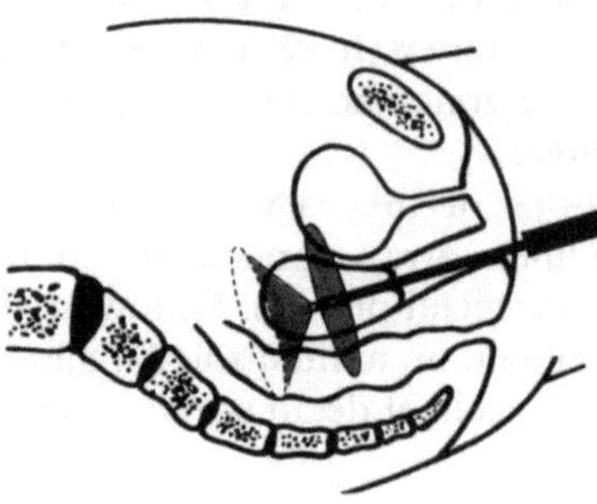

Hysterosonography is generally performed with slender, transaxial rotary scanners with a 360° angle of vision (same type used in cystosonography). To permit evaluation of the fundus, most probes for hysterosonography can switch their scan angle from the standard 90° to 135° forward (Fig. 3). In most scanners the shaft is rigid, and only the transducer rotates at the tip. But it is also possible to have the shaft rotate with the transducer, or to pass the rotating shaft into the uterus through a hysteroscope sheath. A cover for the transducer was originally developed to protect the endometrium but has not

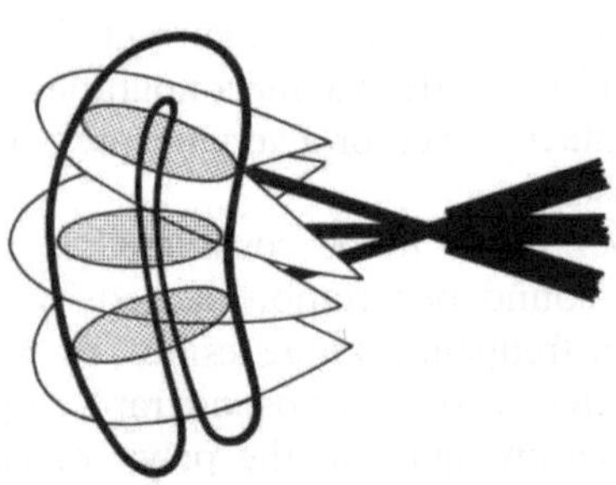

Fig. 2. Serial scans with a front-viewing probe enable a complete transaxial survey of the uterus owing to the empty bladder (comparable to the scans obtained in abdominal sonography)

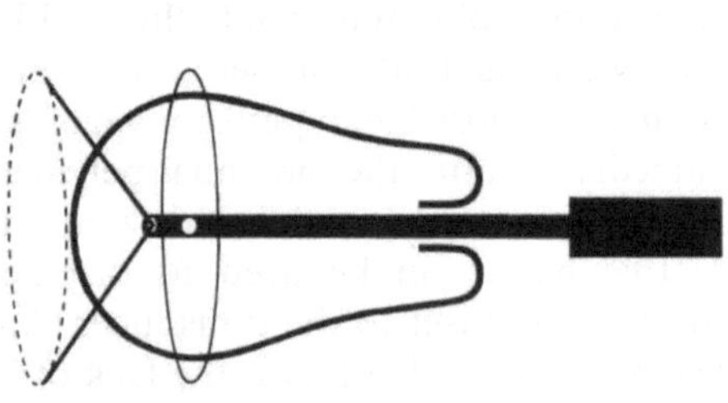

Fig. 3. Hysterosonography is usually performed with a transaxial rotary scanner. Switching to a 135° forward view is useful for evaluating the uterine fundus

proved essential. Neither is it necessary to instill fluid through a resectoscope sheath. Variable transducer frequencies have proven useful due to the relatively strong sound attenuation by the uterine musculature. Thus, very high frequencies can be used to provide optimum detail in the near zone, and lower frequencies can be used to define the boundaries of an enlarged uterus.

Technique

Prior to hysterosonography the bladder should be emptied and the vagina cleaned with an antiseptic. For subsequent insertion of the probe, the anterior lip of the cervix is grasped with a tenaculum and the cervical canal dilated to Hegar 8–10. The general anesthesia required for this dilation is a major disadvantage of hysterosonography compared with all other endosonographic procedures in gynecology. After the uterus has been sounded for length, the previously sterilized probe is passed into the uterine cavity to the maximum determined depth (the depth of insertion is read on a scale on the scanner shaft). Serial transaxial scans of the uterus are taken by gradually withdrawing the probe. The depth of probe insertion relative to the external os uteri is indicated on recorded scans.

Because of the need for general anesthesia, hysterosonography is generally performed in conjunction with diagnostic dilation and curettage.

3 Rectosonography

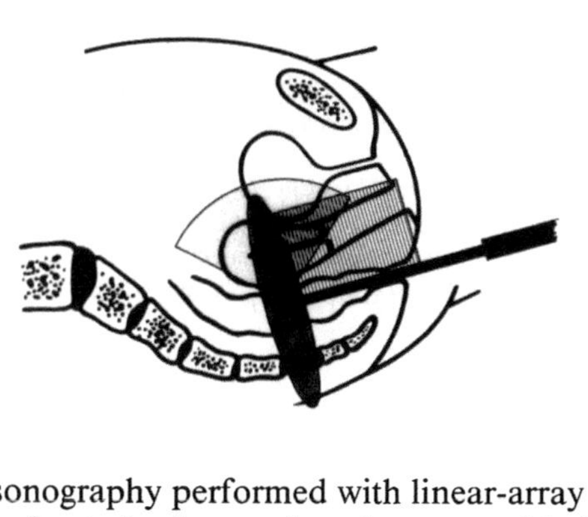

Rectosonography performed with linear-array scanners or front-viewing probes does not add significantly to evaluations of the uterus and vagina compared with the simpler and better-tolerated procedure of vaginosonography. Rectosonography does offer advantages, however, in cases where transverse or longitudinal scans are desired of structures that are not accessible vaginally due to inadequate vaginal length, i. e., structures located higher in the lesser pelvis.

Rectosonography causes approximately the same discomfort as rectal palpation, and almost all patients tolerate the procedure reasonably well after brief counseling. Analogous to a digital examination, rectosonography may be difficult or impossible in the presence of painful hemorrhoids, tight neoplastic strictures, or irritation of the rectal mucosa secondary to irradiation.

Technique

Following bowel preparation as for rectoscopy, approximately 400 ml of physiologic saline is instilled into the urinary bladder if the latter is not already full. This is done both to straighten the normally anteverted uterus and to provide an additional aid to orientation besides the rectal balloon and pelvic wall. As in vaginosonography, the lithotomy position is preferred as it allows the greatest freedom of scanner manipulation. Preliminary rectal palpation is helpful for assessing the course of the rectum and identifying any strictures that are present. The probe is first inserted as far into the rectum as possible, normally to a depth of between 15 and 20 cm. If insertion proves difficult, it may be helpful to insert the probe through a rectoscope sheath. Once the scanner is in place within the rectum, a balloon previously placed over the end of the scanner is filled with about 50 ml of liquid so that all air is expelled. Besides providing a short water-coupling path, the balloon distends the rectal wall slightly and protects it from injury. The tip of the scanner rotates freely in the rectal balloon while the scanner shaft rotates in a stationary sheath, similar to the resectoscope sheath used in cystosonography. In other probes the rotating scanner is enclosed in a fixed sheath that is applied directly to the rectal wall. In these types of probes and others, a condom with a small amount of coupling gel in its tip is slipped over the entire scanner shaft as a hygienic precaution. The scanner is gradually withdrawn while making a series of transaxial scans whose depth relative to the anus is read on the scanner shaft. In this way the pelvis is systematically explored with the filled bladder, rectal balloon, and pelvic wall serving as landmarks.

4 Cystosonography

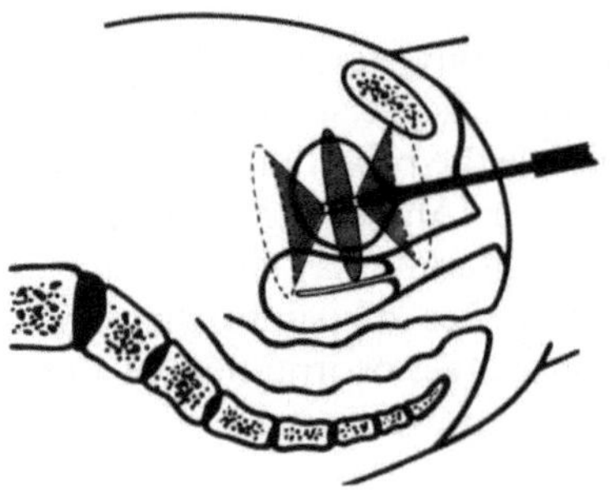

Cystosonography has the least relative importance of the endosonographic procedures in obstetrics and gynecology. It is performed with a transaxial rotary scanner passed into the urinary bladder via a CH 24 resectoscope sheath, through which fluid is subsequently instilled to fill the bladder to the desired extent. For imaging the neck and fundus of the bladder, most vesical probes can switch their scan angle from the standard 90° to 45° backward or 135° forward (Fig. 4). Thus, almost the entire bladder wall can be examined by adding small lateral movements of the scanner shaft. It is important, however, that the beam strike the bladder wall at a perpendicular angle; a tangential incidence makes accurate evaluation difficult. Because the conditions of the examination are more favorable than in abdominal scanning (i. e., water coupling by the full bladder, ability to use higher frequencies due to the small penetration required), cystosonography can provide exceptionally clear views of the entire bladder wall and perivesical region. This is considered a major advantage over traditional cystoscopy, which can only demonstrate mucosal changes.

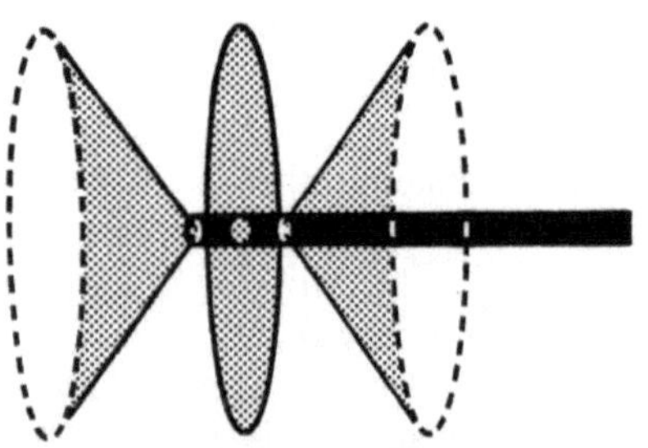

Fig. 4. Cystosonography is usually performed with transaxial rotary scanners. Switching to a 135° forward view or 45° backward view is useful for evaluating the bladder neck and fundus

A limiting factor in the examination is the caliber of currently available probes, which must be introduced through a Ch 24 sheath (though this causes relatively few problems in the female urethra). Moreover, greater attention must be given to antisepsis in cystosonography than in rectosonography or vaginosonography.

References

Deutinger J, Bernaschek G (1989) The endovaginal debate. J Ultrasound Med 8: 817-818

Gammelgaard J, Holm H (1980) Transurethral and rectal ultrasonic scanning in urology. J Urol 124: 863

Holm HH, Northeved A (1974) A transurethral ultrasonic scanner. J Urol 111: 238-241

Lutz H, Rösch W (1976) Transgastroscopic ultrasonography. Endoscopy 8: 203

Reifart N, Strohm WD (1982) Detection of atrial septum defects by transoesophageal two-dimensional echocardiography with a mechanical sector scanner. In: Hanrath P, Bleifeld W, Sougquet J (Eds) Cardiovascular diagnosis by ultrasound. Transoesophageal, computerized, contrast, Doppler echocardiography. Nijhoff, Boston, pp 247-250

Strohm WD, Phillip J, Hagenmüller F, Classen M (1980) Ultrasonic tomography by means of an ultrasonic fiberendoscope. Endoscopy 12: 241

Watanabe H, Igar D, Tanahashi Y (1974) Development and application of new equipment for rectal ultrasonography. J Clin Ultrasound 2: 91-98

Watanabe H, Saitoh M, Mishina T (1977) Mass screening program for prostatic diseases with rectal ultrasonography. J Urol 117: 746

Yamanaka T, Sakai H, Yoshida Y (1982) Ultrasonic endoscopy for the diagnosis of abdominal lesions. I. Clinical evaluation of an ultrasonic gastroendoscope with electronic linear scanning system (prototype). Gastroenterol Endosc 24: 598

Normal Early Pregnancy

1 Chorionic Cavity

1.1 General

It is important to use consistent terms when describing the age of a conceptus. While gestational age in embryology is measured from the time of fertilization, the reference point in obstetrics is the start of the last normal menstrual period. Thus the duration of pregnancy is stated in terms of menstrual age and is measured in weeks. European obstetricians customarily state the current week of fetal development rather than the completed week, so gestational age data tend to be 1 week higher than equivalent data stated in the American and English obstetric literature and in pediatrics. The "5th week of gestation," for example, denotes the period from 4 weeks + 1 day to 4 weeks + 7 days.

For a better understanding of obstetric endosonography and especially, of early pregnancy detection by means of vaginosonography, it is helpful to review some fundamentals of embryology (Langman 1970). By day 17 or 18 of the menstrual cycle, the morula, measuring about 0.1 mm in diameter, enters the uterine cavity in the 16-cell stage and develops into a blastocyst. On about day 19 or 20 the blastocyst attaches to the endometrium, its diameter increasing to 0.2-0.3 mm through fluid absorption. Superficial implantation of the blastocyst is accomplished by day 21 or 22. At this time the trophoblast and embryoblast begin to develop along different lines. The trophoblast embeds more deeply into the endometrium, differentiating into two layers, the syncytiotrophoblast and the cytotrophoblast, while the cells of the embryoblast form the bilaminar germ disk. Implantation is completed by day 24 or 25 of the menstrual cycle, and the epithelial defect in the uterine mucosa closes. Meanwhile the endometrium becomes edematous and thickened. The site of implantation of the blastocyst is almost always in the upper part of the uterine cavity, close to the fundus, and usually to one side of the midline - a circumstance that can be documented sonographically and especially by vaginosonography in early pregnancy. After implantation has occured, the endometrium is referred to as the decidua and consists of three regions: the decidua capsularis, which covers the blastocyst on the side facing the uterine cavity; the decidua basalis, separating the blastocyst from the uterine muscular wall; and the rest of the endometrium, henceforth called the decidua parietalis.

The blastocyst contains two cavities: the true blastocystic cavity, later termed the primary yolk sac, and the much smaller amniotic cavity. The two

Fig. 1. Blastocyst at approximately 12 days post menstruation age. The extraembryonic mesoderm is permeated by lacunas that will coalesce to form the extraembryonic coelom (the future chorionic cavity)

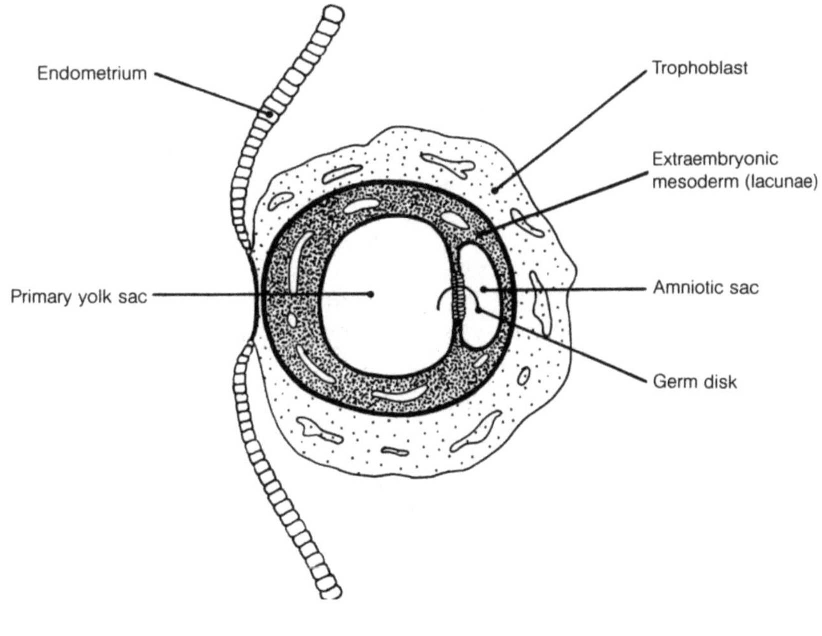

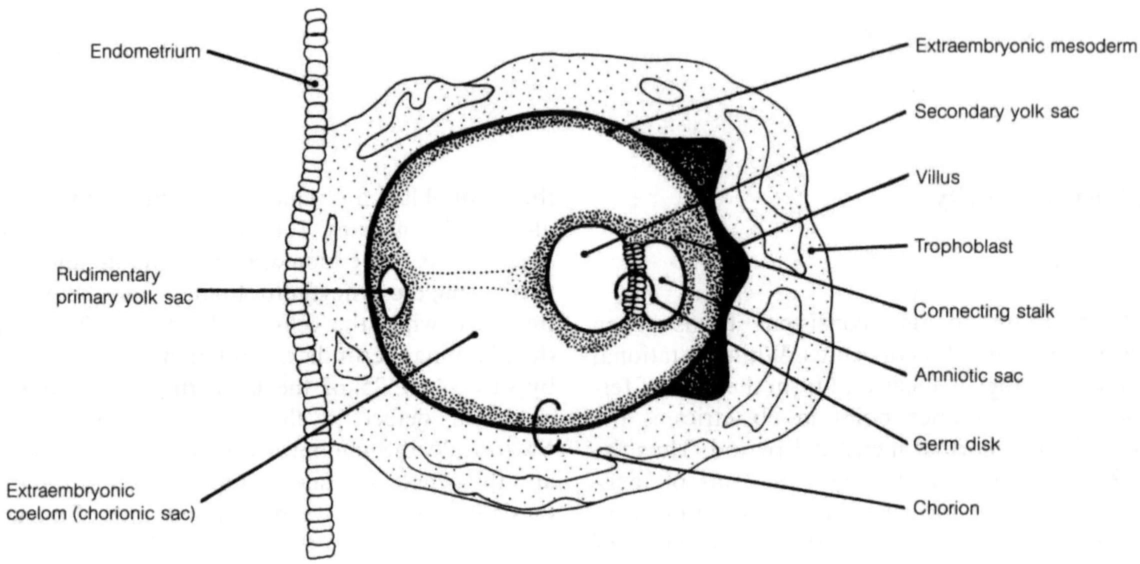

Fig. 2. Blastocyst at approximately 14 days post menstruation with remnants of the primary yolk sac and the newly-formed secondary yolk sac

are separated by the germ disk (Fig. 1). Trophoblastic cells split off to form the extraembryonic mesoderm, which surrounds both the amniotic cavity and the primary yolk sac. The small clefts in the extraembryonic mesoderm coalesce on about day 20 of the menstrual cycle to form the extraembryonic coelom (Fig. 2). While the latter continues to enlarge, the primary yolk sac regresses and is replaced by the smaller secondary yolk sac. The yolk sac and amniotic cavity are joined to the trophoblast by the connecting stalk, which is the precursor of the umbilical cord. The extraembryonic mesoderm and adjacent trophoblasts comprise the chorion and its cavity. By about day 29 or 30 the chorionic cavity reaches a diameter of 1-2 mm and becomes the first structure of early pregnancy that can be visualized with vaginosonography.

With abdominal ultrasound it is sometimes possible (given optimum conditions and up-to-date equipment) to detect the chorionic cavity when it has reached a diameter of 2-3 mm (Hansmann et al. 1985). However, it is generally reported that the chorionic cavity (commonly referred to as the gestational sac) is identified no earlier than the middle of the 7th week of pregnancy. A variety of factors such as obesity, myomatous changes, and especially a retroflexed uterus can make very early pregnancy detection by abdominal ultrasound difficult or impossible.

1.2 Vaginosonography

Vaginosonography offers many advantages over traditional abdominal scanning, making it the method of choice for early detection of the intrauterine chorionic cavity. Owing to the transvaginal approach and the proximity of the transducer to the organ of interest, there is less attenuation of the ultrasound pulse by absorption, specular and nonspecular reflection, scattering, and refraction than with abdominal scanning. Therefore the echoes are technically easier to amplify and process. For example, an abdominally applied 5-MHz linear-array transducer does not provide the same resolution as a vaginally inserted 5-MHz probe, which permits the reliable identification and evaluation of intrauterine structures no larger than 2 mm in diameter.

Because growth curves for the chorionic cavity as measured by abdominal ultrasound usually begin in the 6th or 7th week of gestation, it is crucial that standard vaginosonographic measurements be established from the time the menstrual period is missed. It is also important to acquire quantitative data on the ability of vaginosonography to identify the chorionic cavity, since such data are essential for a meaningful evaluation of the capabilities of the vaginal method. They are also crucial for the correct interpretation of findings in everyday practice.

Our own investigations (Bernaschek et al. 1988) were performed with a 5-MHz vaginal scanner in 24 pregnant women who knew precisely their dates of conception or last menstrual period. The size of

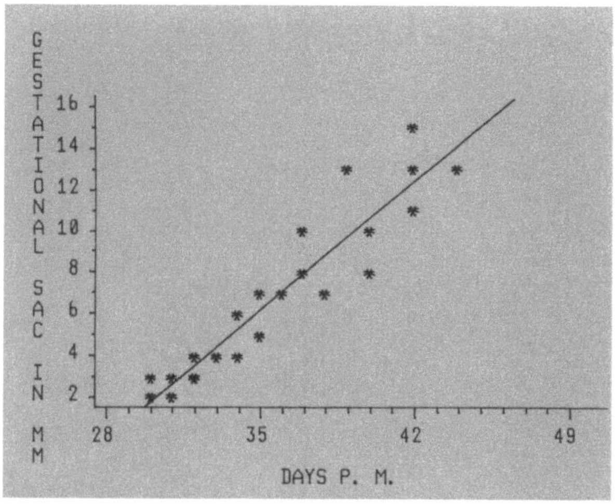

Fig. 3. Relation between gestational sac size and duration of amenorrhea ($y = 0.871x - 24.295$, $r=0.877$, $n=24$). The earliest detection was made on the second day after the missed menstrual period. Days p.m., postmenstrual days

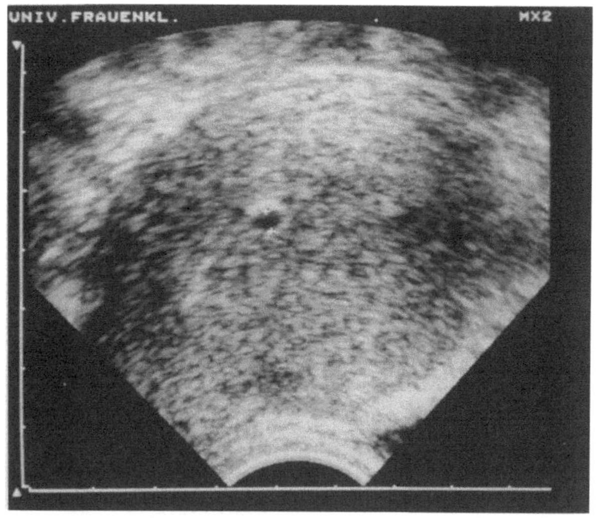

Fig. 5. Endosonographic appearance of the chorionic cavity at 4 weeks + 6 days on a longitudinal scan through the uterus. Diameter of chorionic cavity is 4-5 mm

the chorionic cavity was determined by averaging the largest transverse, anteroposterior, and longitudinal diameters as measured between the inside borders of the trophoblast (chorion). By performing a regression analysis from the vaginosonographically measured diameters of the intrauterine gestational sac and the duration of amenorrhea in women with well-documented dates of conception or last menstruation (from our sterility clinic), we noted a linear relationship between gestational age and the size of the gestational sac (Fig. 3). From this relationship we find that the mean diameter of the gestational sac on the 2nd day after a missed period (i.e., 30 days of amenorrhea) is 2 mm and could serve as a criterion for the early detection of pregnancy (Fig. 4). De Crespigny et al. (1988) also reported the earliest detection of a gestational sac at 4 weeks + 2 days. Reports claiming detection of a 3-mm gestational sac 1 week after conception or 1 week before a missed period (Haid et al. 1985) or a 5-mm chorionic cavity at the time of a missed period (Popp and Lemster 1986) are certainly too optimistic and are not consistent with embryologic principles.

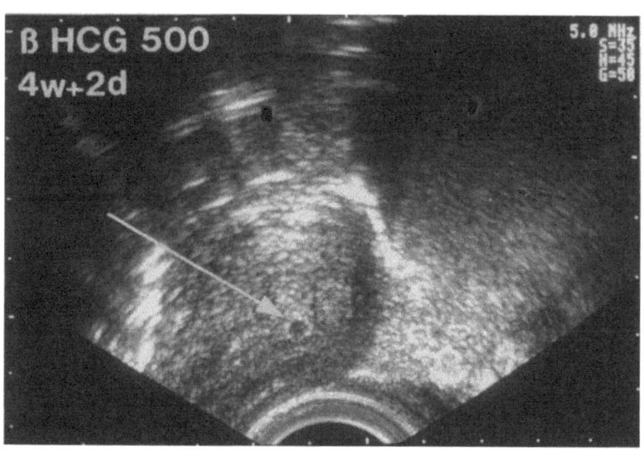

Fig. 4. Endosonographic appearance of the chorionic cavity *(arrow)* 2 days after a missed period (30 days of amenorrhea). Corresponding β-hCG level is 500 mIU/ml

By the end of the 5th week of gestation we were able to detect the chorionic cavity by vaginosonography in all intact intrauterine pregnancies in our series. Similar results are reported by Rempen (1987), Terinde and Kozlowski (1988), and De Crespigny et al. (1988). Thus, the main advantage of vaginosonographic visualization of the chorionic cavity is that it advances pregnancy detection by 1-2 weeks. The chorionic sac size is 4-5 mm by the end of the 5th week of gestation (Fig. 5). At that time the chorionic cavity appears as a usually asymmetric located, ring-shaped structure in the region of the uterine fundus (Fig. 6). Its shape and location serve to distinguish it from the slightly more echogenic, slitlike uterine cavity. The more strongly echogenic trophoblastic rim is initially narrow and of uniform thickness (see Fig. 4). The chorionic cavity is approximately circular up to a diameter of about 1 cm, but as it enlarges at a rate of about 1 mm per day, it soon acquires an oval shape. In the absence of myomas, this distortion is caused entirely by the intrinsic tension of the uterus and the variable pressure from the vaginal probe. Despite an empty bladder, the chorionic cavity may

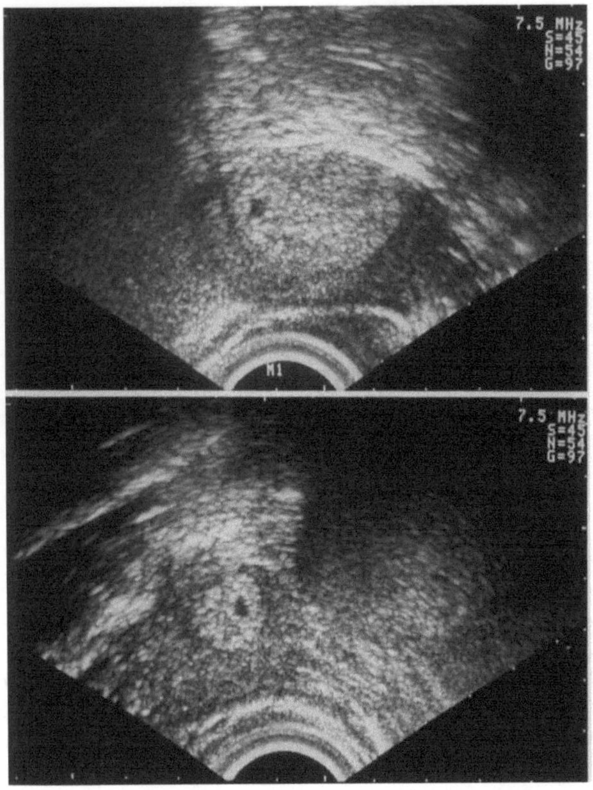

Fig. 6. Pregnancy at 4 weeks + 5 days. *Top:* transverse scan through the uterus in the fundal region. The chorionic cavity is located excentrically near the right tubal outlet. *Bottom:* parasagittal longitudinal scan in the same patient

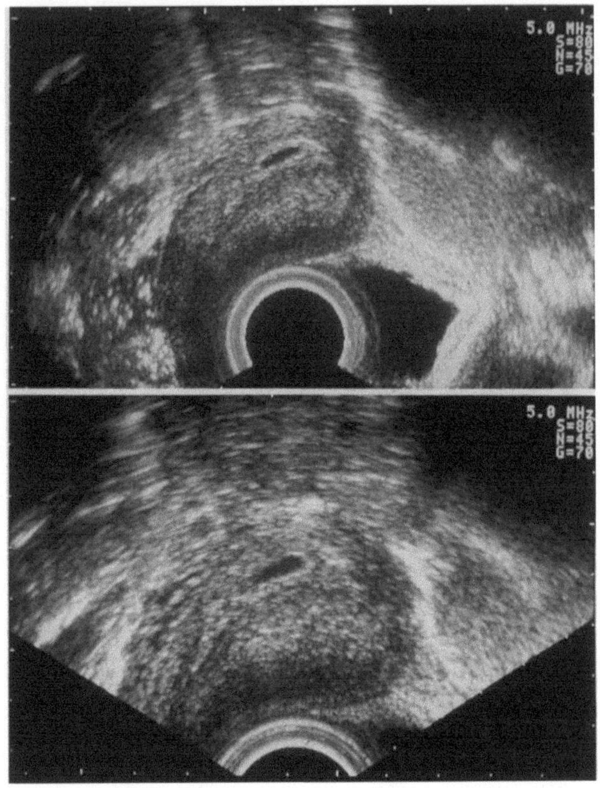

Fig. 7. Pregnancy at 5 weeks + 0 days. *Top:* longitudinal scan through the uterus showing a markedly elongated chorionic cavity. Follow-up a few days later showed the embryo and cardiac activity. *Bottom:* zoom image of the same uterus

display a markedly oval shape that does not necessarily signify an abnormal pregnancy (Fig. 7). By the 6th week of gestation the trophoblast usually shows an asymmetric thickening reflecting the differentiation of the chorion into the chorion laeve and chorion frondosum. This thickening covers about two-thirds the circumference of the gestational sac (Fig. 8).

Currently, vaginosonography seems to have overcome the drawbacks of abdominal scanning with regard to the early and reliable diagnosis of intrauterine pregnancy. Our findings indicate that it is possible to identify the intrauterine gestational sac as early as the 2nd day after a missed menstrual period.

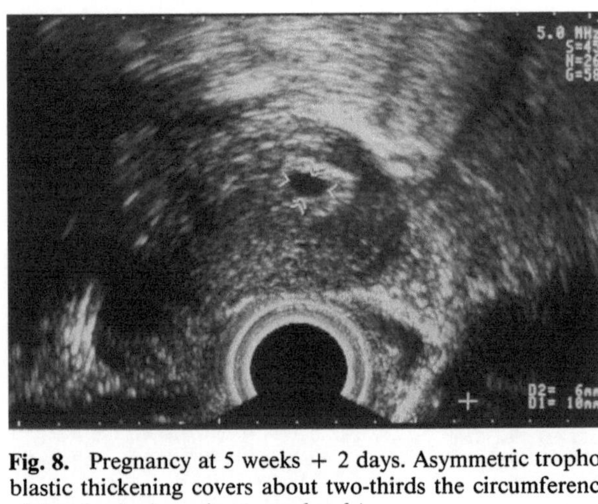

Fig. 8. Pregnancy at 5 weeks + 2 days. Asymmetric trophoblastic thickening covers about two-thirds the circumference of the chorionic cavity *(arrowheads)*

1.3 Recognition of Pseudogestational Sac

Vaginosonography allows a more accurate differentiation between an intact gestational sac and a pseudosac owing to its ability to visualize the endometrium with greater clarity in a higher percentage of cases. The shape and especially the location of a saclike structure can also be determined with greater precision. An intact, early chorionic cavity is usually distinguished by its eccentric location in the endometrium (Fig. 6), predominantly in the region of the fundus and cornua. Also, the intact chorionic cavity can usually be clearly distinguished from the uterine cavity. While the early gestational sac is usually surrounded by a bright, well-defined border of uniform thickness, by the 6th week of gestation the sac shows asymmetric thickening due to differentiation of the chorion. This appears as a wide, echogenic crescent spanning two-thirds the circumference of the gestational sac, which by now is usually oval (Fig. 8). The pseudogestational sac, on the other hand, appears as a symmetrical fluid collection in the uterine cavity and presents an elongated or fusiform shape on longitudinal scans (Fig. 9).

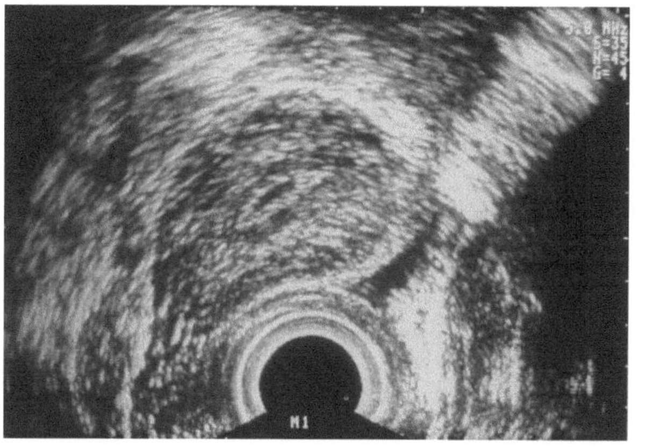

Fig. 9. Longitudinal scan through the uterus showing a pseudogestational sac in the presence of a tubal pregnancy

1.4 Correlation Between Chorionic Cavity and β-hCG

The β-hCG concentration in the venous blood is usually determined by radioimmunoassay. This assay may be calibrated against the Second International Standard for Chorionic Gonadotropin or, less commonly, against the First International Reference Preparation. When a comparison between these international standards is wanted, the conversion factor is 1 mIU/ml of β-hCG (Second International Standard) = 2.2 mIU/ml (First International Reference Preparation). In our study (Bernaschek et al. 1988) we used a commercially available radioimmunoassay (Serono, West Germany), which was calibrated against the Second International Standard. The sensitivity of the assay was 2 mIU/ml.

The radioimmunoassay of the β-subunit of hCG is considered the most sensitive of the endocrine tests currently available. Since the production of β-hCG begins very early in pregnancy and indeed may even precede the time of nidation, this test seems to be the earliest indicator of pregnancy. But despite the early diagnosis, this test gives no information about the location of the pregnancy. Thus, ultrasound evaluation of the early pregnancy may be indicated to provide optimum care for the patient.

Batzer et al. showed in 1983 that abdominal scanners may demonstrate an intrauterine pregnancy when the gestational sac has a mean diameter of 6 mm 20 days after ovulation, and the serum β-hCG has reached a level of 570 mIU/ml. In one case Hansmann et al. (1985) demonstrated a small, well-defined, ringlike structure 2 mm in size with abdominal ultrasound when the β-hCG level was only 216 mIU/ml. Cacciatore et al. (1988) were able to detect a gestational sac on abdominal scans at a β-hCG level of 894 mIU/ml. More recently Nyberg et al. (1985) used abdominal scanners to correlate maternal serum β-hCG levels with sonographic intrauterine findings in 49 normal, early intrauterine pregnancies. The smallest gestational sac size that they were able to detect was 4 mm, and the corresponding β-hCG level was 357 mIU/ml. However, the identification of such a characteristic gestational sac at this early stage represented the exception rather than the rule. A variety of anatomic and physical factors such as obesity, adhesion, a retroverted uterus, an inadequately filled bladder, and external pressure on the uterus from a full bladder prevent a clear identification of this structure. This lack of accuracy did not allow the routine use of abdominal sonography to establish the diagnosis of a very early intrauterine pregnancy.

1.4.1 Earliest Detection with Vaginosonography

The ability to use higher ultrasound frequencies and the resulting improvement in image detail necessarily allow for earlier gestational sac detection with vaginosonography, implying a correlation with lower serum levels of β-hCG.

Our own studies were performed in 40 women randomly selected from our clinical population. Most patients had a workup because of unclear menstrual dates. The lowest serum concentration of β-hCG at which a gestational sac could be plainly identified was 141 mIU/ml (Fig. 10). However, 6 of the 40 pregnancies that were identified by β-hCG determination could not be visualized by vaginosonography because the gestational sac size was below the resolution of the scanner (Fig. 11). Consequently these intrauterine pregnancies with β-hCG levels up to 280 mIU/ml were not diagnosed at this early stage in our study. Nevertheless, we

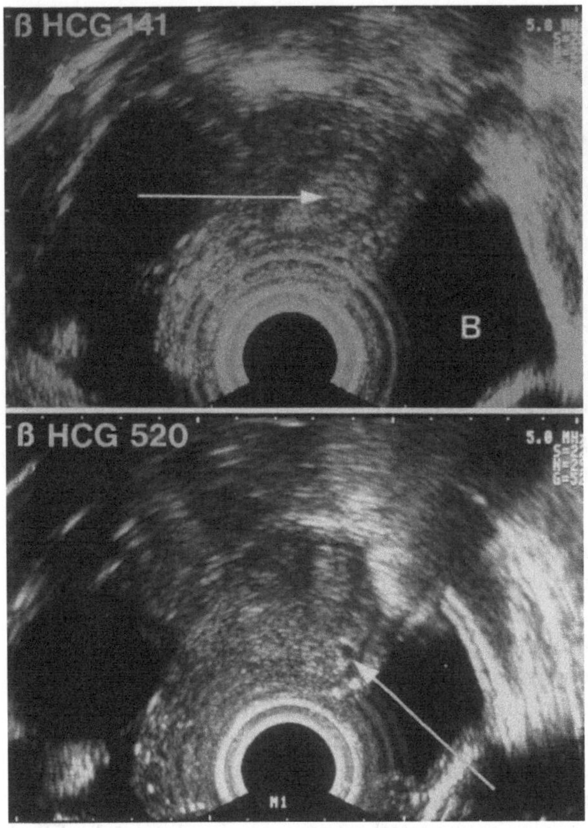

Fig. 10. *Top:* longitudinal scan through the uterus demonstrating a chorionic cavity *(arrow)* only 2 mm in size. Corresponding β-hCG level is 141 mIU/ml. *Bottom:* same case 2 days later. Corresponding β-hCG level is 520 mIU/ml

always correctly identified a pregnancy by vaginosonography when serum β-hCG levels were greater than 300 mIU/ml.

Similarly, Haid and Martin (1980) found that the earliest point at which the gestational sac could be detected was at a β-hCG level of 200 mIU/ml. Nyberg et al. (1988) were able to visualize a gestational sac when the β-hCG level was only 115 mIU/ml. The findings of Blumenfeld et al. (1988) were unusual, in that these authors were unable to detect an intrauterine sac until the β-hCG level reached 420 mIU/ml.

In conclusion, vaginosonography seems to combine the diagnostic advantages of two methods of early pregnancy detection: the sensitivity of serum β-hCG determinations and the accurate and reliable visualization of the conceptus with ultrasound. Although radioimmunoassay remains the most sensitive test, vaginosonography can identify a pregnancy at times when it becomes necessary for the woman to know whether she is pregnant or not.

1.4.2 Discriminatory Limit

The "discriminatory limit" refers to the level of serum β-hCG above which an intrauterine preg-

Fig. 11. Graph showing the correctly identified intrauterine pregnancies and corresponding β-hCG levels. The six filled circles indicate false-negative cases

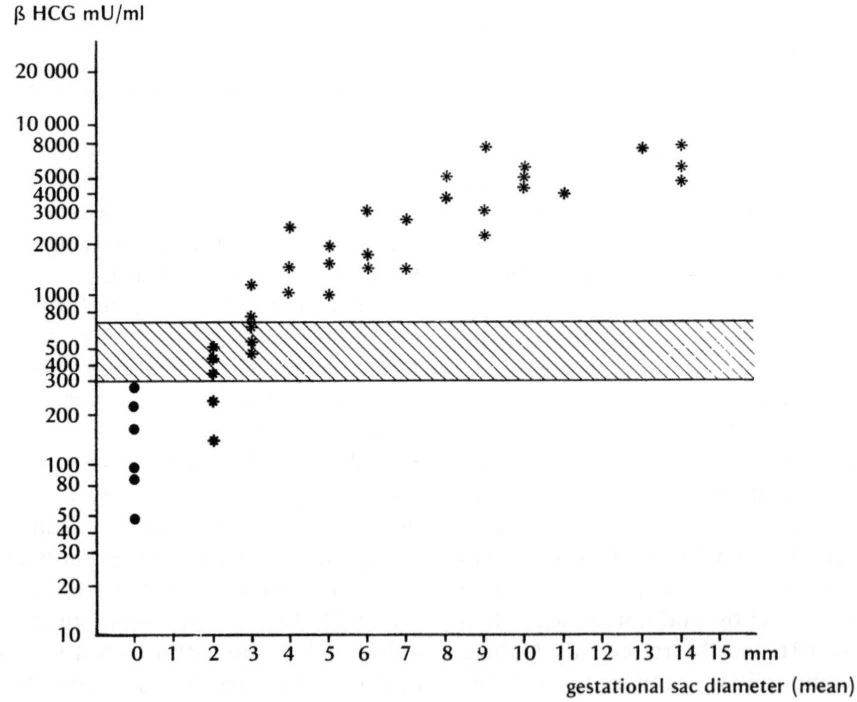

nancy can consistently be detected with ultrasound. Failure to detect an intrauterine gestational sac above the discriminatory limit should always suggest an ectopic pregnancy or recent spontaneous abortion as a differential diagnosis.

By correlating abdominal ultrasound findings with a simultaneous quantitative β-hCG assay, Kadar et al. (1981) reported that a gestational sac should be demonstrated when the β-hCG level exceeds 6500 mIU/ml. This level corresponds to a level of approximately 3250 mIU/ml calibrated against the Second International Standard. More recently, Nyberg et al. (1985) found that gestational sacs are consistently detected when the β-hCG level is 1800 mIU/ml (Second International Standard) or more. In addition, Nyberg et al. (1987) observed a strong correlation ($p < 0.001$) between gestational sac size and the quantitative β-hCG level during the first 8 weeks of pregnancy. An β-hCG level of 1800 mIU/ml (Second International Standard) corresponds to a gestational sac that is approximately 5–6 mm in diameter, while an β-hCG level of 3250 mIU/ml corresponds to an average sac size of approximately 8 mm (Nyberg et al. 1987). Since gestational sac size and quantitative β-hCG levels vary proportionately during early pregnancy, the β-hCG level reported will depend on the resolving capacity of the sonographic equipment and on the scanning technique. This validates the prediction of Nyberg et al. (1985) that "...any apparent β-hCG discriminatory zone will most likely be a function of current or future sonographic technique and resolution. Given the progressive rise in β-hCG levels and normal growth of the gestational sac, it is reasonable to expect that at any given time in the technical advancement of sonographic equipment, there will exist a β-hCG level above which the gestational sac is always visible."

It is in this context that one can appreciate the great advantage of vaginosonography, which can significantly lower the discriminatory limit for pregnancy detection. In a recent study (Bernaschek et al. 1988) we were unable to visualize three-fourths of all intrauterine pregnancies at β-hCG levels below 300 mIU/ml (Fig. 11), but we correctly identified all intrauterine pregnancies when the β-hCG was above 300 mIU/ml. Considering the exponential rise of serum β-hCG by a factor of 1.7 per day

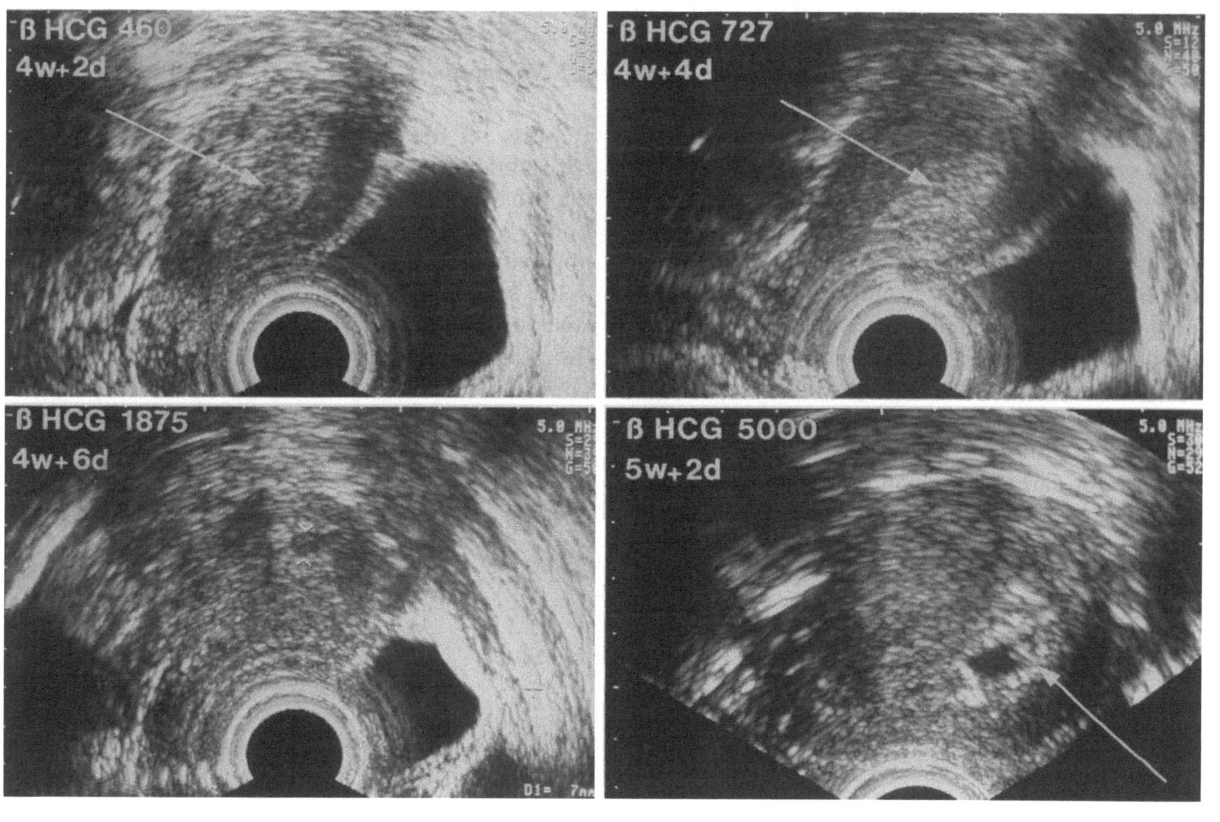

Fig. 12. Four vaginosonographic examinations in the same patient over a 7-day period, demonstrating the growth of the gestational sac. The corresponding β-hCG levels are shown in the upper left corner. Note the initial appearance of the fetal pole *(arrow)* in scan 4

in early pregnancy and the correct identification of gestational sacs at concentrations of 300 mIU/ml in our series, we presume that a gestational sac should not be missed by vaginosonography at β-hCG levels of 750 mIU/ml, even with a retroverted uterus. Because Nyberg et al. had 12 patients and we had only 8 patients at this very early gestational age, we dare not lower the discriminatory limit to 2 mm and 300 mIU/ml of β-hCG. By taking the more conservative value of 750 mIU/ml as the discriminatory limit, we currently should be able to detect any normal intrauterine pregnancy by vaginosonography. Perhaps time and more patients will show which value (300 or 750 mIU/ml) is clinically more useful. Based on our experience to date, if we see a patient with a β-hCG of 300 mIU/ml and no gestational sac, we repeat the vaginal scan after 2 days. We would then expect an increase in gestational sac diameter of about 1 mm/day and a rise in β-hCG above 750 mIU/ml into our discriminatory zone (Fig. 12).

Our results approximate those of Goldstein et al. (1988), who place the discriminatory limit at 1025 mIU/ml in cases where the uterus presents a homogeneous echo structure. If myomas are present, a normal gestational sac may not be visualized by vaginosonography, and diagnosis is possible only at higher β-hCG levels. Similarly, Nyberg et al. (1988) place the discriminatory limit at 1000 mIU/ml (Second International Standard). The gestational sac was detected in only one of five cases when levels were below 500 mIU/ml, and in four of five cases when levels were between 500 and 1000 mIU/ml. Sacks et al. (1988) state that the discriminatory limit for vaginosonography is between 500 and 800 mIU/ml.

Here again, vaginosonography seems to be advantageous. The early exclusion of an intrauterine gestational sac combined with a positive pregnancy test and no clinical evidence of spontaneous abortion, together with the improved evaluation of the adnexal regions and the detection of even small amounts of fluid in the cul de sac, will provide grounds for earlier suspicion of ectopic pregnancy. All this can be accomplished at serum β-hCG levels of 750 mIU/ml or higher. From these indications patients may be selected for laparoscopic confirmation, and necessary surgery may be performed at an early stage applying techniques that will preserve the function of the fallopian tube.

2 Yolk Sac

2.1 General

Early in the course of embryonic development the embryo becomes constricted off from the yolk sac. At the end of the 5th postmenstrual week a portion of the yolk sac is enveloped within the embryo to form the foregut and somewhat later the hindgut. Further folding of the embryo includes another portion of the yolk sac as the midgut (Langmann 1970), the communication between the yolk sac and midgut becoming the yolk duct (omphalomesenteric duct). The yolk sac is the first structure inside the chorionic cavity that can be demonstrated by vaginosonography. It lies outside the amniotic cavity and is joined to the chorionic attachment of the connecting stalk by an elongated yolk stalk, which transmits the yolk duct.

The amnion initially encloses the embryo rather tightly but gradually expands until, by the end of the 10th week, the amnion fills the entire chorionic cavity by virtue of its greater relative growth. As this occurs, the yolk sac becomes caught between the amnion and chorion and diminishes in size. Finally the yolk sac becomes incorporated into the umbilical cord and thereafter exists only as a rudimentary vesicle.

2.2 Vaginosonography

Reports on the earliest detection of the yolk sac with abdominal ultrasound range from the end of the 6th week to the start of the 7th week of gestation. Vaginosonography can demonstrate an approximately circular, well-defined, echo-free area - the secondary yolk sac - by the end of the 5th week or the start of the 6th week. At this early stage the yolk sac occupies approximately one-third of

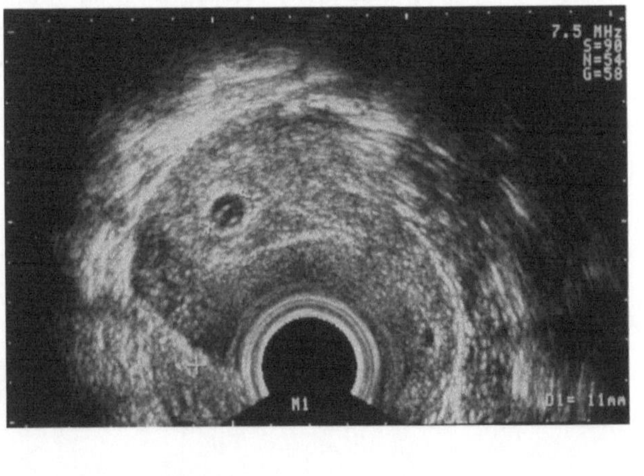

Fig. 13. Longitudinal scan through a retroverted uterus in ▶ the 6th week of gestation. The secondary yolk sac, 4 mm in diameter, is visible within the chorionic cavity

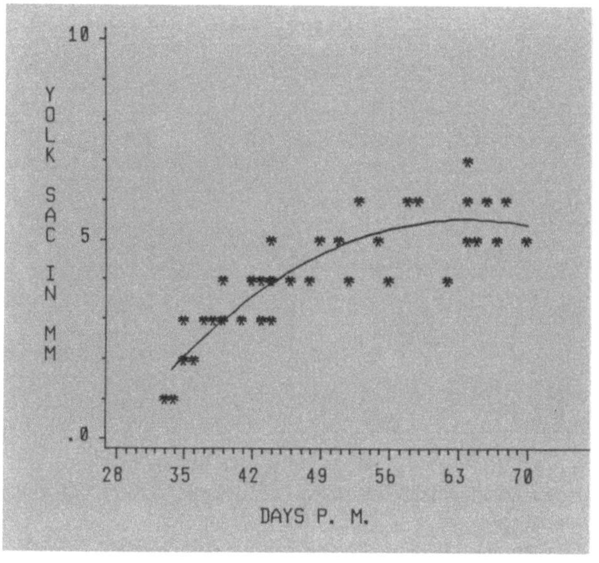

Fig. 14. Relation between yolk sac diameter and duration of amenorrhea in 35 intact pregnancies between the 5th and 10th postmenstrual weeks. The earliest point at which the yolk sac could be demonstrated in the chorionic cavity was the end of the 5th week. Days p.m., postmenstrual days

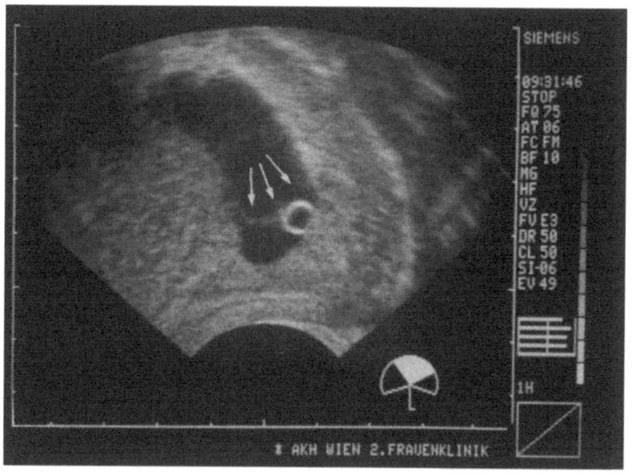

Fig. 15. The yolk sac is pushed against the margin of the chorionic cavity by the expanding amnion *(arrows)*

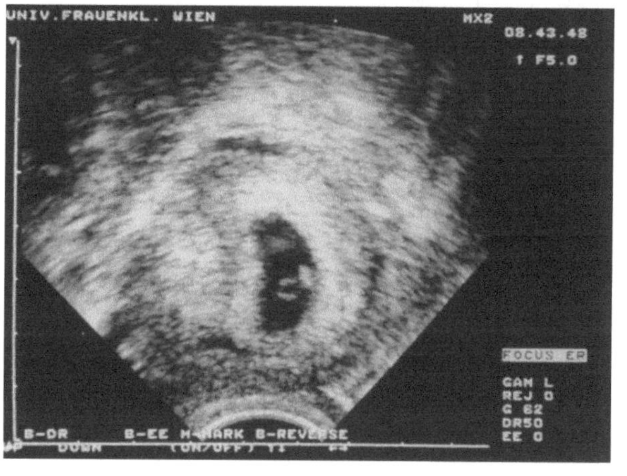

Fig. 16. In the 6th week of gestation the yolk sac is usually adjacent to the embryo and provides an important landmark for locating it

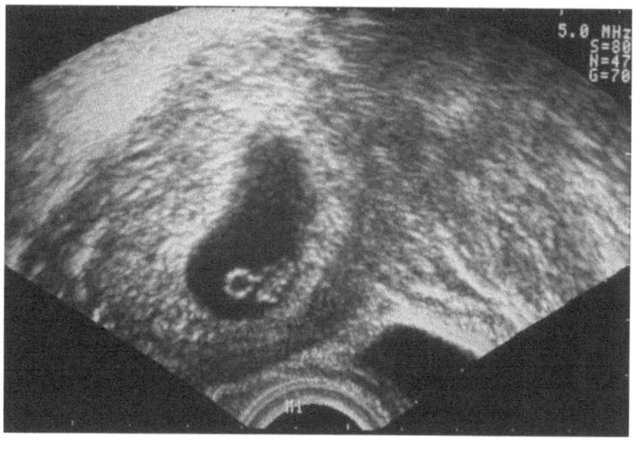

Fig. 17. Yolk sac and yolk stalk in an embryo with a CRL of 9 mm

the chorionic cavity. By the middle of the 6th week (Fig. 13) the yolk sac can be identified in all intact pregnancies, as indicated by the results of our own study (Fig. 14). This is consistent with a study by Levi et al. (1988), who showed that a yolk sac can be consistently visualized in a normal early pregnancy when the chorionic cavity reaches a mean diameter of 8 mm. The importance of identifying the yolk sac lies in the early exclusion of blighted ovum, for a yolk sac does not form in the absence of embryonic development. Additionally, Ferrazzi et al. (1988) describe larger yolk sacs in abnormal pregnancies than in intact pregnancies of equal age. The average diameter increase in intact pregnancies is 3 mm from the end of the 6th to the end of the 10th week (Funk and Fendel 1988). The yolk sac usually reaches its average full diameter of 5 mm in the 7th week, maintaining this size almost until its disappearance at the end of the first trimester (Popp and Lemster 1986). Before that it is pressed against the edge of the chorionic cavity by the rapidly expanding amnion (Fig. 15). The yolk sac remains closely adjacent to the embryo in early gestation (Fig. 16), a circumstance which facilitates recognition of the embryo and its cardiac activity. With further development the yolk sac distances itself from the embryo by the length of the yolk stalk. On a favorable scan plane the yolk sac can be traced along its stalk to the embryonic "abdomen" (Fig. 17).

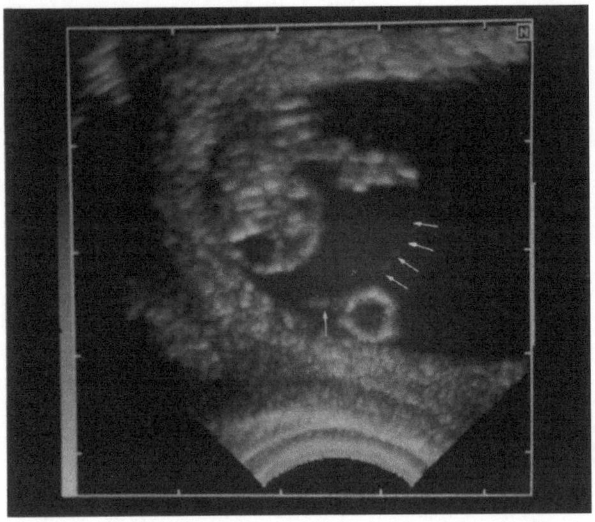

Fig. 18. Extra-amniotic *(arrows)* yolk sac and embryonic head with cerebral vesicles

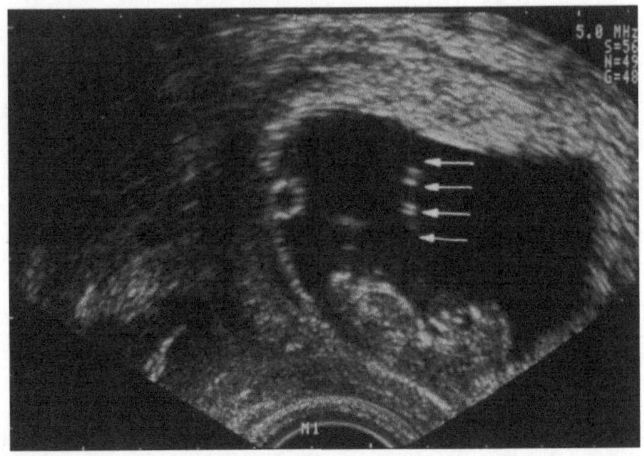

Fig. 19. The yolk sac displaced against the wall of the chorionic cavity; the umbilical cord *(arrows)* points to the chorion frondosum

The circular, thick-walled, echo-free yolk sac is easily distinguished from the thin-walled amniotic cavity, though sometimes it may resemble the fetal head with its cerebral vesicles (Fig. 18). The yolk sac is not useful for locating the chorion frondosum, because the length of its stalk allows it to lie anywhere in the chorionic cavity. The chorionic attachment of the umbilical cord is the most useful guide to identifying the chorion frondosum (Fig. 19).

3 Embryo

3.1 General

The embryo is considered to be present by the 4th week of menstrual age, or the 2nd week of development. The period up to and including the 10th postmenstrual week is called the embryonic period. Generally it may be said that all essential organ systems are formed by the end of the embryonic period, and that growth and maturation of these systems take place in the subsequent fetal period.

The 14-day-old embryo, which at this stage consists of the bilaminar germ disk (ectoderm and

Fig. 20a–e. Schematic drawings of an embryo: **a** on postmenstrual day 42, CRL= 3 mm; **b** on postmenstrual day 49, CRL= 7 mm; **c** on postmenstrual day 56, CRL=13 mm; **d** on postmenstrual day 63, CRL=18 mm; **e** on postmenstrual day 70, CRL=30 mm

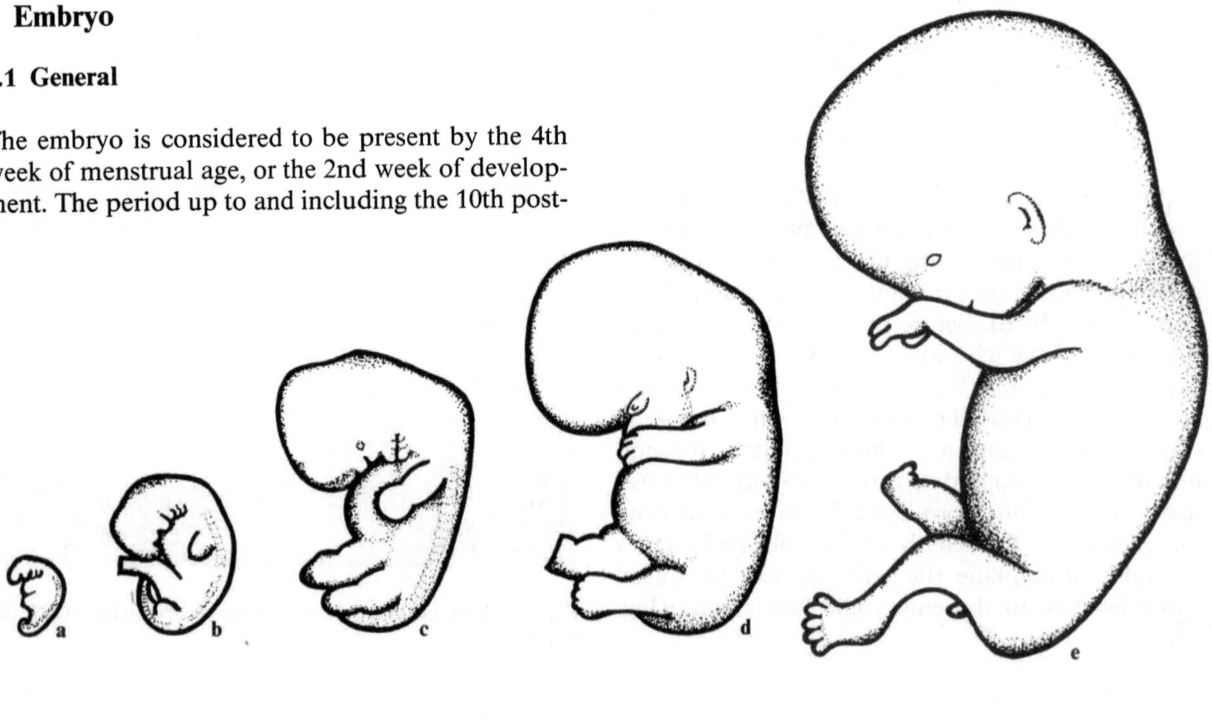

endoderm) and the prechordal plate, is joined to the chorion by the connecting stalk (Fig. 2). On about the 30th postmenstrual day a third embryonic layer begins to develop, the intraembryonic mesoderm. The ectoderm gives rise to the brain, spinal cord, peripheral nervous system, the epithelial portions of the sensory organs, the skin and its appendages, etc. The mesoderm gives rise to the bones, cartilage, connective tissue, muscles, heart, vessels, kidneys, gonads, blood, etc. The endoderm supplies the epithelial portions of the digestive and respiratory tract, the liver, the pancreas, and other organs.

The neural tube that develops from the ectoderm closes between postmenstrual days 39 and 41 (6th week). At this time the curved embryo has an overall length of 2.5-3 mm and thus can be demonstrated by vaginosonography. The mesoderm becomes subdivided on each side of the neural tube into individual segments composed of epithelioid cells; the segments thus formed are known as somites. Some 42-45 somite pairs are present by the 45th day of menstrual age (7th week). They are used in embryology to determine the age of the embryo during this stage, but they cannot be resolved sonographically. Here we are concerned only with the general body contours of the embryo that can be delineated with ultrasound. They are illustrated in Fig. 20 for embryos of varying age and size. The gross external morphology of the embryo shows a subdivision into head, neck, body, and limbs. The upper limbs appear as paddle-shaped buds at about the beginning of the 7th postmenstrual week (Langman 1970), and the lower limbs develop shortly thereafter (Fig. 21). The crown-rump length (CRL) at this time is approximately 7 mm.

Robinson was able in 1973 to demonstrate the embryo and measure the CRL in the 7th week of

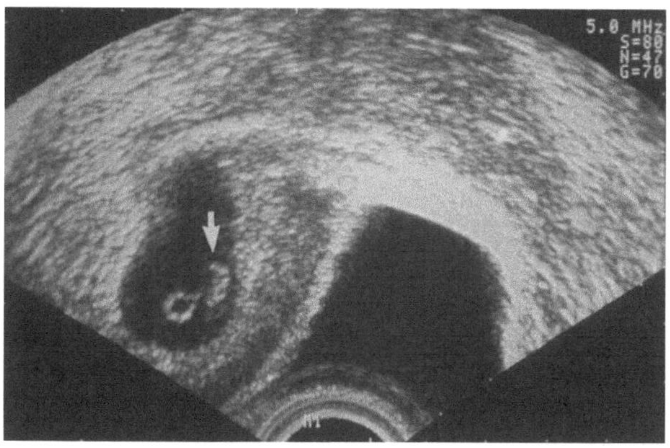

Fig. 21. Embryo with a CRL of 8 mm and a paddle-shaped lower limb *(arrow)*

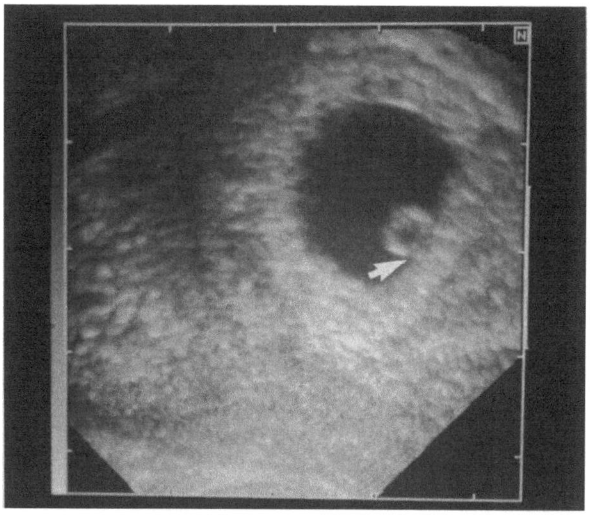

Fig. 22. Yolk sac and embryonic pole *(arrow)* at about 6 weeks gestation (5 weeks + 6 days). Heart activity can be demonstrated

gestation using transabdominal ultrasound. Hackelöer and Hansmann (1976) also cited the 7th week as the earliest time for detecting the embryo with abdominal ultrasound, noting that detection was consistently possible by the 9th week. Even then, however, maternal obesity or a retroflexed uterus can make visualization of the embryo difficult. An embryo "crouched" in the corner of the gestational sac is also easily missed.

3.2 Vaginosonography

The earliest point at which an embryo can be demonstrated by vaginosonography is the beginning of the 6th week of gestation, when the embryo has an overall length of 2-3 mm. In the early stages the embryonic pole lies close to the connecting stalk, directly adjacent to the yolk sac (Fig. 22). Early identification of the embryo may be hampered by its proximity to the echogenic trophoblast, although the rhythmic pulsations of the cardiac tube offer an effective means of identification. The embryonic pole can be visualized in only about 20% of cases at the beginning of the 6th week of gestation, in about 75% at the beginning of the 7th week (Fig. 23), and consistently by the 8th week (Fig. 24) (Rempen 1987). Similar results are reported by Levi et al. (1988), who were able to identify an embryo in all normal early pregnancies when the gestational sac reached an average diameter of 16 mm. We performed our own CRL measurements in 55 pregnant women with well-documented dates of last men-

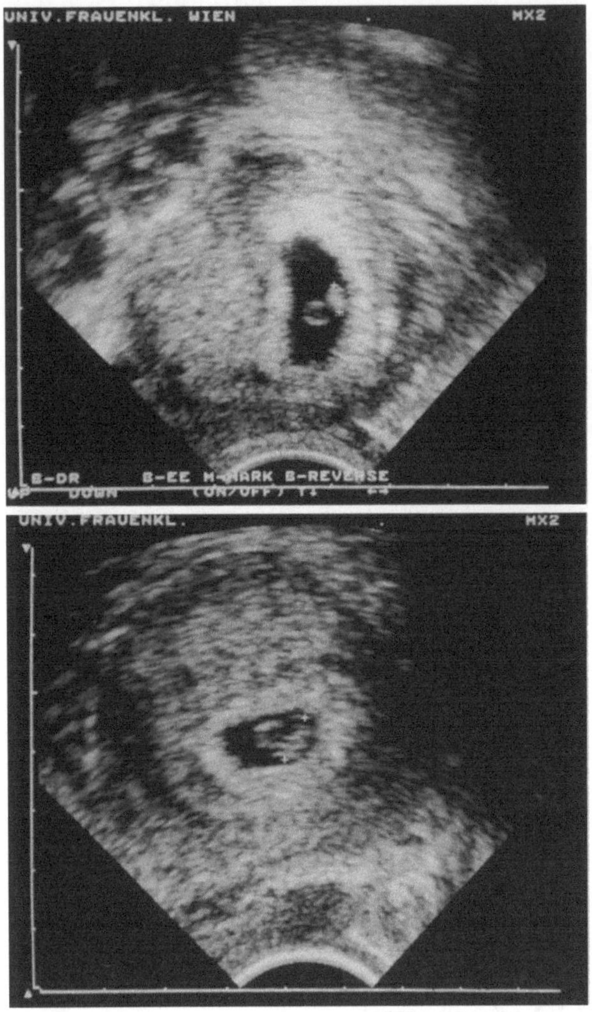

Fig. 23. *Top:* yolk sac and embryo early in the 7th week of gestation (6 weeks + 1 day) with a CRL of 5 mm. *Bottom:* yolk sac and embryo at the middle of the 7th week with a CRL of 8 mm

struation (Fig. 25). The values correlate closely with the data of Rempen (1987) and of Funk and Fendel (1988): CRL=2.5 mm (standard deviation 0.7) at 5 weeks gestation, 5.3 mm (1.8) at 6 weeks, 10.2 mm (4.2) at 7 weeks, 18.5 mm (4.7) at 8 weeks, 27.9 mm (3.1) at 9 weeks, and 38.2 mm (7.0) at 10 weeks. Very early CRL measurements are the least certain due to difficulties in accurately defining the cranial and caudal ends of an embryo positioned close to the wall. Comparisons of vaginosonographic CRL measurements with embryologic data indicate that the ultrasound measurements frequently underestimate embryonic size at very early stages (6th–7th week of gestation) (Rempen 1987). This may relate to the flexed embryonic posture imposed by the close confines of the amniotic sac. For this reason

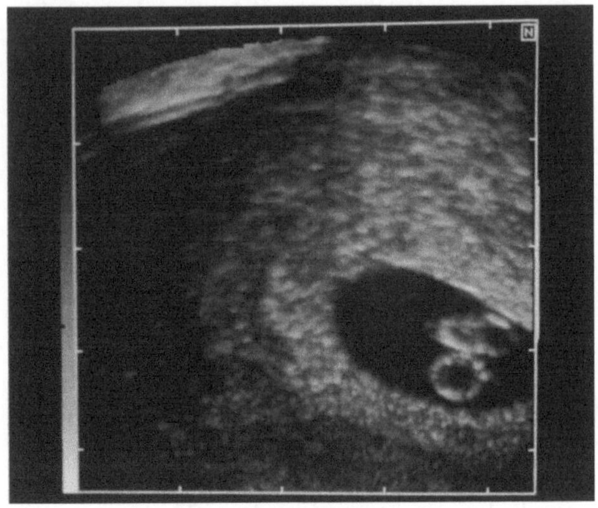

Fig. 24. Yolk sac and embryo at almost 7 weeks gestation (6 weeks + 6 days) with a CRL of 9 mm. The echo-free skull still exhibits a smaller diameter than the yolk sac

and because of the relatively small increase in CRL compared with the growth of the chorionic cavity in the 6th and 7th weeks, it appears that gestational age in this period can be more reliably estimated from the average gestational sac size. From the end of the 7th week the CRL increases by approximately 8–11 mm per week; this is the main reason why estimation of gestational age on the basis of CRL is more accurate from the 8th week of pregnancy (Figs. 26, 27).

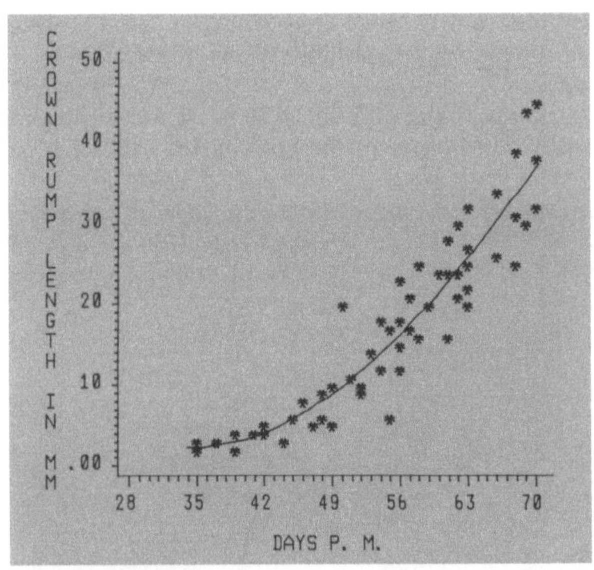

Fig. 25. Relation between embryonic CRL and duration of amenorrhea. The earliest detection of the embryo was made at 5 weeks + 0 days. days p.m., postmenstrual days

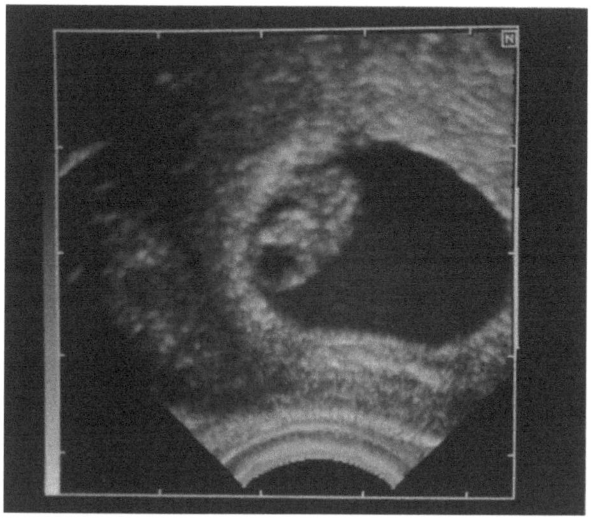

Fig. 26. Embryo with a CRL of 15 mm (7 weeks + 5 days). A portion of the neural tube is visible adjacent to the cerebral vesicles

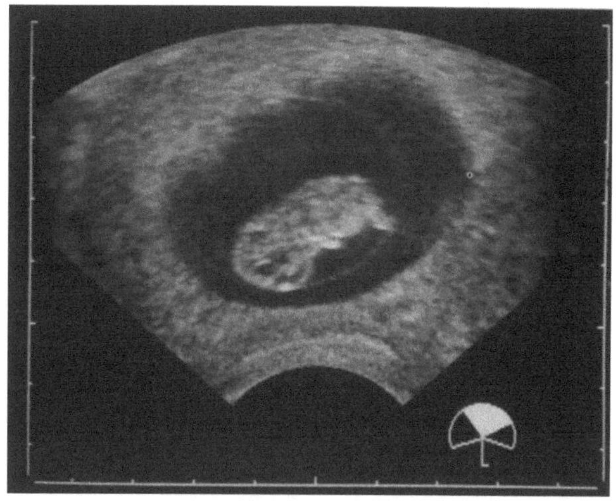

Fig. 27. Embryo with a CRL of 25 mm at the beginning of the 10th week of gestation. Rudimentary upper and lower limbs are seen alongside the cerebral vesicles

4 Cardiac Activity

4.1 General

The vascular system of the human embryo develops around the middle of the 5th postmenstrual week, when the embryo can no longer meet its nutritional needs by diffusion alone. At this time mesenchymal cells proliferate in the chorion, the connecting stalk, the wall of the yolk sac, and shortly thereafter in the embryonic mesoderm, forming separate cellular aggregations from which the extra- and intraembryonic vascular systems are derived. Approximately 3 days elapse between the initial appearance of intraembryonic blood vessels and the formation of the cardiac tube. Thus the rudimentary heart is present by the end of the 5th week, consisting first of a pair of tubes that soon establish contact with the embryonic blood vessels and fetal membranes. Cardiac pulsations commence at about the beginning of the 6th week of gestation, when the embryo is approximately 2 mm long. Thus, assuming that conception occurred on day 14 of the menstrual cycle, the detection of embryonic life is not possible before the 36th postmenstrual day.

Pregnancy may be diagnosed as intact only if the viability of embryonic structures is confirmed. The earliest point at which Hackelöer and Hansmann (1976) were able to detect heart activity with abdominal ultrasound was on day 42 of the menstrual cycle; Robinson (1973) recognized heart activity on day 45. The average embryonic heart rate reported in the 7th week of gestation was only 123 beats per minute (bpm). By the 9th week it rises to approximately 171 bpm and thereafter slowly declines to the normal fetal range of 120–160 bpm. While heart activity can occasionally be detected with abdominal ultrasound in the 7th week of gestation, reliable confirmation is usually not obtained until the 9th week (Hansmann et al. 1985; Terinde and Kozlowski 1988).

4.2 Vaginosonography

Using a vaginal probe, Kratochwil and Eisenhut in 1967 were able to detect embryonic heart activity no earlier than day 46 of the menstrual cycle. However, only pulsed-echo and A-mode techniques were available at that time. The vaginal route remains the most favorable approach for sonographic evaluations in early pregnancy, and tremendous progress has been made in the design of vaginal probes and of the ultrasound unit itself. It is now possible, under favorable examining conditions, to demonstrate heart activity at the same time embryonic structures are first visualized. Indeed, in the early stages the embryonic pole is usually identified by the presence of cardiac activity, which appears as a rhythmic "blinking" on the real-time scans. Heart activity can be recorded and documented using the time-motion technique. Haid et al. (1985) documented heart activity with a vaginal probe on day 38, and thus only 2 days after the onset of cardiac pulsations. Similarly, Levi et al. (1988) identified cardiac pulsations in gestational

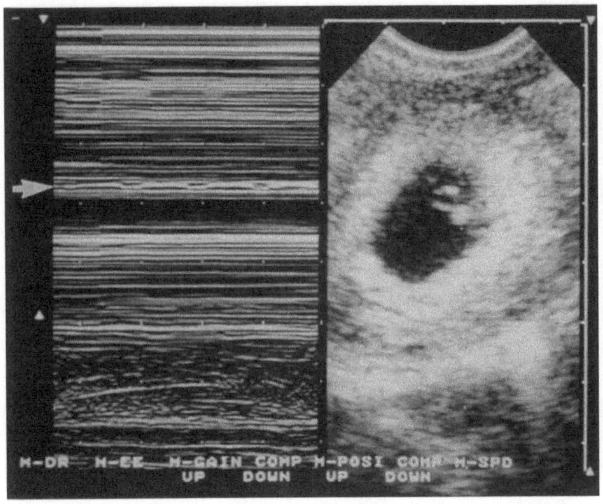

Fig. 28. Time-motion display of embryonic heart activity *(arrow)* on the 43rd postmenstrual day (heart rate 60 bpm)

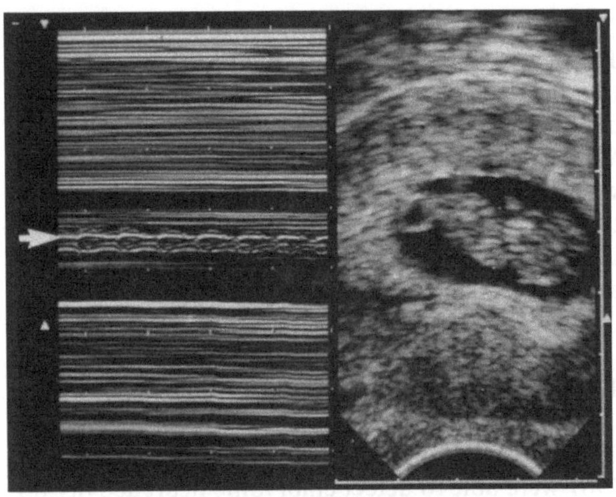

Fig. 29. Time-motion display of embryonic heart activity *(arrow)* on the 63rd postmenstrual day (heart rate 115 bpm)

sacs with mean diameters as small as 9.5 mm. The earliest point at which Blumenfeld et al. (1988) were able to detect heart activity by vaginosonography was the 40th postmenstrual day. It should be noted, however, that the maternal heartbeat can mask embryonic pulsations in the 6th and 7th postmenstrual weeks, and that passive movements of the embryo in synchrony with the material pulse can mimic cardiac activity. Vaginosonography can demonstrate embryonic heart activity, and thus confirm a viable pregnancy, by the 8th week of gestation in all cases, and it can do so by the 7th week in the majority of cases (Funk and Fendel 1987; Rempen 1987; Terinde and Kozlowski 1988). Reports claiming 100% detection of heart activity in the 6th week of gestation (Hogaki et al. 1987) appear too optimistic. However, vaginosonography can advance the detection of a viable embryo by an average of 1 to 1 1/2 weeks compared with abdominal ultrasound.

It is reported that the rate of pulsation of the cardiac tube is 60-70 bpm in the 6th week of gestation (Haid and Martin 1986), increases to 75-100 bpm by the 7th week, and thereafter attains the rates demonstrable with abdominal ultrasound. These findings are consistent with our own research (Figs. 28, 29). Pulsations of the cardiac tube appear in early stages as wavelike movements in the embryonic thoracic region that resemble peristaltic waves (Schaaps et al. 1986).

5 Amniotic Cavity

5.1 General

In the course of its development from the germ disk, the embryo lengthens and develops folds which constrict it off from the yolk sac and cause it to be increasingly enveloped within the amniotic cavity. The amnion initially forms a relatively tight enclosing membrane with only a thin fluid layer separating the embryo from the thin amniotic epithelium. This fluid layer cannot be demonstrated even by vaginosonography. But as the amniotic cavity continues to expand, the amniotic epithelium, which consists of only five cell layers, can be visualized. Near the end of the first trimester the expanding amniotic cavity completely fills the chorionic

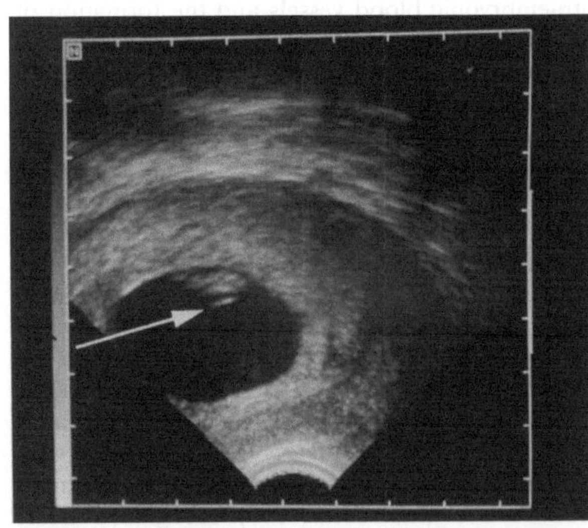

Fig. 30. The amnion usually appears only as a point or streaklike echo *(arrow)* in the 6th and 7th weeks of gestation

cavity due to its greater relative growth, so that the amnion apposes to the chorion. At that time the membranes can no longer be defined as separate structures with ultrasound.

The extremely thin amniotic epithelium can be demonstrated by high-resolution abdominal scanning (Jeanty and Romero 1984; Hansmann et al. 1985), although most authors claim that detection is possible only in some cases by the 7th to 8th week of gestation. The sound waves must strike the extremely thin amnion at a perpendicular angle and must be reflected without scatter in order to demonstrate the membrane. Since the amniotic cavity is usually spherical, this condition is met only when the beam passes through the center of the sphere. Even then, however, only two echoes are obtained.

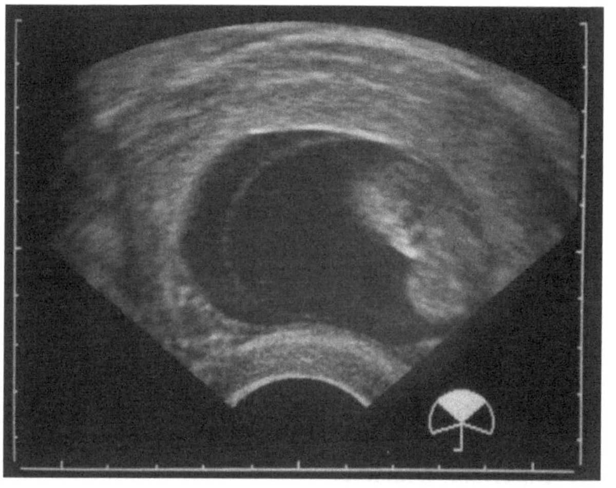

Fig. 32. Here the amnion is delineated more clearly by the slightly contrasting densities of the chorionic and amniotic fluid

5.2 Vaginosonography

The first detailed vaginosonographic studies of the amnion were performed by Rempen (1987) and Funk and Fendel (1988). They report that the amniotic cavity is consistently demonstrated by the 9th week of gestation and is occasionally seen as early as the 6th week. At that time the amnion appears as a point or streaklike echo flanking the embryo; it can be measured on only one plane (Fig. 30). With increasing expansion of the amnion, however, almost the entire circumference can be visualized (Fig. 31). Chorioamniotic discrimination is further aided by an increasing contrast in the echogenicity of the amniotic and chorionic fluid (Fig. 32). While the amniotic cavity appears devoid of echoes, the

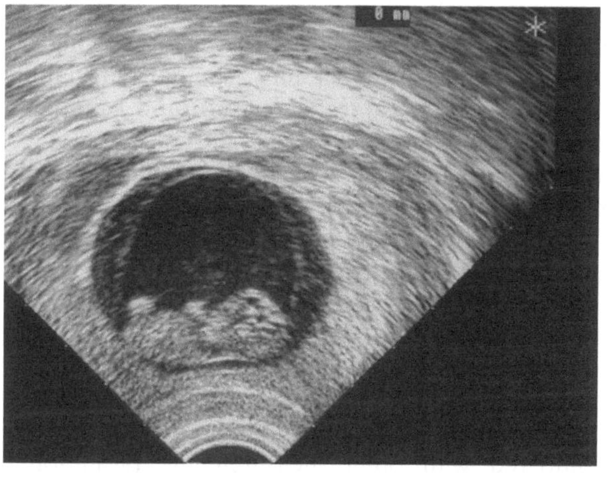

Fig. 33. Increased echogenicity of the chorionic fluid toward the end of the first trimester

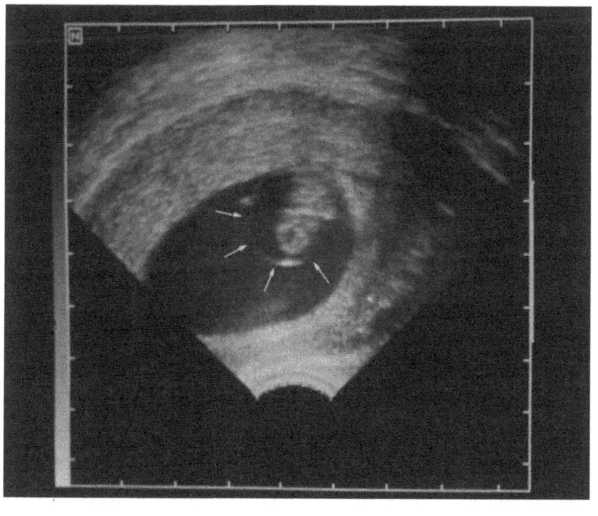

Fig. 31. The whole circumference of the amniotic sac *(arrows)* is visualized

chorionic cavity becomes filled in with low-level echoes that are best appreciated at high gain settings (Fig. 33). This is believed to be due to differences in the composition of the fluids, the chorionic fluid apparently acquiring a greater density with passage of time. In about 70% of cases the amniotic cavity is defined by a sharp boundary echo, and in about 30% of cases it is recognized by its contrasting density (Funk and Fendel 1988). It is possible, therefore, to survey the entire amniotic cavity on multiple planes. Protected as it is by the chorionic fluid, the amniotic sac retains its spherical shape for some time. The amniotic sac assumes the shape of the chorionic cavity only when the sac has expanded sufficiently to fill it. By the end of the first trimester on average the amnion apposes to the chorion, and the two membranes can no longer be separated by vaginosonography, although small areas of chorioamniotic separation may still be present. The diameter of the amniotic cavity at 7 weeks gestation is 13 mm (standard deviation 4.5), 20 mm (SD 5.8) at 8 weeks, 33 mm (SD 4.0) at 9 weeks, and 40 mm (SD 6.0) at the end of 10 weeks (Rempen 1987). Thus, the development of the amnion closely parallels the increase in the CRL.

6 Other Biometric Data in the First Trimester

6.1 Biparietal Diameter

The cephalic pole of the embryo can be identified and measured at an early stage by vaginosonography. Rempen (1987) was able to detect the cephalic pole as early as the end of the 7th week of gestation. At that time the cranial pole can be measured in approximately 40% of cases, showing an average diameter of 5.4 mm (standard deviation 1.0). The average values increase from 8.1 mm at 8 weeks to 12.2 mm at 9 weeks and 15.4 mm at 10 weeks. By the end of 9 weeks the cephalic pole is consistently delineated with vaginal ultrasound (Fig. 34). Popp and Lemster (1986) report slightly lower average values in their study. Thus, the values measured by vaginosonography can supplement the growth curves for biparietal diameter (BPD) by adding new and earlier data. Although early biometry of the cephalic pole has assumed secondary importance in determining gestational age when compared with other measurements (gestational sac, CRL), it could acquire significant future value in the early detection of gross fetal anomalies such as anencephaly.

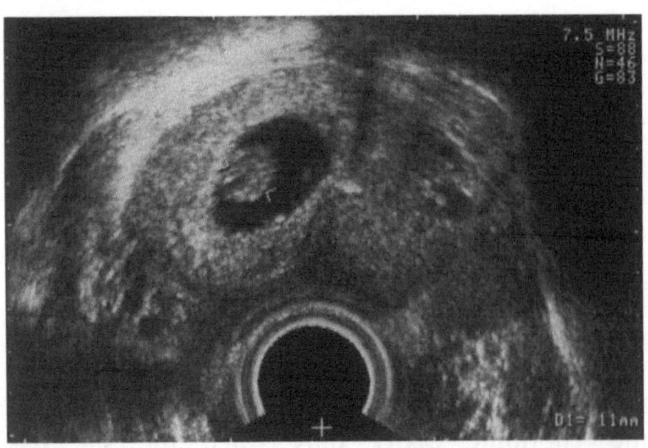

Fig. 34. Transverse scan through the cephalic pole *(arrowheads)* in the 9th week of gestation

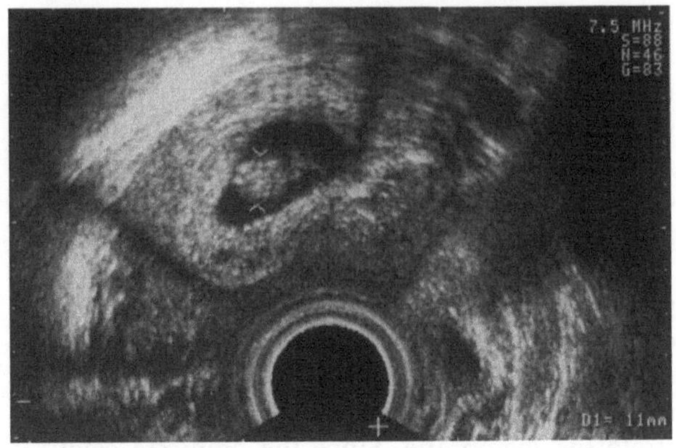

Fig. 35. Transverse scan through the abdomen in the 10th week of gestation. *Arrowheads* show fetal trunk diameter

6.2 Trunk Diameter

Popp and Lemster (1986) report that the average fetal trunk diameter is 2 mm at the beginning of the 7th week of gestation. This value shows an almost linear increase (Fig. 35) until the end of the 10th postmenstrual week, when it averages 11 mm. The relatively large scatter in the data of Popp and Lemster (1986) may be due to the presence of a physiologic omphalocele in some embryos. Green and Hobbins (1988) report finding this condition in approximately 20% of fetuses at the end of the first trimester.

6.3 Umbilical Cord and Placenta

The earliest point at which the umbilical cord can be detected is the end of the 8th week. The cord usually appears as a linear feature (Fig. 36) extending from its origin at the umbilicus to its insertion on the chorion frondosum (Fig. 37). The umbilical cord is, in fact, the best landmark for locating the chorion frondosum - a circumstance that has particular importance in ultrasound-guided chorionic biopsy.

Among the first structures demonstrated by vaginosonography in the developing placental region are myometrial blood vessels approximately 2 mm in diameter. While these vessels may initially be found at some distance from the chorionic cavity, they move progressively closer to the trophoblast as gestation proceeds. They eventually form a network of blood vessels outside the echogenic zone of the trophoblast. Rhythmic movements synchronous with the maternal pulse can be observed within the blood vessels (Schaaps et al. 1986). Blumenfeld et

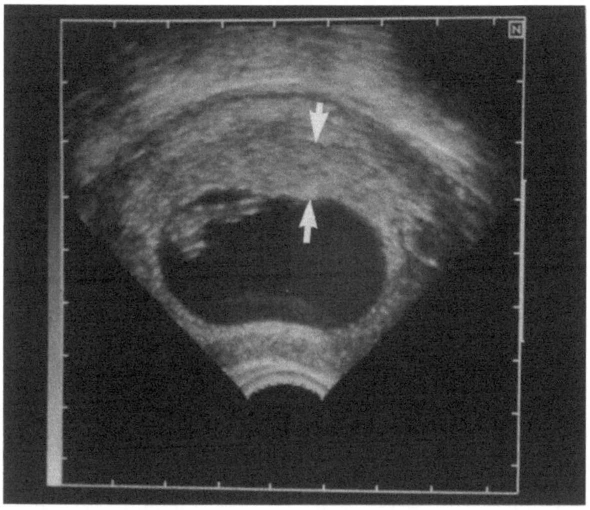

Fig. 36. Appearance of the umbilical cord at the end of the 9th week gestation. At this time the cord forms almost a straight line between the umbilicus and the placental insertion of the cord. The arrows indicate the thickness of the trophoblast

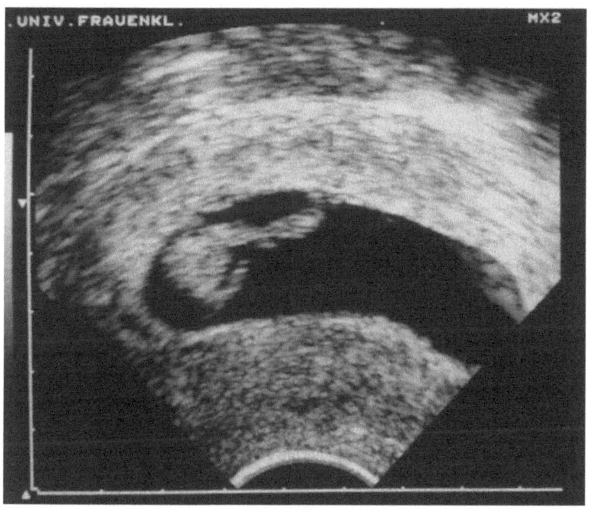

Fig. 37. Scan in the 11th week showing the umbilical cord extending from the umbilicus to the placental insertion of the cord

al. (1988) also believe that these vessels are precursors of the maternal placental circulation. The trophoblast itself, which appears more echogenic, is clearly visible by the 8th week of gestation and has a thickness of about one-third to one-fifth the diameter of the chorionic cavity. Contrary to the recommendation of Blumenfeld et al. (1988), however, the trophoblast should not be included in gestational sac measurement, as this may lead to errors in estimating gestational age. The trophoblastic rim becomes relatively narrower in the 9th week of gestation, its thickness diminishing to only one-fourth to one-sixth the chorionic diameter (Fig. 36). The placenta is delineated more clearly at this stage, so its position relative to the uterine wall can be assessed with greater accuracy. This means that ultrasound-guided chorionic biopsy can be safely performed in the 9th week of gestation.

7 Multiple Pregnancy

Multiple pregnancies may result from the simultaneous fertilization of more than one ova or from the division of a fertilized ovum in the course of its development. The early sonographic diagnosis of multiple pregnancy is based on the demonstration of two or more gestational sacs and is possible as early as 5 weeks' gestation (Fig. 38). Abdominal ultrasound studies (Levi 1976) have shown that approximately 70% of multiple pregnancies are either reduced to a single gestational sac or lost altogether in the first trimester (Fig. 39). In the interest of the patient and ultimately of the examiner, a multiple pregnancy generally is not diag-

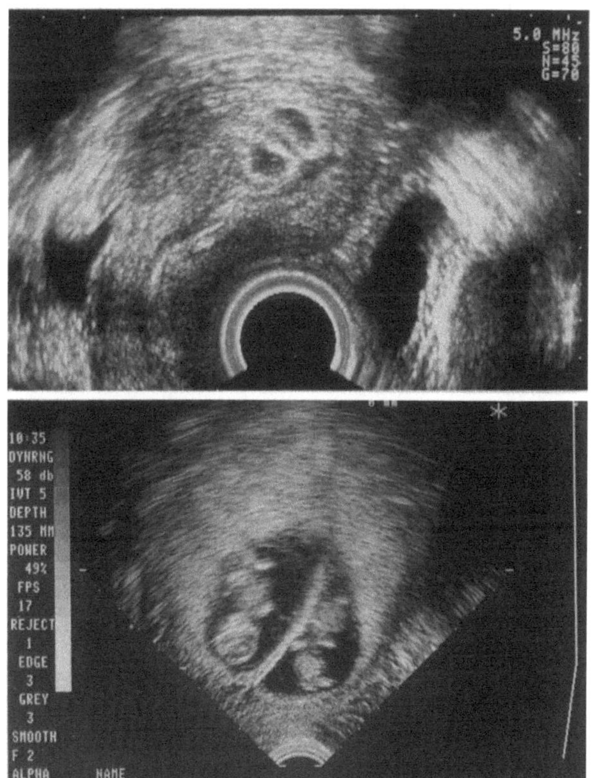

Fig. 38. A twin pregnancy in the 5th week of gestation *(top)* and in the 12th week *(bottom)*

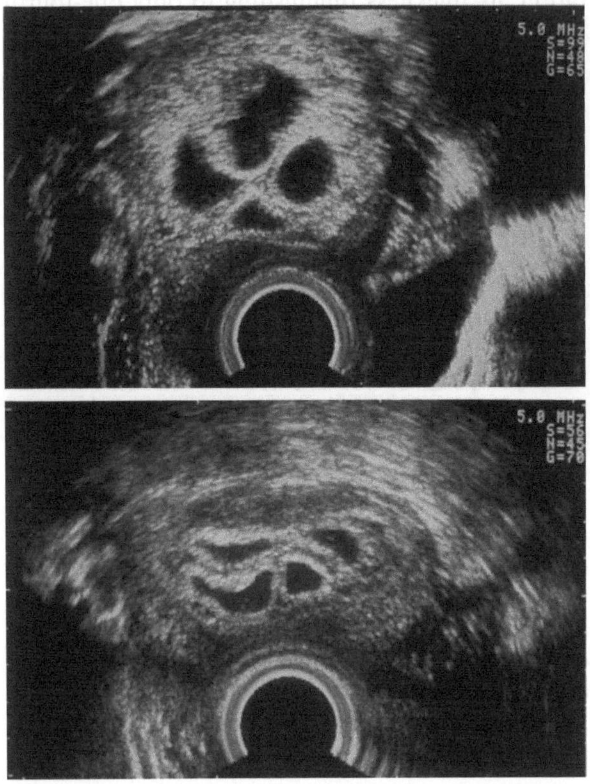

Fig. 39. Quadruplets. *Top:* longitudinal scan through the uterus in the 6th week of gestation. *Bottom:* transverse scan through the uterus in a different pregnancy

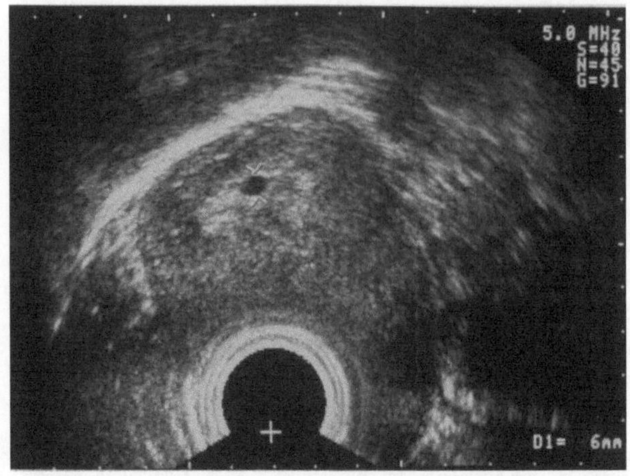

Fig. 40. Longitudinal transvaginal scan through a retroverted uterus at 4 weeks + 6 days. The gestational sac could not be visualized with abdominal ultrasound

nosed until viability of all the fetuses has been established. This is an area where vaginosonography is of particular value, for it can demonstrate two or more yolk sacs and confirm embryonic heart activity at least 1 week earlier than is possible with abdominal ultrasound.

8 Summary

Here it is helpful to recall the definition of "week of gestation" as it is customarily used by European authors. The "6th week of gestation," for example, refers to the period between 5 complete weeks (i.e., 5 weeks + 0 days) and 6 complete weeks (i.e., 5 weeks + 7 days). The following statements should be interpreted with that definition in mind.

With modern ultrasound equipment, it is occasionally possible to detect an intrauterine pregnancy, define the embryonic pole, and confirm viability at a very early stage even when abdominal scanning is employed (Green and Hobbins 1988). However, when we relate the percentage detection rate to the corresponding week of gestation, we find that vaginosonography offers significant advantages. Various obstacles to abdominal scanning such as obesity, an inadequately filled bladder, and a retroverted uterus (Fig. 40) significantly postpone reliable detection compared with vaginosonography, which can demonstrate the most important parameters of early pregnancy 1-2 weeks earlier. The chart in Fig. 41 is intended to answer the following two questions:

1. What can be seen with vaginosonography?
2. What must be seen with vaginosonography?

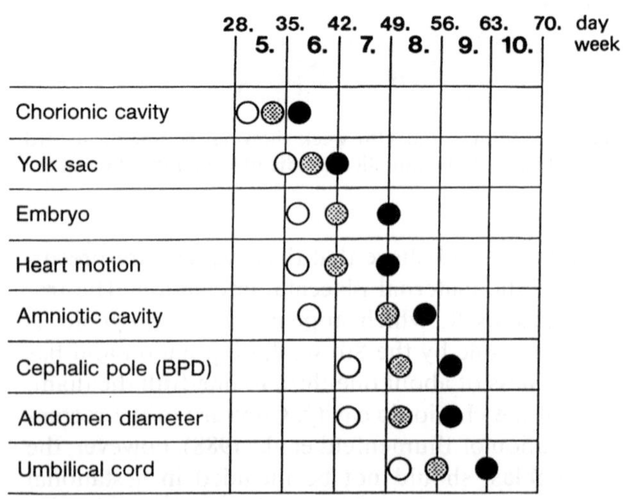

Fig. 41. Detectability of various parameters in early pregnancy by vaginosonography. ○ earliest detection; ◍ usually detected; ● always detected; p.m. post menstruation

References

Batzer F, Weiner ST, Corson ST, Schlaff S, Otis CH (1983) Landmarks during the first forty-two days of gestation demonstrated by the β-subunit of human chorionic gonadotropin and ultrasound. AM J Obstet Gynecol 146: 973-979

Bernaschek G, Rudelstorfer R, Csaicsich P (1988) Vaginosonography versus serum human chorionic gonadotropin in early detection of pregnancy. AM J Obstet Gynecol 158: 608-612

Blumenfeld Z, Rottem S, Elgali S, Timor-Tritsch I (1988) Transvaginal sonographic assessment of early embryological development. In: Timor-Tritsch IE, Rottem S (eds) Transvaginal sonography. Heinemann, London, pp 87-108

Cacciatore B, Ylöstalo P, Stenman U, Widholm O (1988) Suspected ectopic pregnancy: ultrasound finding and HCG levels assessed by an immunofluorometric assay. Br J Obstet Gynaecol 95: 497-502

De Crespigny LC, Cooper D, McKenna M (1988) Early detection of intrauterine pregnancy with ultrasound. J Ultrasound Med 7: 7-10

Ferrazzi E, Brambati B, Lanzani A, Oldrini A, Stripparo L, Guerneri S, Makowski EL (1988) The yolk sac in early pregnancy failure. AM J Obstet Gynecol 158: 137-142

Funk A, Fendel H (1988) Ultraschallechographische Darstellbarkeit und Messung der Amnionhöhle und des Dottersacks in der frühen Schwangerschaft: vergleichende Untersuchung von intakten und gestörten Schwangerschaften. Z Geburtshilfe Perinatol 192: 59-66

Goldstein SR, Snyder JR, Watson C, Danon M (1988) Very early pregnancy detection with endovaginal ultrasound. Obstet Gynecol 72: 200-204

Green JJ, Hobbins JC (1988) Abdominal ultrasound examination of the first-trimester fetus. AM J Obstet Gynecol 159: 165-175

Hackelöer BJ, Hansmann M (1976) Ultraschalldiagnostik in der Frühschwangerschaft. Gynakologe 9: 108

Haid C, Martin J (1986) Frühschwangerschaft und Extrauteringravidität aus der Sicht eines intravaginalen Linear-array-Scanners. In: Popp LW (ed) Gynäkologische Endosonographie. Klemke, Quickborn, pp 101-105

Haid C, Zech H, Martin J (1985) Verbesserte Frühdiagnose der intrauterinen Schwangerschaft durch Ultraschall-Vaginalsonde. Geburtshilfe Frauenheilkd 45: 371-375

Hansmann M, Hackelöer BJ, Staudach A (1985) Ultraschall in Geburtshilfe und Gynäkologie. Springer, Berlin Heidelberg New York

Hogaki M, Ohkawa T, Takizawa M, Takeuchi Y (1987) Transvaginal detection of fetal signals by doppler method. J Perinat Med 15 Suppl 1: 161

Jeanty P, Romero R (1984) Obstetrical ultrasound. McGraw-Hill, New York, p 37

Kadar N, Romero R (1986) Abnormal pregnancy: early diagnosis by US and serum chorionic gonadotropin levels (letter). Radiology 161: 854-857

Kadar N, De Vre G, Romero R (1981) Discriminatory hCG zone: its use in sonographic evaluation for ectopic pregnancy. Obstet Gynecol 58: 156-161

Kratochwil A, Eisenhut L (1967) Der früheste Nachweis der fetalen Herzaktion durch Ultraschall. Geburtshilfe Frauenheilkd 27: 176-180

Langmann J (1970) Medizinische Embryologie. Thieme, Stuttgart

Levi S (1976) Ultrasonic assessment of the high rate of human multiple pregnancy in the first trimester. J Clin Ultrasound 3-7

Levi S, Lyons EA, Lindsay DJ (1988) Early diagnosis of non-viable pregnancy with endovaginal US. Radiology 167: 383-385

Nyberg DA, Filly RA, Mahony BS, Monroe S, Laining FC, Jeffrey RB (1985) Early gestation: correlation of HCG levels and sonographic identification. Am J Roentgenol 144: 951-954

Nyberg DA, Laing FC, Filly RA, Mack LA, Zarutskie PW (1987) Ectopic pregnancy: diagnosis by sonography correlated with quantitative hCG levels. J Ultrasound Med 6: 145-150

Nyberg DA, Mack LA, Laing FC, Jeffrey RB (1988) Early pregnancy complications: endovaginal sonographic findings correlated with human chorionic gonadotropin levels. Radiology 167: 619-622

Popp LW, Lemster S (1986) Vaginosonographische Detailbefunde der Frühschwangerschaft: erlebte Embryologie, klinische Anwendung. In: Popp LW (ed) Gynäkologische Endosonographie. Klemke, Quickborn, pp 107-126

Rempen A (1987) Vaginale Sonographie der intakten Gravidität im ersten Trimenon. Geburtshilfe Frauenheilkd 47: 477-482

Robinson HP (1973) Sonar measurements of fetal crown-rump-length as means of assessing maturity in the first trimester of pregnancy. Br Med J 4: 28

Schaaps JP, Mustin J, Lambotte R (1986) Vaginosonographische Aspekte der Uterus-Trophoblastzirkulation. In: Popp LW (ed) Gynäkologische Endosonographie. Klempke, Quickborn, pp 127-132

Sacks GA, Fleischer AC, Kepple DM (1988) Clinical applications of transvaginal sonography in obstetrics and gynecology. Med Rev 23: 15-24

Terinde R, Kozlowski P (1988) Ultraschalldiagnostik der gestörten Frühgravidität. Gynäkologe 21: 210-219

Disorders of Early Pregnancy

1 General

Bleeding and cramping are among the most fequent complications of pregnancy in the first trimester. Since clinical and laboratory studies are unable to establish the integrity of the pregnancy or the viability of the fetus with absolute confidence, sonography provides the only means of assigning a specific diagnosis to the initially nonspecific symptoms. Moreover, the prognostic value of ultrasound in this situation can be a great comfort to the patient who has learned little from hormone tests and pelvic examinations but is alarmed and distressed by her symptoms. Given the lack of acceptance of early ultrasound screening at many centers, it is rare for an abnormal pregnancy to be recognized in the first trimester before clinical symptoms have appeared.

By its ability to demonstrate the chorionic cavity, yolk sac, fetal pole, and cardiac activity at a significantly earlier stage than abdominal ultrasound, vaginosonography, has revolutionized the diagnosis of the abnormal early pregnancy. Another advantage of vaginal scanning over abdominal scanning is its superior resolution, which is of great value in differentiating between an intact and abnormal pregnancy. The following sonographic parameters - their time of appearance, size, shape, boundaries, and intra- or extrauterine location - are important in recognizing disorders of early pregnancy:

1. Intrauterine ring
2. Yolk sac
3. Embryo
4. Detection of fetal life
5. Topography of the uterus and adnexa and fluid in the cul-de-sac
6. Correlation with β-hCG levels

The goals of the evaluation are twofold: to detect the abnormal pregnancy as early as possible in order to shorten the period of uncertainty, and to avoid premature intervention in a putatively nonviable pregnancy.

2 Threatened Abortion

Almost all patients with early pregnancy complications who are referred for ultrasound evaluation have bleeding as their major symptom and are placed in the clinical category of "threatened abortion." However, the subsequent ultrasound examination confirms this diagnosis in only about half of cases. In cases where a viable fetus is demonstrated, there is only about a 10% likelihood that the pregnancy will end in abortion (Anderson 1980; Hackelöer 1985).

Even with vaginosonography it is rarely possible to determine the etiology of vaginal bleeding in an intact pregnancy. It is not uncommon for hematomas to form as a result of the hemorrhage. Most are located above the internal cervical os (Fig. 1). Abdominal sonography usually displays the bleeding site as an echo-free zone within the uterus that may be indistinguishable from a second gestational sac in an abnormal twin pregnancy. However, vaginosonography, with its superior resolution, can always demonstrate minute echoes even in a fresh hemorrhage, furnishing an effective criterion for differentiation.

As the hematoma becomes more organized, the echoes become more distinct while the area dimin-

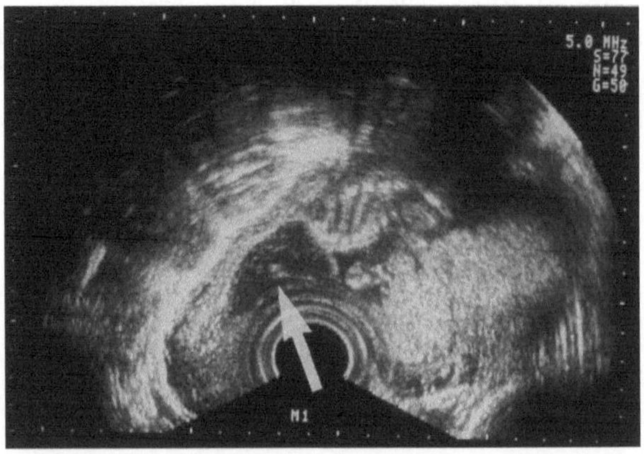

Fig. 1. Longitudinal vaginal scan through the uterus. An area of hemorrhage *(arrow)* is visible above the internal os

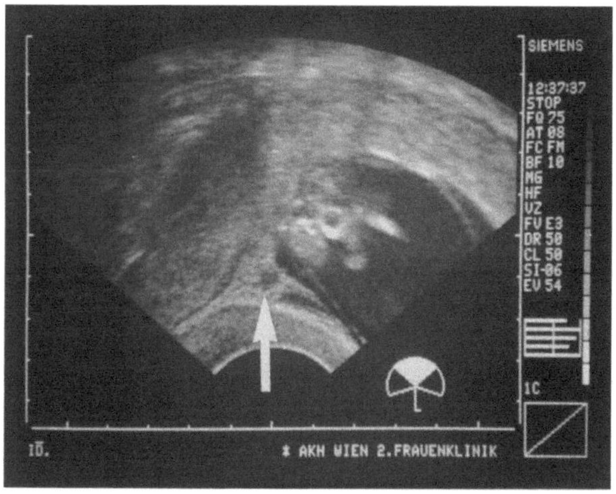

Fig. 2. Longitudinal scan through the uterus with an "older" hemorrhage above the internal os *(arrow)*

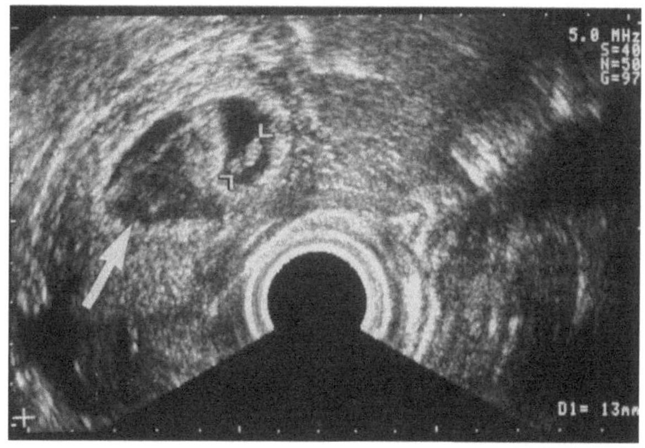

Fig. 4. Transverse scan through the uterus showing a retrochoroidal hematoma *(arrow)* adjacent to an intact pregnancy *(arrowheads)*

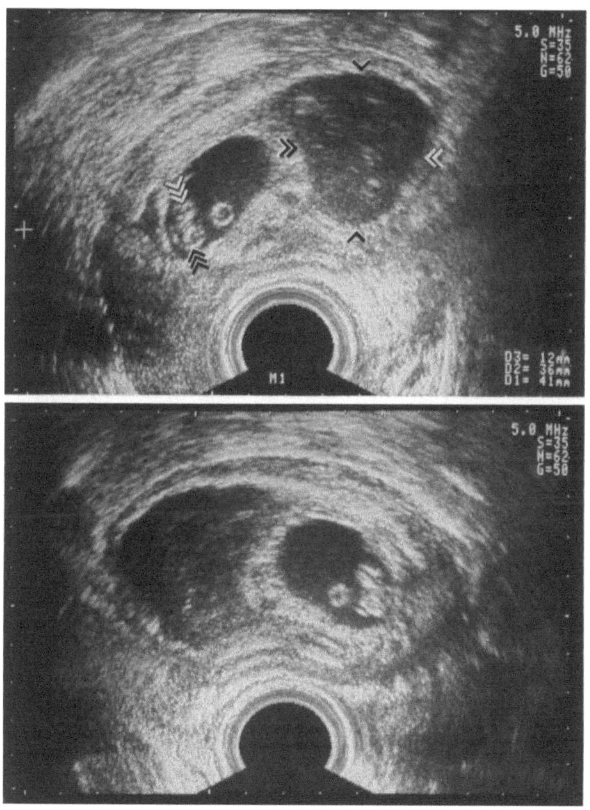

Fig. 3. Retrochoroidal hematoma in the region of the uterine fundus. *Top:* longitudinal scan through the uterus; *bottom:* horizontal scan in the same case. *Single arrowheads,* anteroposterior diameter of hematoma; *double arrowheads,* longitudinal diameter of hematoma; *triple arrowheads,* crown-rump length

ishes in size and acquires a flattened shape (Fig. 2). Chorionic bleeding can also lead to hematomas in the fundal region (Fig. 3) or to retrochoroidal pooling of blood adjacent to an intact pregnancy (Fig. 4). The prognosis in such cases depends on the size of the hematoma but tends to be poor. Sauerbrei and Dieu (1986) state that the outcome of the pregnancy is generally poor when the hematoma volume exceeds 60 ml. As a rule, however, chorionic hemorrhage has a reasonably good prognosis. Goldstein et al. (1983) and Jouppila (1985) found that 80% of pregnancies with sonographically demonstrable hematomas were viable.

Differentiation between an intact pregnancy and a blighted ovum, missed abortion, or incomplete abortion cannot be accomplished right away by clinical means and must rely on ultrasound. Given the symptoms of bleeding and cramping vaginosonography is the method of choice for deciding whether conservative or operative treatment is indicated.

3 Blighted Ovum

A blighted ovum presents sonographically as an echo-free sac, usually small for gestational age, that shows minimal growth on subsequent scans and may even diminish in size. A fetal pole or yolk sac is not seen. Differential diagnosis is based on the exclusion of a very early intact pregnancy in which the embryo is still too small to be resolved. The earliest point at which an embryo can consistently

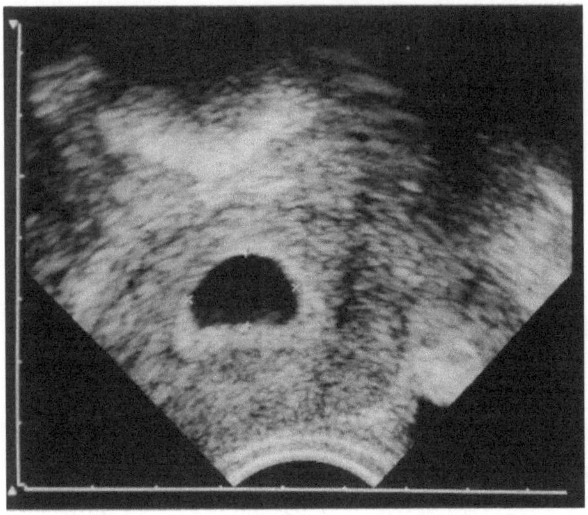

◀ **Fig. 5.** Longitudinal scan through the uterus in an anembryonic gestation. The gestational sac has a mean diameter of 15 mm, and there is no evidence of a yolk sac or embryo

be discerned in relation to gestational sac size has changed markedly in recent years. Thus, it has become possible to make a confident diagnosis of blighted ovum at smaller and smaller mean gestational sac sizes. Romero et al. (1984) and Hackelöer (1985) state a mean sac diameter of 30 mm as the cutoff point when abdominal scanning is used. Nyberg et al. (1986) and Scott et al. (1987) place the limits respectively at 25 mm and 26 mm. Vaginosonography has improved this situation considerably. The superior resolution of vaginal scans, even in a retroverted uterus, has made it possible to confirm an anembryonic pregnancy at a significantly earlier stage. For example, Bernaschek et al. (1988) and De Crespigny (1988) were able to confirm a blighted ovum at mean sac diameters of only 15 mm and 12 mm, respectively (Fig. 5). Similarly, Levi et al. (1988) could predict a nonviable pregnancy from failure to demonstrate an embryo at a sac size of 16 mm. Another criterion for nonviable pregnancy is the absence of a yolk sac, which generally serves as a guide in vaginosonography for locating the embryo in the chorionic cavity (Fig. 6). Levi et al. were able to correctly diagnose anembryonic gestation in all cases where a yolk sac was absent at a mean gestational sac size of 8 mm. Other criteria for blighted ovum such as a misshapen gestational sac (Fig. 7) with indistinct contours (Fig. 8) and a discrepancy between sac size and uterine size are obsolete, given the capability for early fetal delineation with vaginosonography.

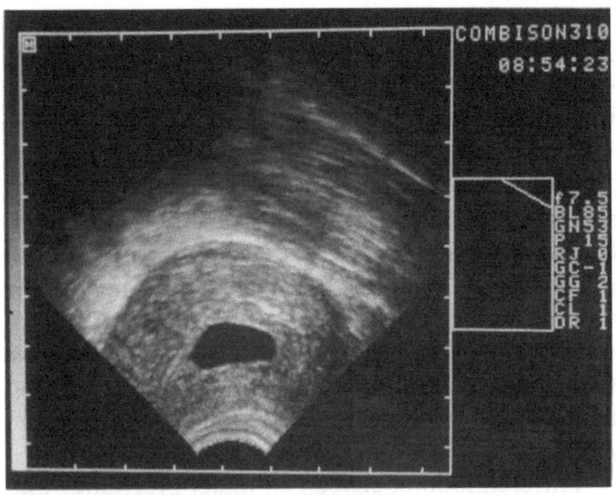

Fig. 6. Longitudinal scan through the uterus in an anembryonic gestation. Mean gestational sac diameter is 16 mm

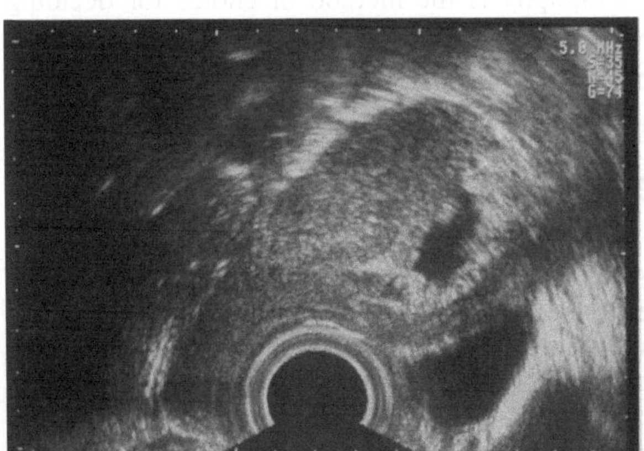

Fig. 7. Longitudinal scan through the uterus in an anembryonic gestation

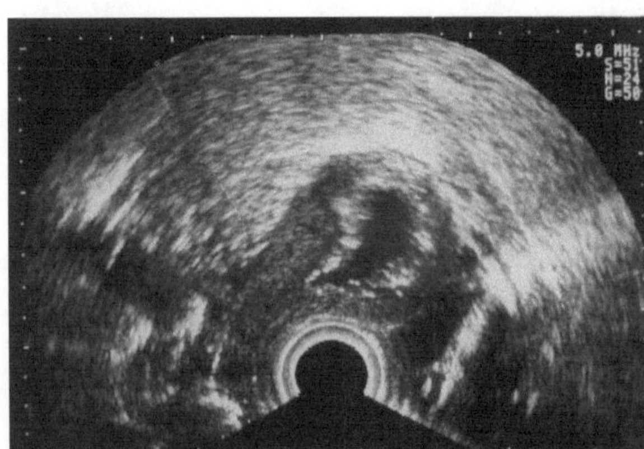

Fig. 8. Longitudinal scan through the uterus in an anembryonic gestation

Despite the capabilities of this method, it is not always prudent to make a final diagnosis at the earliest possible time. Usually it is best to verify a diagnosis of blighted ovum by follow-up scans, especially if uncertainty exists. Given the rapid growth rate of the early gestational sac and the early visualization of the yolk sac and embryo, we would advise performing a repeat scan 4-5 days later. In any case the menstrual history and clinical findings should be disregarded, and the final diagnosis of blighted ovum should rely entirely on vaginosonographic findings.

4 Missed Abortion

The diagnosis of missed abortion depends on the point at which fetal viability can be sonographically demonstrated. In abdominal ultrasound, an absence of cardiac activity does not establish the diagnosis of missed abortion until the fetus has reached a CRL of 10 mm or more (Longley and Sabbagha 1987). Vaginosonography, on the other hand, can confirm missed abortion when the CRL is only 5 mm, since pulsations of the cardiac tube can always be demonstrated at the time the embryo is seen. A CRL of 5 mm corresponds to a mean gestational sac diameter of 18 mm. By the 9th week of pregnancy, fetal musculature has developed to the point where body movements can be perceived. Other criteria, such as a size discrepancy between the gestational sac and embryo (Fig. 9), indistinct embryonic structures (Fig. 10), and passive swaying movements of the embryo in response to percussion have little or no importance in vaginosonography.

Evaluation of the yolk sac provides an additional early criterion for the diagnosis of missed abortion. The diameter of the yolk sac has been found to be larger in nonviable pregnancies (Fig. 11) than in viable pregnancies of equal age (Ferrazzi et al. 1988). In 44% of cases these authors noted a distinctive type of missed abortion in which a yolk sac was present but a fetal pole could not be seen. Bernard and Cooperberg (1985) and Hurwitz (1986), using abdominal scanning, similarly noted that an isolated yolk sac in a gestational sac larger than 25 mm should raise suspicion of missed abortion. With vaginosonography, this should be possible at a gestational sac size of 15-20 mm.

Funk and Fendel (1988 b) directed attention to the amniotic cavity in their vaginosonographic studies of missed abortion. They showed that the CRL and amniotic cavity both fail to increase in

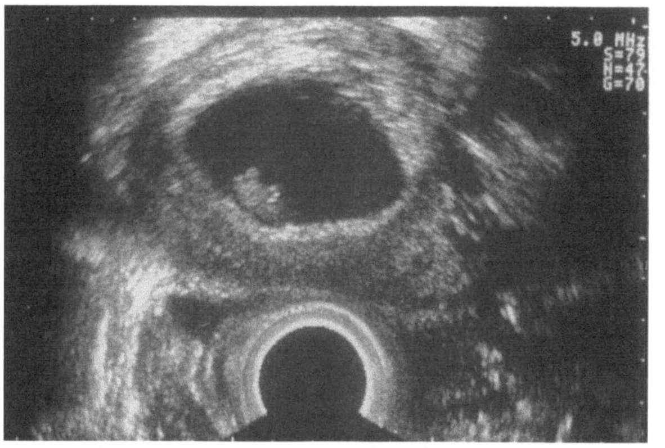

Fig. 9. Transverse scan through the uterus in a missed abortion, showing a discrepancy between gestational sac size and CRL

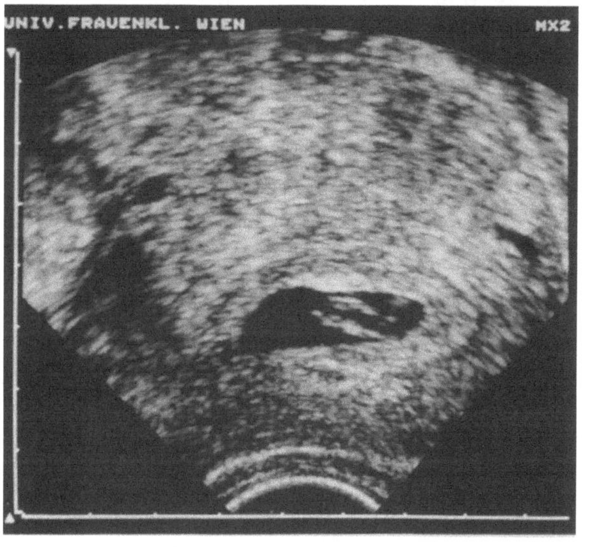

Fig. 10. Longitudinal scan through the uterus showing a missed abortion with indistinct embryonic structures

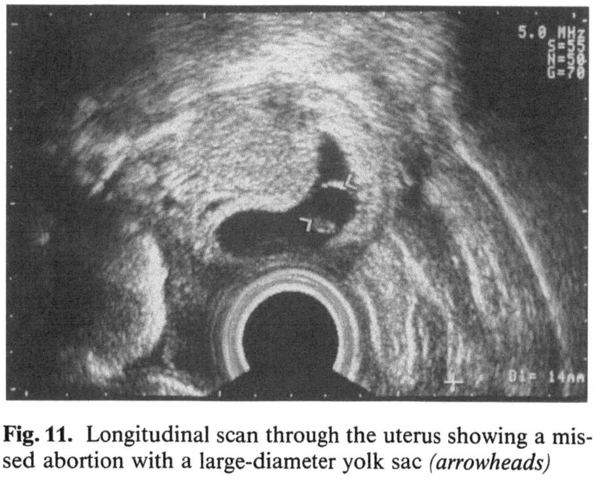

Fig. 11. Longitudinal scan through the uterus showing a missed abortion with a large-diameter yolk sac *(arrowheads)*

size in missed abortion while the chorionic cavity continues to enlarge as in a viable pregnancy.

Generally, however, we caution against premature intervention when a missed abortion is diagnosed. It seems prudent to perform a repeat scan 3 days later, especially if doubt exists. This will decrease the risk of error while still saving some of the time gained by the use of vaginosonography.

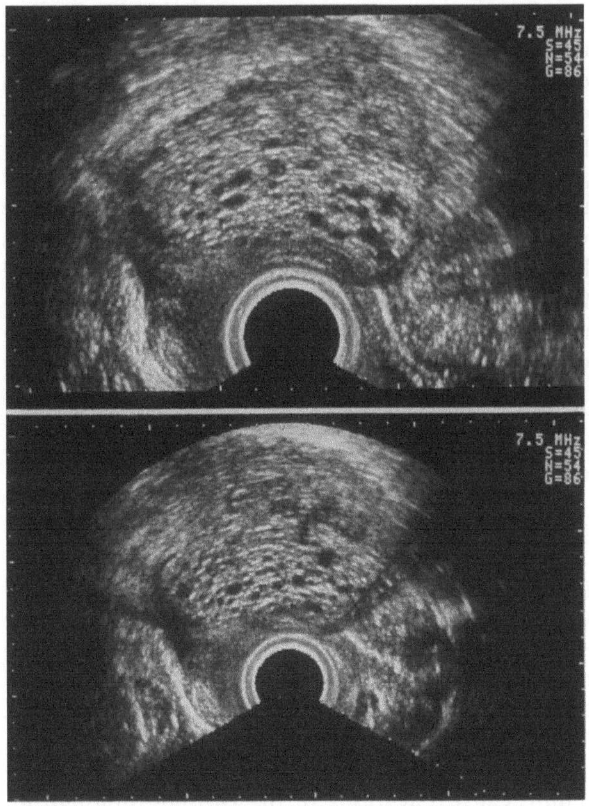

Fig. 12. Vaginal scans of a hydatidiform mole. *Top:* longitudinal scan through the uterus; *bottom:* transverse scan

5 Incomplete Abortion

Incomplete abortion is diagnosed mainly from the history and clinical findings. Nevertheless, sonography is useful for making a definitive diagnosis, especially when clinical findings are inconclusive. The sonographic features of incomplete abortion are variable. The placental remnants left in the uterine cavity and cervical region usually produce dense echoes. Occasionally there may be an irregular, partially collapsed gestational sac filled with nonhomogeneous echoes. Retrochorioidal hematomas may also be present. Except in cases of marked obesity or a retroflexed uterus, vaginosonography can add little to the results of transabdominal scanning.

Complete abortion is characterized by an empty uterus whose cavity appears as an echogenic line less than 4 mm wide. While this condition can be detected by abdominal scanning, it can be diagnosed with much greater confidence by vaginosonography. By confirming a complete abortion, sonography can obviate the need for curettage in many patients. Serial β-hCG assays are useful for confirming the absence of active trophoblastic tissue in cases managed conservatively.

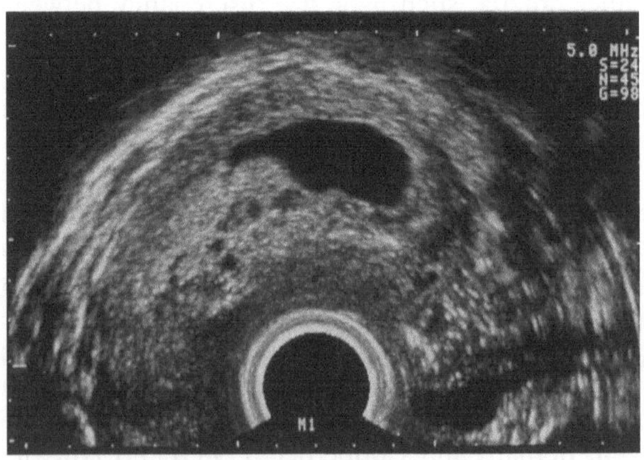

Fig. 13. Longitudinal scan through the uterus showing circumscribed molar degeneration of the placenta

6 Hydatidiform Mole

Hydatidiform mole is easily recognized by its characteristic "snowstorm"-like sonographic appearance, first described by Donald in 1965. However, greatly enlarged uteri filled with small cystic spaces (Fig. 12) are rarely seen nowadays owing to advances in antenatal diagnosis. It is more common to encounter early forms showing molar degeneration in circumscribed areas of the placenta or partial molar degeneration with or without a demonstrable embryo. Equivocal sonographic findings may be clarified by a β-hCG assay. However, a hydatidiform mole is not always associated with an elevated hCG titer. Vaginosonography is most rewarding in very obese patients and in early cases showing only small, circumscribed cystic changes (Fig. 13).

7 Ectopic Pregnancy

7.1 General

The incidence of ectopic pregnancy has increased markedly in the last two decades. This is attributed mainly to the rising incidence of adnexitis and the more widespread use of intrauterine contraceptive devices. Recurrence of ectopic pregnancy following tube-conserving operative treatment is also a significant factor.

Ectopic pregnancy is difficult to diagnose because of the variability of symptoms and morphologic presentations. Although abdominal sonography has become the most reliable tool for evaluating other early pregnancy disorders, it has proved disappointing in the diagnosis of ectopic pregnancy. While endosonography cannot eliminate all problems, vaginal scanning does offer some advantages in the diagnosis of extrauterine pregnancy, and these will be discussed below.

7.2 Exclusion of Intrauterine Pregnancy

This question is paramount, because concurrent intrauterine and extrauterine pregnancies occur in only 1 of 30000 pregnancies, so the discovery of an intrauterine pregnancy virtually excludes an ectopic gestation. The sonographic detection of a viable early pregnancy presents no difficulties. Problems can arise in the differentiation between gestational and pseudogestational sac within the uterus. The combination of β-hCG determination and ultrasound investigation is essential. Vaginosonography is of particular value in this regard owing to its consistent ability to demonstrate a gestational sac immediately after a missed menstrual period.

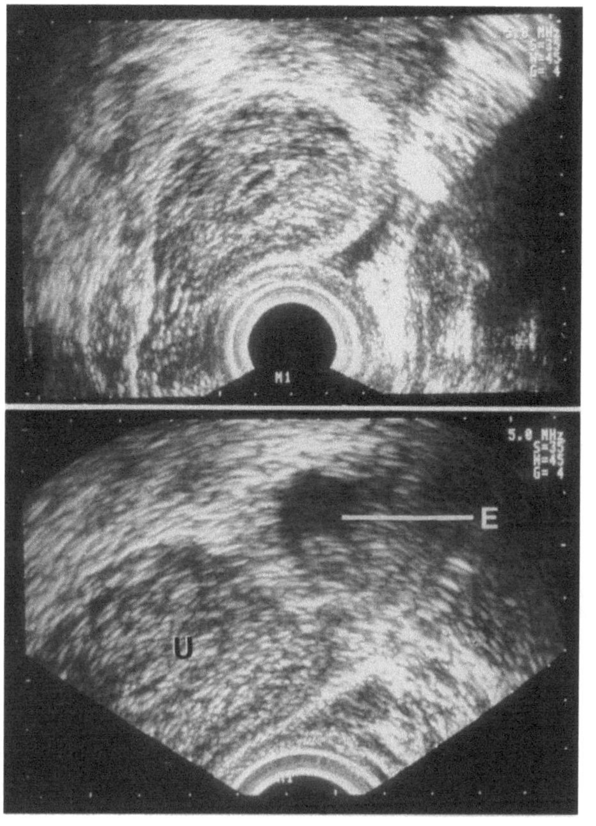

Fig. 14. Ectopic pregnancy. *Top:* longitudinal scan through the uterus with a small central fluid collection. *Bottom:* an intact ectopic pregnancy *(E)* is seen to the left of the uterus *(U)*

7.2.1 Recognition of a Pseudogestational Sac

The endometrial thickening caused by hormonal stimulation can produce an intrauterine decidual ring that is considered a suggestive sign of ectopic pregnancy. The ultrasound studies of Abramovici et al. (1983) indicate that a pseudogestational sac mimicking a true gestational sac due to blood collection and endometrial proliferation is present in up to 20% of patients with ectopic pregnancy. However, it is likely that the earlier diagnosis provided by vaginosonography will significantly reduce the frequency of demonstrable pseudogestational sacs (Fig. 14). In particular, the large pseudogestational sacs that may be mistaken for a blighted ovum could become less common (Fig. 15). Differentiating features from an early intrauterine pregnancy are described in Sect. 1.3 of the chapter "Normal Early Pregnancy."

7.2.2 Discriminatory Limit

Though the diagnosis of intrauterine pregnancy appears simple, problems arises when it is necessary to exclude intrauterine pregnancy in the face of a suspected ectopic gestation. The question, reduced to its simplest form, is: What is the earliest point at which an intrauterine pregnancy can be confidently diagnosed with ultrasound? Because information on menstrual dates cannot be considered reliable, this question can be addressed only by correlating sonographic findings with measurements of β-hCG levels. The discriminatory limit is the β-hCG level above which an intrauterine pregnancy can consis-

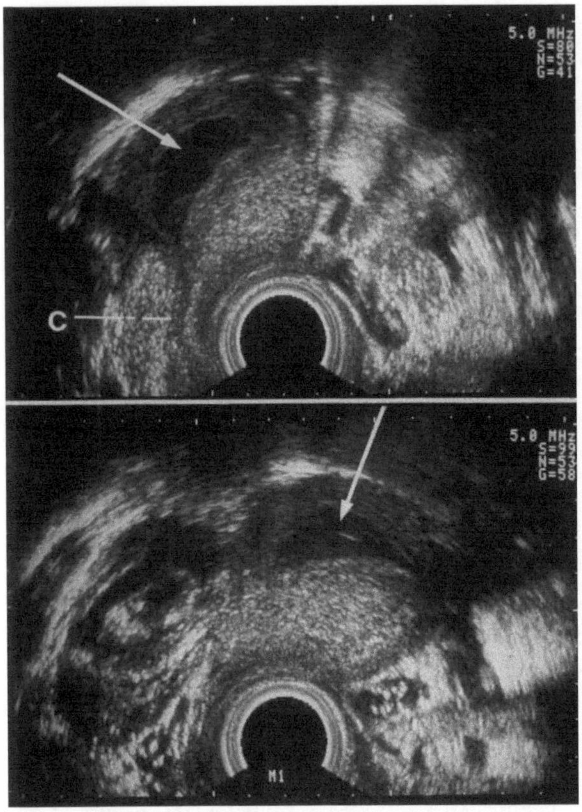

Fig. 15. Ectopic pregnancy. *Top:* longitudinal scan through the uterus showing a large pseudogestational sac *(arrow)* and the cervical canal *(C)*. *Bottom:* transverse scan through the uterus showing the pseudogestational sac *(arrow)* in the same case. A partly anechoic, partly echogenic area consistent with ectopic pregnancy appears in the right adnexal region

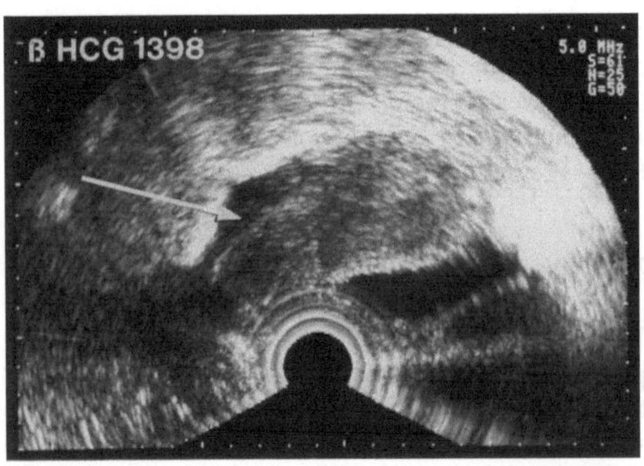

Fig. 16. Longitudinal scan through the lesser pelvis in ectopic pregnancy. The uterus appears empty. The corresponding β-hCG level is 1398 mIU/ml, and there are small amounts of free fluid in the cul-de-sac *(arrow)*

Fig. 17. Adjacent to the uterus is a large, intact ectopic conceptus with a CRL of 43 mm *(arrowheads)* and positive cardiac activity

tently be detected with ultrasound. The improved resolution of vaginosonography has made it possible to lower this limit to between 500 and 1000 mIU/ml (Fig. 16). Details on the significance of the discriminatory limit in normal pregnancies are given in Sect. 1.4 of the chapter "Normal Early Pregnancy."

Approximately 65% of all ectopic pregnancies show β-hCG levels greater than 1000 mIU/ml (Nyberg et al. 1988). In almost all such cases vaginosonography will demonstrate a suspicious area in the adnexal region consistent with ectopic gestation. When the β-hCG level is less than 1000 mIU/ml, this is possible in only one-third of cases.

7.3 Demonstration of an Intact Ectopic Pregnancy in the Adnexal Region

An ectopic pregnancy is rarely demonstrated directly by abdominal ultrasound (1%-5%), because the products of conception tend to deteriorate quickly outside the uterus and may present only as a nonspecific mass having both cystic and solid features. Thus, the demonstration of a small echo-free area in the adnexal region is not sufficient grounds for diagnosing an intact ectopic pregnancy, especially if the β-hCG levels are below normal or show little tendency to rise on serial measurements. However, vaginosonography can sometimes demonstrate a yolk sac within the ectopic gestational sac as early as 6 weeks gestation. Using vaginosonography, Funk and Fendel (1988 a) reported a considerable increase in the rate with which ectopic pregnancies could be directly visualized and evaluated (Fig. 17). They were able to demonstrate an ectopic pregnancy in 13 of 16 cases examined. Rempen (1988) also found an increased detection rate of intact ectopic

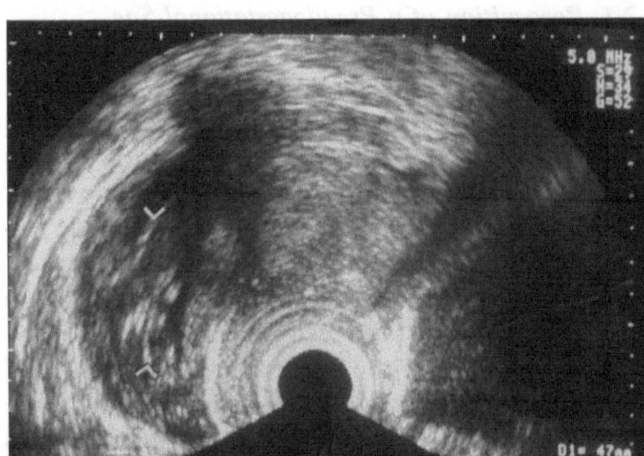

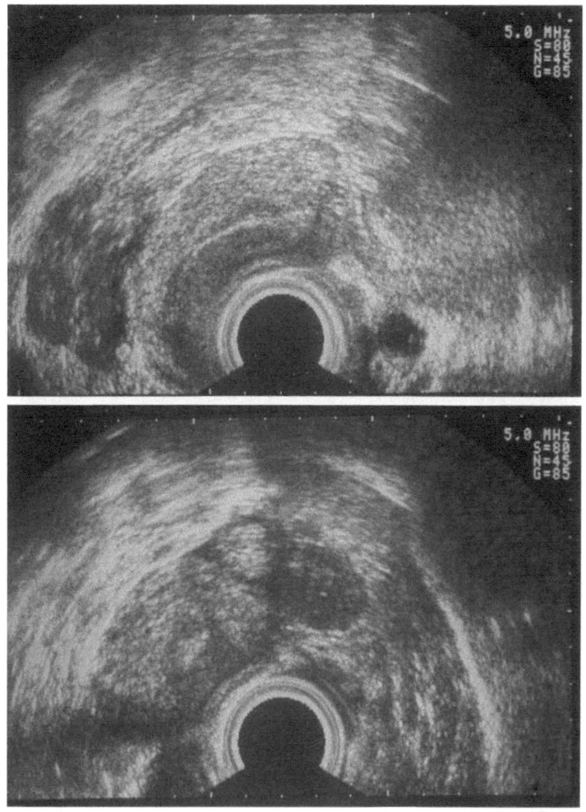

Fig. 18. Ectopic pregnancy. *Top:* longitudinal scan through the lesser pelvis. The uterus appears empty, and there is a nonspecific hypoechoic area in the cul-de-sac consistent with tubal pregnancy. *Bottom:* transverse scan in the same case, showing the suspicious area on the left side of the empty uterus

pregnancies with vaginal ultasound (71%) and even detected fetal life in 24%.

The early, reliable detection of an extrauterine gestational sac is important in selecting patients for "conservative" treatment (see the chapter "Endosonographically Guided Punctures") and for the planning of tube-conserving surgical procedures. Once an intact ectopic pregnancy has been diagnosed by vaginosonography, the patient should be referred for surgical treatment without delay. Waiting for symptoms to appear will only worsen the patient's chances.

7.4 Nonspecific Masses in the Adnexal Region

Cystic/solid masses associated with ectopic pregnancy tend to have a more elongated shape than corpus luteum cysts and contain more echogenic areas. They may be seen in cases of tubal rupture as well as tubal abortion. However, even the fine resolution provided by vaginosonography cannot distinguish these masses from inflammatory tubal changes based on sonographic findings alone (Fig. 18).

One advantage of vaginosonography is that the probe can be used to "palpate" the lesser pelvis under vision to allow the more selective examination of sites that are exceptionally painful. In some cases secondary sonographic criteria will have to be combined with clinical findings in order to localize the ectopic pregnancy. It is noteworthy that Rempen (1988) observed abnormal adnexal findings in 90% of all ectopic pregnancies evaluated by vaginosonography.

7.5 Free Fluid in the Abdomen

Following a tubal rupture, large amounts of free fluid may collect in the lesser pelvis even when clinical symptoms are mild (Fig. 19). Smaller amounts

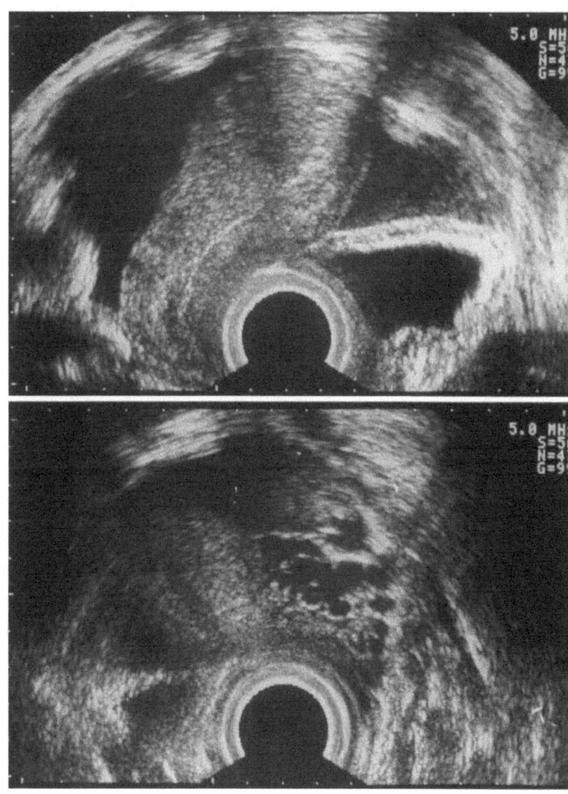

Fig. 19. Ruptured ectopic pregnancy. *Top:* longitudinal scan through the lesser pelvis. The uterus is empty, and there is copious free fluid in the cul-de-sac. *Bottom:* transverse scan in the same case showing a nonspecific, partly anechoic area on the left side

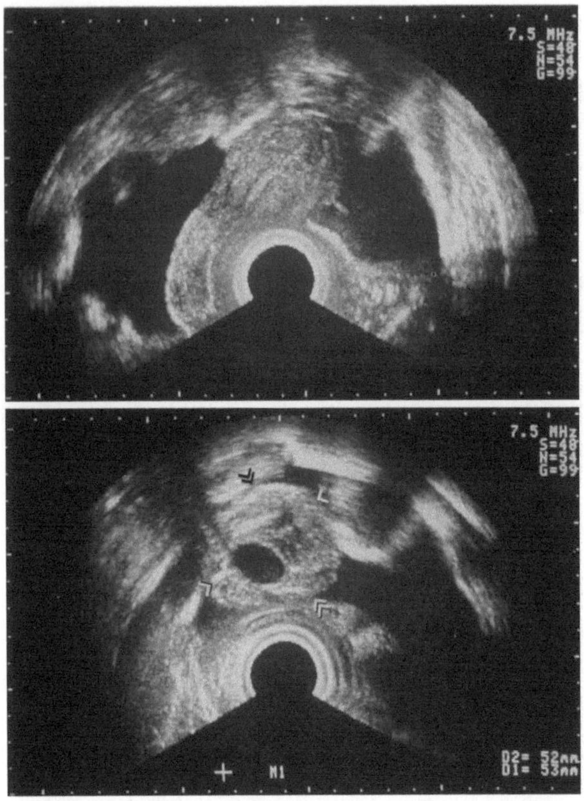

Fig. 20. Ruptured ectopic pregnancy. *Top:* longitudinal scan through the lesser pelvis. The uterus is empty, and there are large amounts of free fluid in the cul-de-sac. *Bottom:* the tubal pregnancy appears as an echo-free area with a wide rim *(arrowheads)* in the right adnexal region

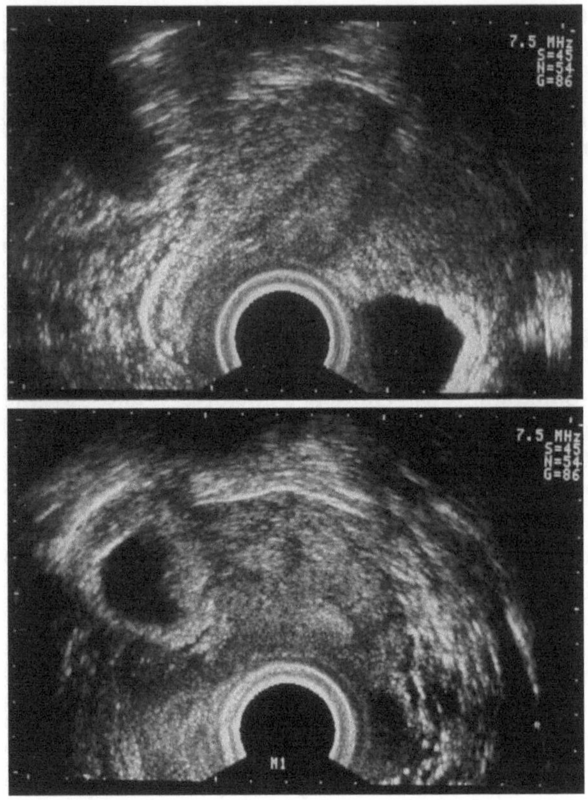

Fig. 21. Right cornual pregnancy. *Top:* longitudinal scan through the uterus showing an empty uterine cavity. *Bottom:* transverse scan through the uterus. A gestational sac appears to the right of the empty uterine cavity. (Sonographic diagnosis confirmed at operation)

of fluid may be seen in the cul-de-sac following a tubal abortion. Vaginosonography is the modality of choice for demonstrating even small amounts of cul-de-sac fluid. It should be noted, however, that small fluid collections may be found even in healthy, nonpregnant women, especially in the period immediately following ovulation. Thus, some experience with vaginosonography is needed in order to interpret correctly the finding of increased cul-de-sac fluid.

Vaginosonography demonstrates a fluid collection in the cul-de-sac (Fig. 20) in 70%–80% of all ectopic pregnancies (Funk and Fendel 1988a; Rempen 1988). Even if free intra-abdominal fluid on vaginal scans serves only as a suggestive sign, it can still aid in directing further diagnostic procedures such as culdocentesis (which is performed optimally under vaginosonographic guidance). Conversely, failure to detect free cul-de-sac fluid on vaginal scans will eliminate the need for culdocentesis.

Vaginosonography has proved especially valuable in acute cases of ectopic pregnancy, because it can be performed quickly and with an empty bladder. This is a particular advantage when general anesthesia is planned, in which case retrograde filling of the bladder might prove necessary if abdominal sonography were used.

7.6 Uncommon Sites of Ectopic Pregnancy

So far there have been no published reports on the endosonographic diagnosis of ovarian and abdominal pregnancies because of their very low incidence, and we have no personal experience with these entities. However, vaginosonography has demonstrated great value in the diagnosis of interstitial and cornual pregnancies. By simultaneous palpation with the free hand, it is possible to follow the synchronous movements of the eccentrically located gestational sac with the empty uterine cavity. In addition, the improved resolution of vaginosonography can demonstrate the thinned uterine wall over the gestational sac (Fig. 21).

7.7 Summary

1. Vaginosonography allows more confident differentiation between a true gestational sac and a pseudogestational sac.
2. An intrauterine pregnancy can be confirmed or excluded at low β-hCG levels.
3. Vaginosonography provides a substantially higher detection rate of ectopic pregnancy.
4. All other sonographic findings, such as endometrial thickening, cystic/solid adnexal masses, and free fluid in the cul-de-sac, are delineated better by vaginosonography, but still they are only suggestive signs that contribute in varying degrees to establishing or excluding an ectopic pregnancy.

References

Abramovici H, Auslender R, Lewin A, Faktor JH (1983) Gestational-pseudogestational sac: a new ultrasonic criterion for differential diagnosis. Am J Obstet Gynecol 145: 377-379

Anderson G (1980) Management of threatened abortion with real-time sonography. Obstet Gynecol 55: 259-262

Bernard KG, Cooperberg PL (1985) Sonographic differentiation between blighted ovum and early viable pregnancy. AJR 144: 597-602

Bernaschek G, Rudelstorfer R, Kölbl H (1988) The contribution of vaginal endosonography in the diagnosis of bleeding in pregnancy. JEMU 9: 204-206

De Crespigny LC (1988) Early diagnosis of pregnancy failure with vaginal ultrasound. Am J Obstet Gynecol 159: 408-409

Donald I (1965) Ultrasonic echo sounding in obstetrical and gynecological diagnosis. Am J Obstet Gynecol 93: 935-941

Ferrazzi E, Brambati B, Lanzani A, Oldrini A, Stripparo L, Guerneri S, Makowski EL (1988) The yolk sac in early pregnancy failure. Am J Obstet Gynecol 158: 137-142

Funk A, Fendel H (1988a) Verbesserte Diagnostik der Extrauteringravidität durch die Endosonographie. Z Geburtshilfe Perinatol 192: 49-53

Funk A, Fendel H (1988b) Ultraschallechographische Darstellbarkeit und Messung der Amnionhöhle und des Dottersackes in der frühen Schwangerschaft: Vergleichende Untersuchungen von intakten und gestörten Schwangerschaften. Z Geburtshilfe Perinatol 192: 59-66

Goldstein SR, Subramanyam BR, Raghavendra BN, Horii SC, Hilton S (1983) Subchorionic bleeding in threatened abortion: sonographic findings and significance. AJR 141: 975-978

Hackelöer BJ (1985) Gestörte Entwicklung. In: Hansmann M, Hackelöer BJ, Staudach A (eds) Ultraschall in Geburtshilfe und Gynäkologie. Springer, Berlin Heidelberg New York, p 57

Hurwitz SR (1986) Yolk sac sign: sonographic appearance of the fetal yolk sac in missed abortion. J Ultrasound Med 5: 435-438

Jouppila P (1985) Clinical consequences after ultrasonic diagnosis of intrauterine hematoma in threatened abortion. J Clin Ultrasound 13: 107-111

Levi CS, Lyons EA, Lindsay DJ (1988) Early diagnosis of nonviable pregnancy with endovaginal US. Radiology 167: 383-385

Longley JV, Sabbagha RE (1987) Abnormal early pregnancy. In: Sabbagha RE (ed) Diagnostic Ultrasound applied to obstetries and gynecology, 2nd edn. Lippincott, Philadelphia, pp 464-473

Nyberg DA, Laing FC, Filly RA (1986) Threatened abortion: sonographic distinction of normal and abnormal gestation sacs. Radiology 158: 397-400

Nyberg DA, Mack LA, Laing FC, Jeffrey RB (1988) Early pregnancy complications: endovaginal sonographic findings correlated with human chorionic gonadotropin levels. Radiology 167: 619-622

Rempen A (1988) Vaginal sonography in ectopic pregnancy. J Ultrasound Med 7: 381-387

Romero R, Jeanty P, Hobbins JC (1984) Diagnostic ultrasound in the first trimester of pregnancy. Clin Obstet Gynecol 27: 286-291

Sauerbrei EE, Dieu HP (1986) Placental abruption and subchorionic hemorrhage in the first half of pregnancy: US appearance and clinical outcome. Radiology 16: 109-112

Scott RF, Featherstone T, Hussey JK (1987) Ultrasound of the empty gestation sac in threatened abortion. Clin Radiol 38: 127-131

Vaginosonographic Examination of the Fetus

1 General

Transabdominal ultrasound, with its large field of view, can demonstrate the presenting fetal part very easily and usually can depict numerous anatomic details. However, this can be difficult if the presenting part is already low in the pelvis when the study is performed. Even in the second trimester it can be difficult to visualize certain fetal structures if the presenting part is shadowed by the fetal limbs. Cases of this kind require either a different examination technique or a different route of approach.

Vaginal sonography was initially believed to have its greatest value in early pregnancy. At that time vaginosonography can play a crucial role in identifying normal and abnormal pregnancies, providing earlier detection of fetal life, and detecting ectopic gestation. The value of vaginosonography in detecting a low-lying placenta is discussed in the chapter "Placenta Previa." But the advantages of this modality (transducer close to the organ of interest, higher transducer frequencies, superior resolution) are leading to increasing application in later stages of pregnancy as well.

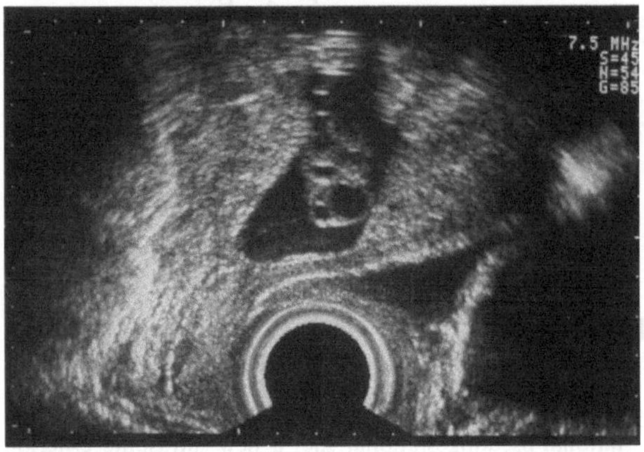

Fig. 1. Vaginal scan of the forelying umbilical cord. All three vessels can be identified

2 Indications

2.1 Examination of the Fetus in the First Half of Pregnancy

2.1.1 Biometry

The late first trimester and early second trimester are particularly well suited for vaginosonographic evaluations of the fetus. At that time it is possible to image the entire fetus, depending on the scanning frequency and the available field of vision. It is advantageous to use a scanner with variable frequencies, for otherwise the presentation may allow only portions of the fetus to be delineated. By the 12th week of gestation the presence of fetal organs can be confirmed, and the entire spine can be visualized. At any stage of pregnancy it may be possible to demonstrate portions of the umbilical cord by vaginosonography (Fig. 1).

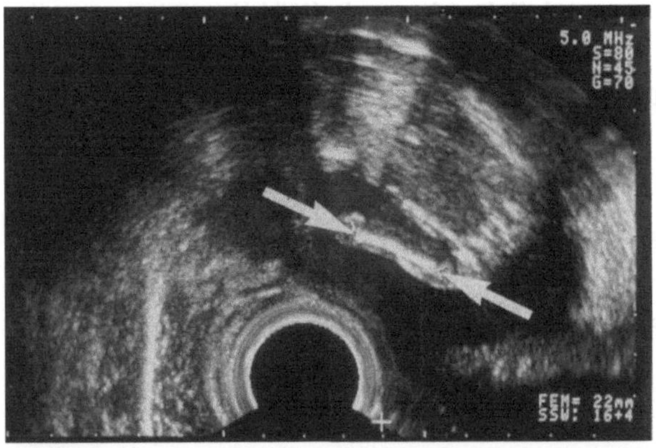

Fig. 2. Vaginal fetal biometry: measurement of femur length in the 16th week of pregnancy *(arrows)*. Transabdominal biometry was not possible in this case due to bowel overlying the uterus

The advantages of vaginosonography for fetal examination include the capacity for earlier biometric evaluation and a more accurate prediction of term, especially when results from gestational sac biometry are not available (Fig. 2).

2.1.2 Visualization of Fetal Organs

The earliest point at which fetal organs can be depicted with ultrasound depends on many factors such as the experience of the examiner, the quality of the equipment, the thickness of the maternal abdominal wall, the amniotic fluid volume, and fetal position. However, an even more important consideration is the week of precnancy in which the fetal organs in general can be demonstrated clearly enough to permit the detection or exclusion of anomalies. Generally it may be said that vaginosonography can demonstrate all fetal organs 1-4 weeks earlier than is possible with transabdominal scanning (Warren et al. 1989). When organ development is evaluated at such an early stage of pregnancy, it must be understood that some structures have not yet reached the stage of development that is traditionally demonstrated by abdominal ultrasound. Thus, while the falx cerebri is not seen when the fetal skull is examined in the 9th week of pregnancy, a scan taken in the 12th week will disclose the midline echo of the falx (Fig. 3).

From a sonographic standpoint, the dominant process in early embryonic development is the formation and differentiation of the central nervous system. Various stages can be recognized in the development of the embryonic brain. Ideally, a caudal and a rostral pole can be distinguished as early as the 6th week of pregnancy (Fig. 4). At the caudal pole are the three cerebral vesicles consisting of the prosencephalon, mesencephalon, and rhombencephalon (Fig. 5) (Wisser et al. 1988). With failure of neural tube closure in the brain region, a diagnosis of anencephaly can be made even at this very early stage. One week later the brain is seen to consist of five parts: the cerebrum, the diencephalon, the mesencephalon, the metencephalon, and the myelencephalon, from which the medulla oblongata is derived. This is the stage at which holoprosencephaly may develop, characterized by facial dysmorphias, arrhinencephaly, and cerebellar aplasia. Additional details appear in the cranium by the 8th week of gestation, and the orbits, thalamus, choroid plexus, and cerebellum can be discerned by the 9th or 10th week. The developing ventricular system and choroid plexus appear to occupy most of the cranial volume at this stage (Fig. 6). Cysts of the choroid plexus usually have no pathologic significance (Green and Hobbins 1988). The maxilla and mandi-

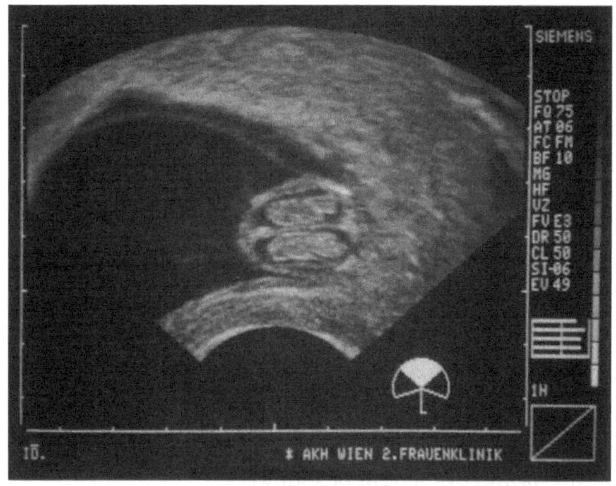

Fig. 3. Vaginal scan in the 12th week already demonstrates the intracranial midline echo, allowing accurate fetal cephalometry

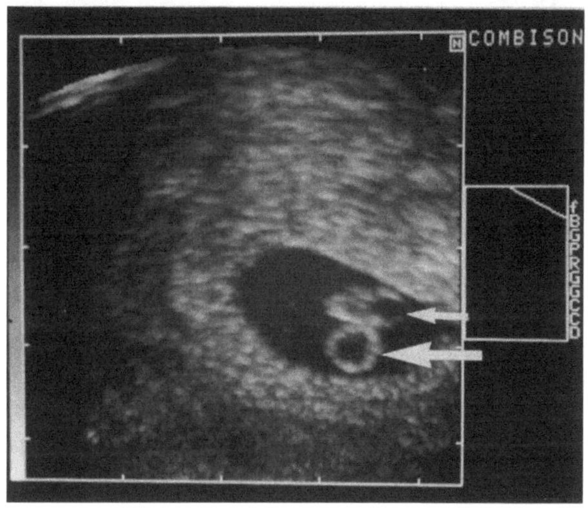

Fig. 4. The caudal and rostral poles can be distinguished as early as 6 weeks' gestation. The *small arrow* indicates the rostral pole, the *large arrow* the yolk sac

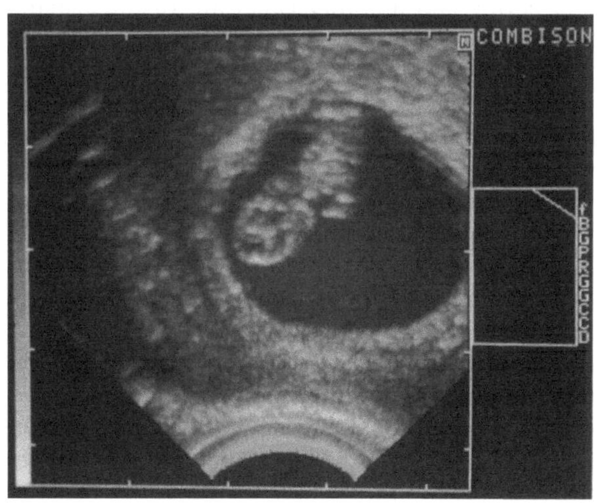

Fig. 5. The three cerebral vesicles consisting of the prosencephalon, mesencephalon, and rhombencephalon can be identified as early as the 6th or 7th week ▶

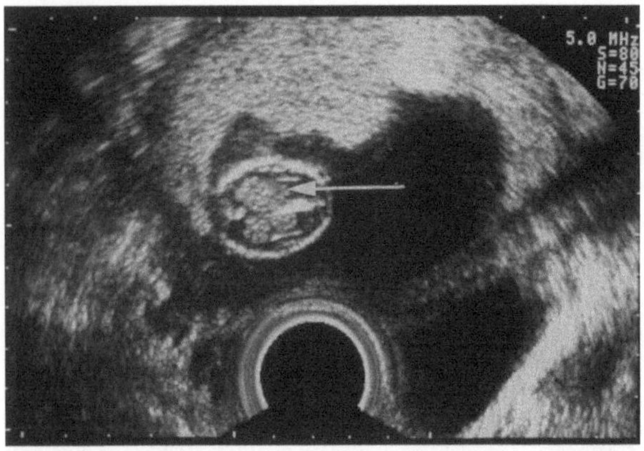

Fig. 6. Most of the cranium appears occupied by choroid plexus *(arrow)* by the 12th or 13th week

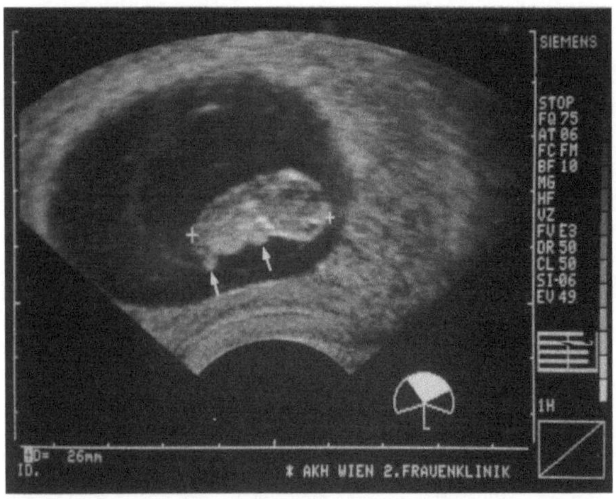

Fig. 7. Fin-like limbs are visible *(arrows)* at a CRL of only 26 mm

ble can also be differentiated at this time. The cerebral cortex and auditory canal are distinguishable by the 11th or 12th week of pregnancy.

In two-thirds of cases, transabdominal ultrasound can demonstrate the fetal head with a midline echo and two large lateral ventricles in the 14th week of pregnancy (Zador et al. 1988). This can be done as early as the 12th week using vaginal ultrasound. However, because the lateral ventricles appear disproportionately large in early pregnancy, hydrocephalus should not be diagnosed before the 16th week of pregnancy.

Identification of the fetal kidneys is also of great importance in early pregnancy. Abdominal scanning can rarely demonstrate both kidneys before the 20th week of pregnancy (Zador et al. 1988), whereas vaginosonography can do so by the 14th week. Consistent delineation of the fetal bladder should also be possible at that time. Cardiac anatomy also can be evaluated by obtaining the four-chamber view (Warren et al. 1989).

The fetal stomach can be visualized in 93% of cases by the 12th week of gestation (Green and Hobbins 1988). A "physiologic diaphragmatic hernia" is still present in 20% of fetuses by the 13th week. This accounts for the great uncertainty of abdominal biometry in early pregnancy (Green of Hobbins 1988).

The limbs can be sonographically identified as early as the 9th week of gestation (Fig. 7). At that time they still appear finlike, and reproducible biometric data cannot be obtained until the bones become mineralized in the 12th week.

Besides allowing earlier visualization of the fetal head and kidneys, vaginosonography can provide earlier and more detailed images of the fetal spine, heart, stomach, bladder, and umbilical cord, depending on the fetal position. We would recommend that a vaginal ultrasound examination be performed routinely in every pregnancy between the 12th and 16th weeks. After 16 weeks, vaginosonography serves as an adjunct to abdominal ultrasound and, depending on the fetal position, can help to substantiate a tentative diagnosis based on transabdominal findings.

2.1.3 Screening for Anomalies

The advancement of prenatal diagnosis into the first trimester offers significant humanitarian, social, and economic benefits (Green and Hobbins 1988). For the experienced examiner performing vaginosonography at this early stage of pregnancy, the question is no longer "What can I see under optimum conditions?" but rather "What structures should be visible in a normal pregnancy, and at what stages?"

With vaginosonography it is possible to make a morphologic assessment of major fetal structures as early as the 12th week of pregnancy. Organ biometry at this stage provides a very effective screening tool that is unequaled by any other method at such an early stage. In a series of 400 examinations performed between the 12th and 15th weeks of pregnancy, 10 major fetal anomalies could be diagnosed by vaginosonography: two cases each of ventriculomegaly and polycystic kidneys, one cystic hygroma, one case each of polydactyly and spina bifida, and several cardiac malformations (Bronshtein et al. 1988 a).

An enlarged nuchal fold recognized at this stage may signify the presence of a chromosome abnormality, which can then be confirmed or excluded by karyotyping.

Before vaginosonography is applied more generally as a screening tool for fetal malformations, additional data are required on the normal sonographic fetal anatomy at this early stage of pregnancy. At present, one should be careful about drawing conclusions from pathologic findings until other studies have been done to confirm the diagnosis.

2.1.4 *Anhydramnios*

Amniotic fluid in normal quantities makes it considerably easier to distinguish fetal parts with ultrasound. In the presence of oligohydramnios or anhydramnios, evaluation of the fetal kidneys in early pregnancy is essential yet is made difficult by the paucity of amniotic fluid. If the problem is compounded by maternal obesity, an accurate ultrasound examination may be impossible. Vaginosonography facilitates evaluation of the fetal organs in these cases as well. This method can demonstrate the kidneys and also provide a four-chamber view of the heart as early as the 14th week of pregnancy - an especially helpful study in the presence of oligohydramnios (Timor-Tritsch 1988).

The advantages of vaginosonography are obvious in cases of marked obesity, oligo- or anhydramnios, or when abnormalities have been noted on transabdominal scans (Figs. 8-12). If additionally the fetal structure of interest is within the lesser pelvis, it may not be possible to obtain satisfactory re-

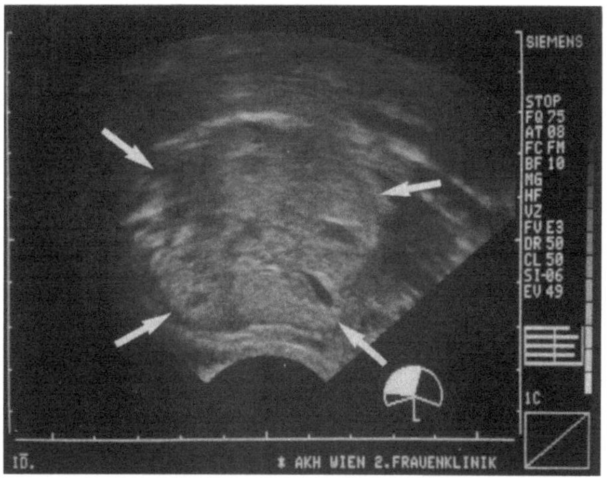

Fig. 9. Transverse scan of the abdomen (*arrows:* abdominal wall)

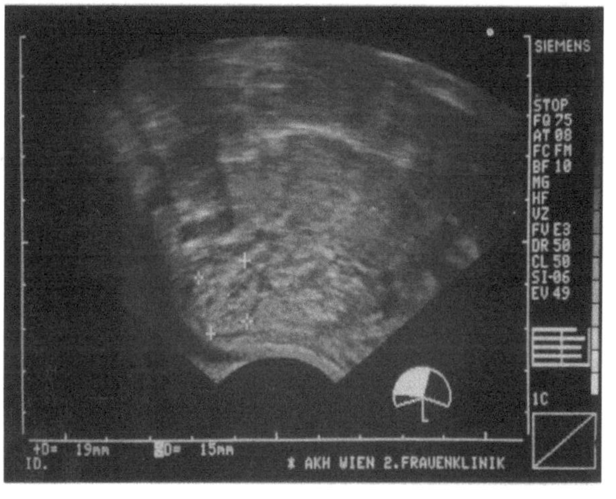

Fig. 10. Transverse scan of the lesser pelvis demonstrating one kidney *(crosses)*

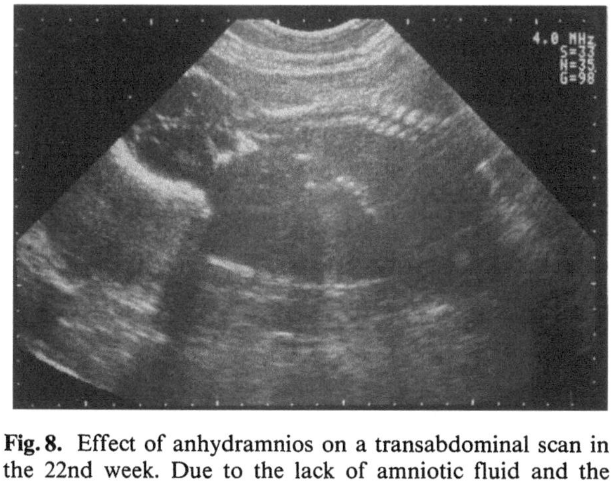

Fig. 8. Effect of anhydramnios on a transabdominal scan in the 22nd week. Due to the lack of amniotic fluid and the acoustic shadow from the spinal column, intra-abdominal structures cannot be evaluated. Transvaginal scanning in this case gave a much clearer picture of structural details

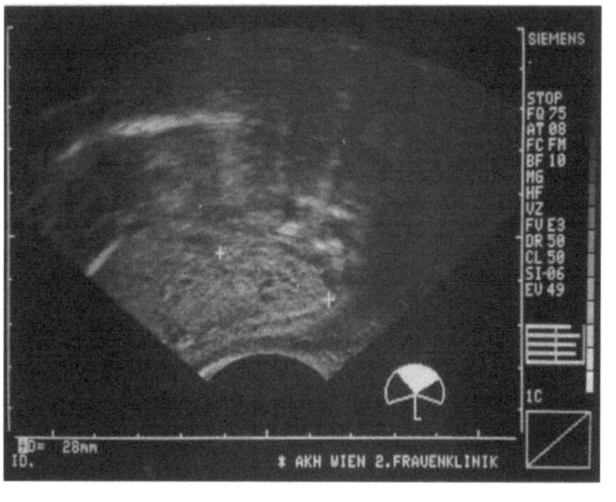

Fig. 11. Longitudinal scan of the lesser pelvis demonstrating one kidney *(crosses)*

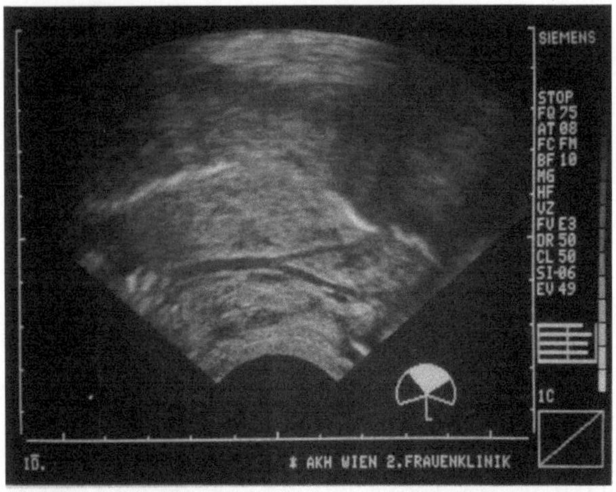

Fig. 12. Longitudinal scan of the trunk showing the aortic bifurcation

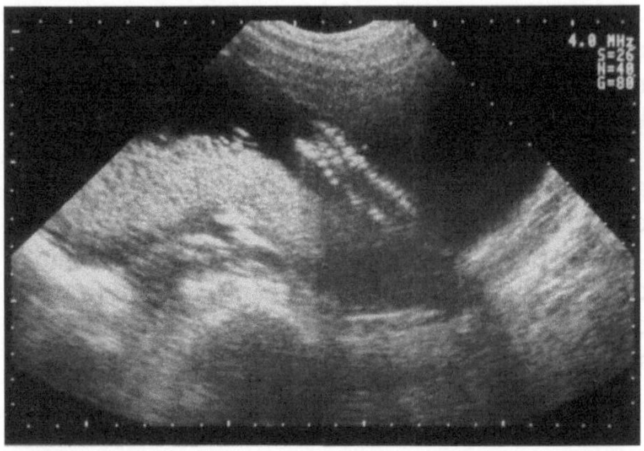

Fig. 13. Transabdominal scanning does not always allow visualization of the fetal spine, even in a frontal plane

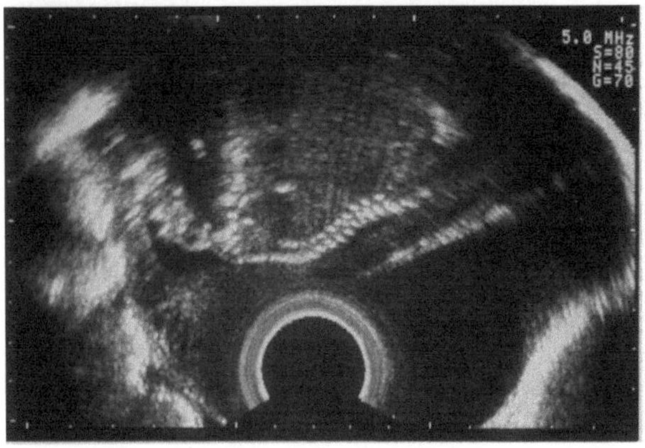

Fig. 14. Vaginal sonogram of an anencephalic fetus in the 16th week of pregnancy. The longitudinal scan can demonstrate the fetal spine but no normal head contour

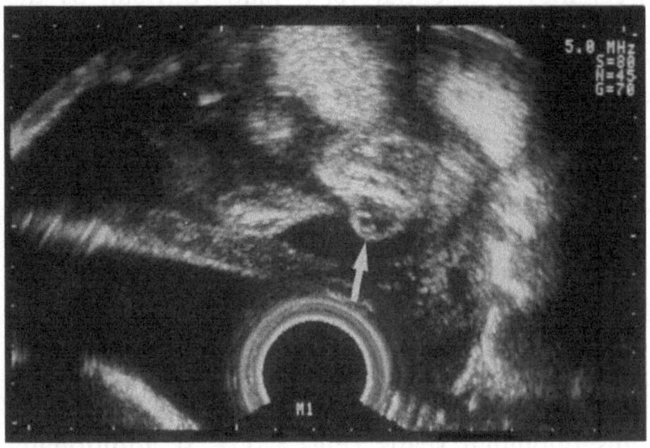

Fig. 15. Vaginosonography was able to demonstrate protrusio bulbi *(arrow)* in case of anencephaly (same case as in Fig. 14)

sults with abdominal scanning. Thus, abdominal scans may be unable to identify an anencephalic fetus in the vertex presentation, even on the sagittal plane (Figs. 13–15).

2.1.5 Sex Determination

In 60% of cases in the second trimester of pregnancy the fetus is in a breech or labile presentation (Staudach 1986), permitting the early determination of fetal sex with ultrasound. This determination should be made no earlier than the 14th week, for direct examinations of aborted fetuses at this stage have shown that sex cannot be determined even morphologically in 20% of cases (Timor-Tritsch 1988). Sex determination is possible in 88% of fetuses examined by vaginosonography between the 15th and 16th weeks. Interestingly, a correct assignment was made in 96% of males and in 98% of females (Bronshtein et al. 1988 b).

Early sonographic sex determination is not considered an essential study, however. If there is risk or suspicion of chromosome-linked disease (e. g., x-recessive hemophilia), karyotyping and subsequent DNA analysis are performed prior to the 12th week of pregnancy, and thus before vaginosonographic sex determination could be applied.

2.2 Examination of the Fetus in the Second Half of Pregnancy

2.2.1 Examination of the Presenting Part

Vaginosonography in the third trimester is limited to examinations of the presenting fetal part or even portions of it due to the limited range of the sound field (depending on the frequency used). Since more than 90% of fetuses have a cephalic presentation in the last trimester and just before term, vaginosonography is especially well suited for evaluating the craniocerebral anatomy at this time. While abdominal sonography can easily provide transverse scans of the fetal head, sagittal and frontal scans are difficult to obtain (Duda et al. 1987). But because cerebral sonograms in newborns are taken through the acoustic window of the fontanelles using sagittal and frontal scan planes, the results of prenatal and postnatal cerebral sonograms have been difficult to compare. Vaginosonography, by providing sagittal and frontal scans of the fetal head, can help to bridge the gap in terminology, scan planes, and documentation between the obstetrician and pediatrician (Fig. 16). Low-frequency transducers (e. g., 3 MHz) are advantageous for this application.

The practical advantages of vaginal fetal brain scanning include the improved delineation of intracranial pathology (Fig. 17) and the detection of cerebral hemorrhage developing in the immediate prepartal period (Duda et al. 1987).

If hydrocephalus detectable at this stage threatens to obstruct the delivery, the situation can be remedied by aspirating cerebrospinal fluid with an endovaginal needle passed under vaginosonographic guidance (see the chapter "Endosonographically Guided Punctures") (Chayen and Rifkin 1987).

In breech presentations, vaginosonography can be used to visualize either the buttocks or the feet (Fig. 18). Even after the onset of labor while the os uteri is still closed, vaginosonography can help to establish the type of breech presentation and direct obstetric management. In breech presentations with a sacrococcygeal teratoma, the relative proportion of cystic and solid components can be assessed more accurately with vaginal ultrasound, although larger tumors may extend beyond the range of the beam.

In summary, transabdominal sonography continues to be the mainstay for fetal evaluation in the second half of pregnancy, but there are cases in which vaginosonography is extremely helpful in providing a more detailed view of the presenting fetal part.

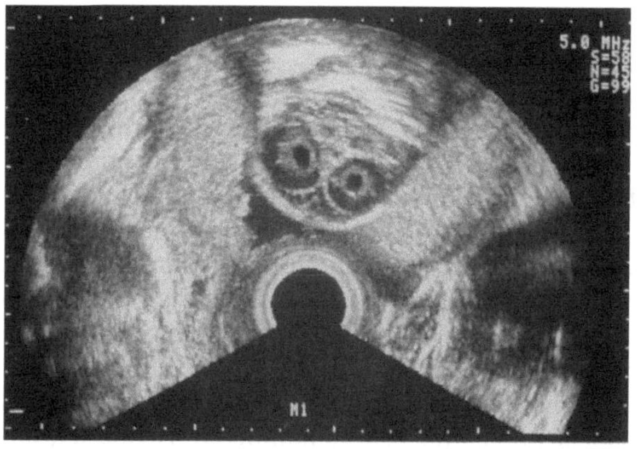

Fig. 16. Frontal scan through the fetal head in the 25th week with an image of the lateral horns

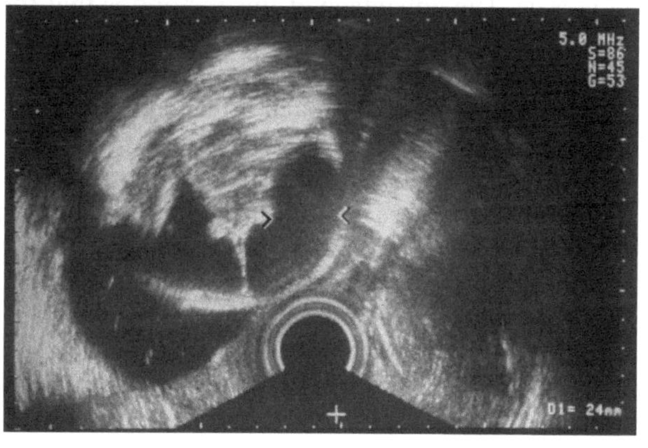

Fig. 17. Vaginosonogram in hydrocephalus. The lateral horn *(arrowheads)* is expanded to 24 mm

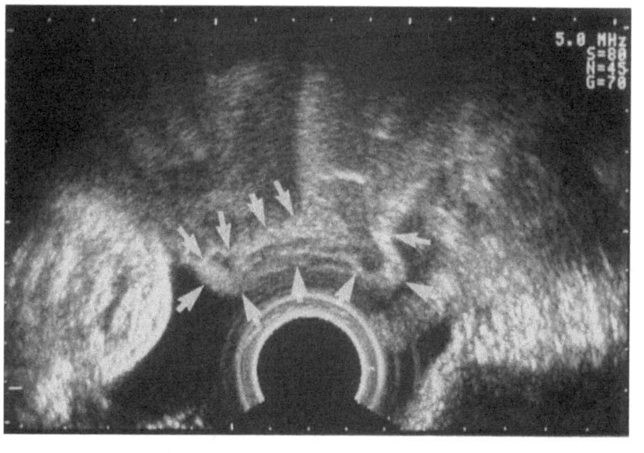

Fig. 18. Vaginosonogram in a breech presentation immediately before delivery. Only the uterine wall separates the foot *(arrows)* from the vaginal scanner

2.2.2 *Vaginal Pulsed Doppler Examination of Fetal Arteries*

Fetal heart activity can be detected earlier in the first trimester when the vaginal route is employed. Doppler ultrasound or combined Doppler and M-mode scanning offer additional advantages (Ikegawa et al. 1987), although Doppler scanning should not be performed routinely at this early stage due to the high energy output of the scanner (S. Campbell, personal communication).

Transabdominal Doppler scanning in the second half of pregnancy can be hampered by obesity, oligo- or anhydramnios, a forelying cord, or an unfavorable position of the fetal skull. Vaginal pulsed Doppler velocimetry of maternal vessels represents a new technique for evaluating blood flow in the fetomaternal system (Deutinger et al. 1988). The same vaginal scanner can be used to measure flow in fetal vessels. The advantage of this method is that the transducer can be placed in very close proximity to the presenting fetal part and thus to the vessels of interest. The vaginal approach offers significant advantages in selected cases (Deutinger et al. 1989).

When transabdominal scanning is used, significant maternal obesity combined with a forelying cord can make it difficult or impossible to obtain a useful Doppler signal from the umbilical artery. However, this is easily accomplished when the vaginal approach is used (Fig. 19).

Coexisting maternal obesity and anhydramnios likewise hamper the transabdominal Doppler examination of fetal arteries. Vaginal scanning in these cases makes it easier to obtain signals from the fetal aorta (Fig. 20).

Similarly, an occiput anterior position of the fetal head may prevent signal acquisition from the internal carotid artery when the transabdominal approach is used (Wladimiroff et al. 1986). The vaginal approach is particularly advantageous in this situation (Fig. 21).

Thus, while the transabdominal approach is generally favored for the Doppler examination of fetal blood vessels, vaginal pulsed Doppler ultrasound can supplement or replace abdominal scanning in the situations described above.

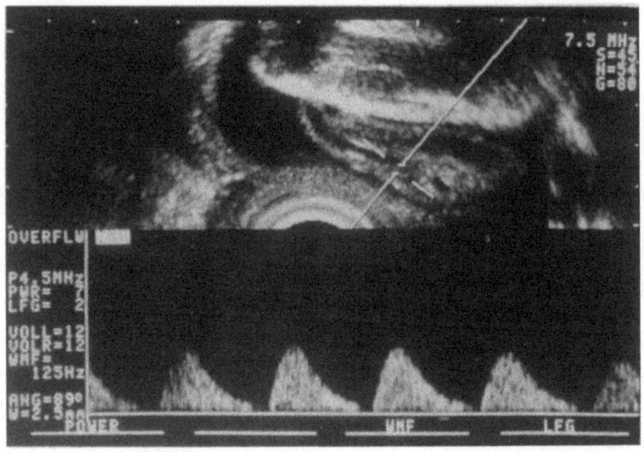

Fig. 19. Vaginal pulsed Doppler examination of the umbilical artery in the 31st week with a forelying cord. The absence of end-diastolic blood flow signifies a deficiency of fetal nutrition

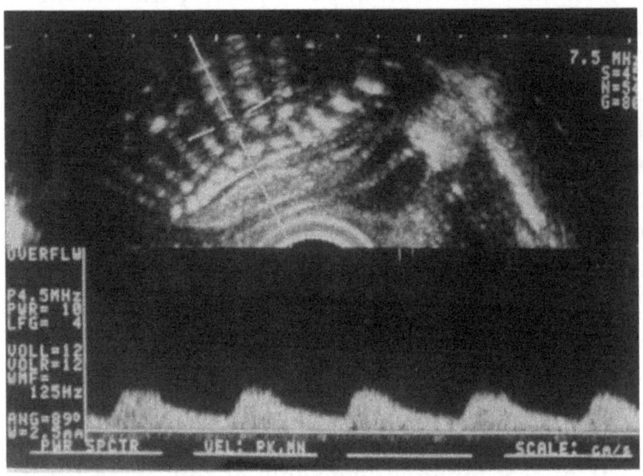

Fig. 20. Vaginal pulsed Doppler examination is the fetal aorta in the 27th week. Signals of acceptable quality could not be obtained by transabdominal scanning because of maternal obesity and anhydramnios

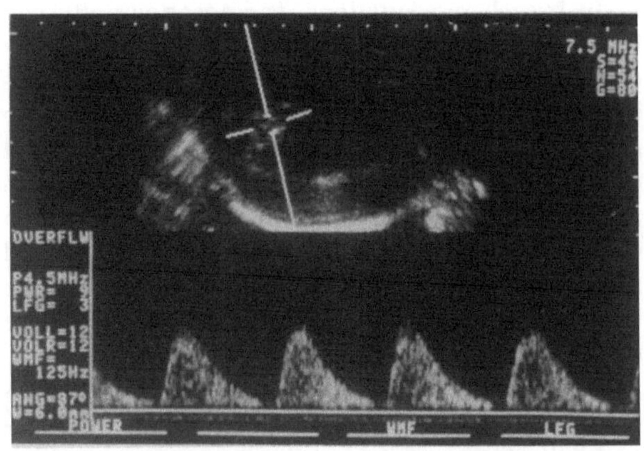

Fig. 21. Vaginal pulsed Doppler examination of the internal ▶ carotid artery in the 36th week. The occipitoanterior position of the fetal skull prohibited a transabdominal approach

References

Bronshtein M, Yoffe N, Brandes JM (1988 a) Transvaginal sonography: a new dimension in early pregnancy evaluation. In: First World Congress on Vaginosonography in Gynecology, 9-12 June 1988, Washington. Syllabus, p 2

Bronshtein M, Yoffe N, Rottem S, Brandes JM (1988 b) Early determination of fetal sex using transvaginal sonography. In: First World Congress on Vaginosonography in Gynecology, 9-12 June 1988, Washington. Syllabus, p 3

Chayen B, Rifkin MD (1987) Cephalocentesis. Guidance with an endovaginal probe and endovaginal needle placement. J Ultrasound Med 6: 221-223

Deutinger J, Rudelstorfer R, Bernaschek G (1988) Vaginosonographic velocimetry of both main uterine arteries by visual vessel recognition and pulsed Doppler method during pregnancy. Am J Obstet Gynecol 159: 1072-1076

Deutinger J, Rudelstorfer R, Häusler P, Bernaschek G (1989) Transvaginal pulsed Doppler velocimetry in fetal arteries. Arch Gynecol Obstet 246: 57-60

Duda V, Czekelius P, Schulz KD (1987) Vaginosonographie: fetaler Cerebralultraschall im letzten Trimenon. Ultraschall Klin Prax [Suppl 1]: 17

Green JJ, Hobbins JC (1988) Abdominal ultrasound examination of the first-trimester fetus. Am J Obstet Gynecol 159: 165-175

Ikegawa A, Ohkawa T, Hogake M, Takeuchi Y (1987) Data processing for detection of embryonal heart rate from linear scan ultrasonography. J Perinat Med 15 [Suppl 1]: 51

Staudach A (1986) Fetale Anatomie im Ultraschall. Springer, Berlin Heidelberg New York, p 33

Timor-Tritsch IE (1988) Transvaginal sonography in early pregnancy. In: First World Congress on Vaginosonography in Gynecology, 9-12 June 1988, Washington. Syllabus, p 29

Warren WB, Timor-Tritsch I, Peisner DB, Raju S, Rosen MG (1989) Dating the early pregnancy by sequential appearance of embryonic structures. Am J Obstet Gynecol 161: 747-753

Wisser J, Krone S, Strowitzki T, Knitza R (1988) Ultraschallembryologie des Zentralnervensystems als Grundlage der Sonopathologie. Ber Gynäkol Geburtshilfe 125 (7-8): 561-562

Wladimiroff JW, Tonge HM, Stewart PA (1986) Doppler ultrasound of cerebral blood flow in the human fetus. Br J Obstet Gynaecol 93: 471-475

Zador IE, Bottoms SF, Tse GM, Brindley BA, Sokol RJ (1988) Nomogram for ultrasound visualization of fetal organs. J Ultrasound Med 7: 197-201

Evaluation of the Cervix

1 General

The diagnosis of cervical incompetence is generally based on a prior history of late abortions or preterm deliveries. In the absence of such antecedents or in primigravidae, about the only way to diagnose cervical incompetence in the past was by vaginal examination. However, this examination provides no objective information on cervical shortening. Only about half the true cervical length is accessible to the palpating finger, and the interpretation of possible cervical shortening is highly subjective. Moreover, dilatation of the cervical canal is detectable by palpation only if it extends to the external os. Thus, palpation can detect an incompetent cervix only at a relatively late stage, and treatment is correspondingly delayed.

Initial reports on ultrasonography of the cervix were published in the late 1970s (Sarti et al. 1979; Christopher et al. 1979; Mahran 1980). Besides providing an objective measurement of cervical length, abdominal sonography was able for the first time to detect dilatations of the cervical canal affecting the internal os. The cervical length in a normal second-trimester pregnancy is 2.5-6 cm (Zemlyn 1981; Michaels et al. 1986; Varma et al. 1987). It is our experience and that of other authors that cervical lengths less than 2.5 cm signify incompetence, although far greater emphasis should be placed on serial measurements than on individual determinations. The process of cervical dilatation begins at the internal os and progresses outward, culminating in prolapse of the gestational sac into the vagina. Even then transabdominal sonography improves evaluation of the actual degree of cervical dilatation and aids in deciding whether it is appropriate to undertake cerclage.

In the final weeks of pregnancy there is a general "maturation" of the lower uterine segment characterized by a shortening of the cervix and a gradual opening of the os (Hendricks et al. 1970; Parikh and Mehta 1961). This process can at best be evaluated subjectively by vaginal palpation. Here again, ultrasound studies have provided the first normal values for cervical length in the final weeks of pregnancy (Pfersmann et al. 1986). Studies indicate significant cervical shortening from an average length of 4 cm to 2.5 cm between 30 and 40 weeks gestation (Fig. 1). The primiparous cervix shortens by an average of 34.4% during this period, and the multiparous cervix by 41.9% (Fig. 2). Accordingly, sonography can be a valuable tool even in the third trimester, especially in cases of impending preterm delivery and in the supervision of patients on tocolytic therapy. Another value of transabdominal ultrasound is that it reduces the need for vaginal examinations and thus eliminates the risk of infection - an especially vital consideration following premature rupture of the membranes.

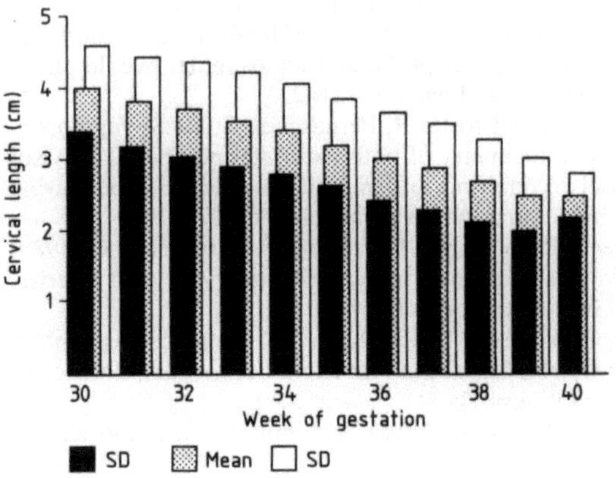

Fig. 1. Cervical shortening between 30 and 40 weeks of pregnancy (means and standard deviations)

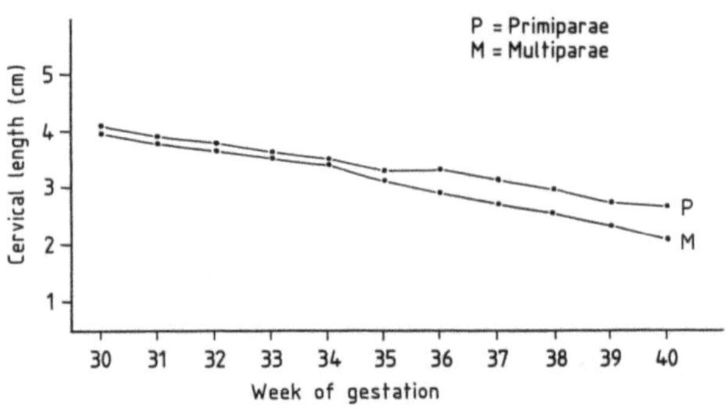

Fig. 2. Comparison of cervical shortening in primiparae and multiparae between 30 and 40 weeks of pregnancy

2 Vaginosonography

Vaginosonographic evaluation of the cervix eliminates many problems of transabdominal scanning. Especially as term approaches, the low cervical position can hamper evaluation due to sound absorption by the symphysis. Besides eliminating this potential barrier, vaginosonography always provides optimum delineation of the cervical canal over its entire length, even when the course of the canal deviates from the sagittal plane, owing to the maneuverability of the vaginal probe. This allows an accurate measurement of the distance between the internal and external os (Fig. 3).

Another advantage of vaginosonography over abdominal ultrasound in evaluating the incompetent cervix is the absence of a distended bladder. Besides altering topographic relationships, a full bladder may exert too much pressure on the lower uterine segment and obscure dilatation of the internal os. While the normal Y-shaped appearance of the lower uterine segment on longitudinal scans remains largely unaffected (Fig. 4), V-shaped (Fig. 5) or U-shaped dilatations of the cervical canal (Fig. 6) may go undetected on abdominal scans (Fig. 7).

Vaginosonography can eliminate this problem (Fig. 6) because it does not alter normal topographic relationships. What is more, the urgency of a situation can be more accurately assessed by endosonography than with the palpating finger, which can perceive the external os as being closed, as in the case illustrated (Fig. 6), but cannot detect more proximal dilatation.

Vaginosonography is also superior for evaluating "dynamic cervical incompetence" in which the degree of dilatation is variable and may even change during the ultrasound examination, probably as a function of uterine tension (Parulekar and Kiwi 1988). The main advantage of vaginosonography in this situation is that the competence of the internal os can be tested provocatively by pressing gently on the abdominal wall with the free hand (Fig. 8).

The horizontal scan plane is best for demonstrating the width of the cervical canal, which should be measured at the level of the internal os. This and the longitudinal plane are best for checking the placement of cervical cerclage and assessing the relation of the band to the internal os.

The first study comparing transabdominal and vaginal sonography for evaluating the cervix demonstrated the superiority of the endosonographic technique (Brown et al. 1986). Besides the advantages stated above, the study confirmed the ability of vaginosonography to adequately demonstrate the

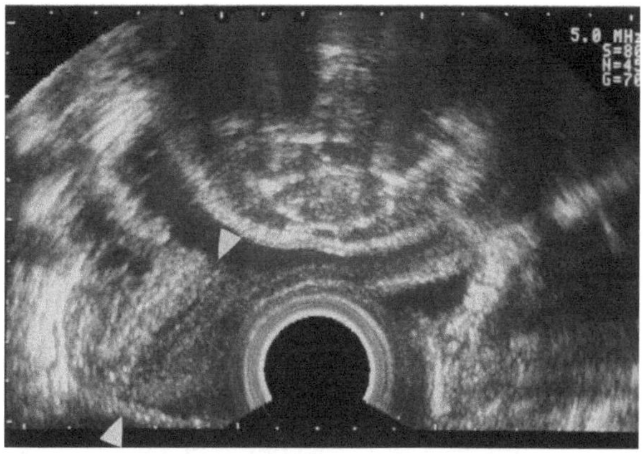

Fig. 3. Longitudinal scan through the lower uterine segment demonstrating the cervical length *(arrowheads)*

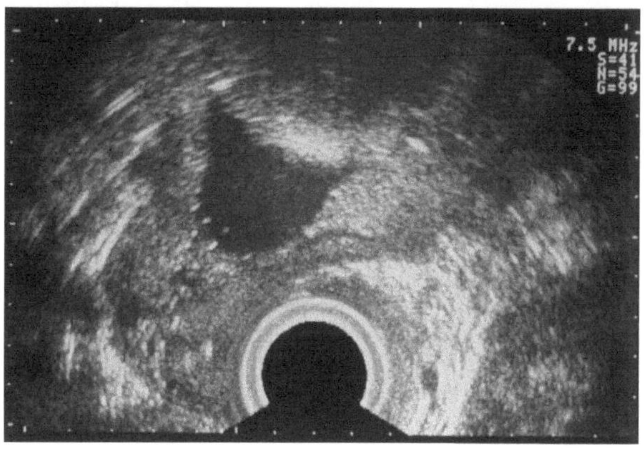

Fig. 4. Longitudinal scan normally demonstrates a Y-shaped lower uterine segment

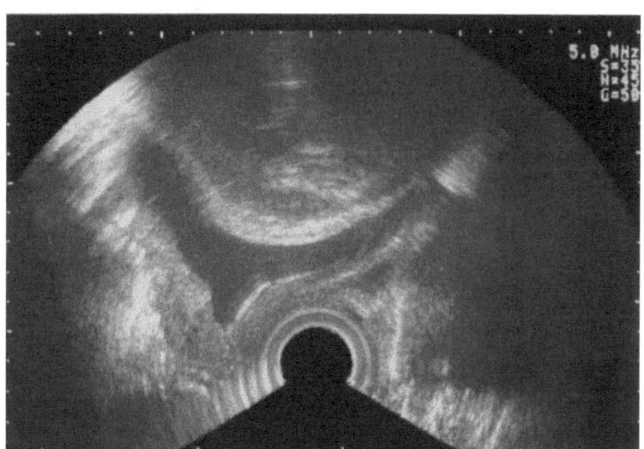

Fig. 5. Scan showing V- or funnel-shaped dilatation of the internal os. The external os is closed

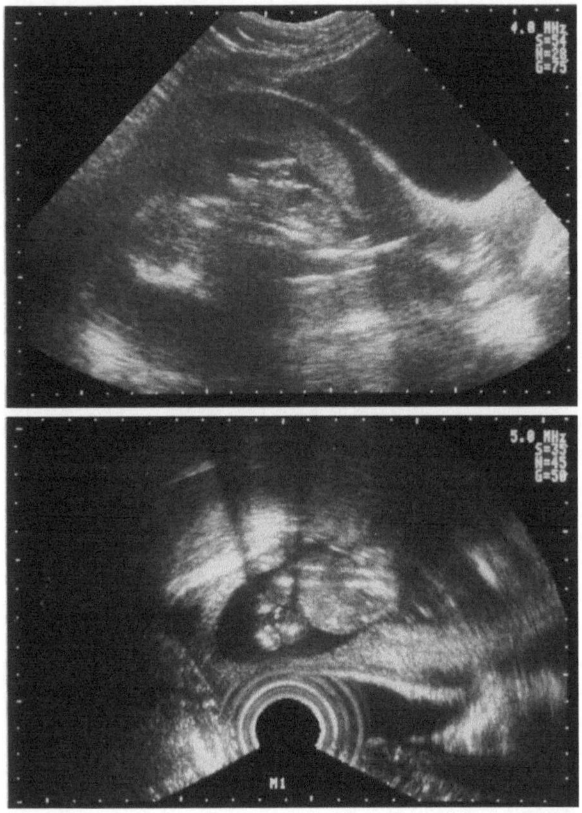

Fig. 6. *Top:* transabdominal scan with a full bladder. The fetal presentation is breech, and the lower uterine segment is compressed. *Bottom:* postvoiding vaginal scan showing a U-shaped (ballooned) lower uterine segment. There is considerable cervical shortening, but the external os is still closed

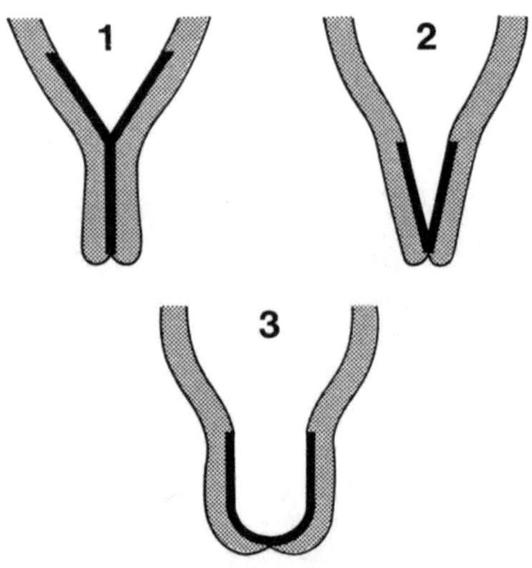

Fig. 7. Sonographic appearances of the lower uterine segment (schematic representation). *1* Normal Y-shaped segment with a closed cervical canal; *2* V-shaped (funneled) lower uterine segment; *3* U-shaped (ballooned) lower uterine segment

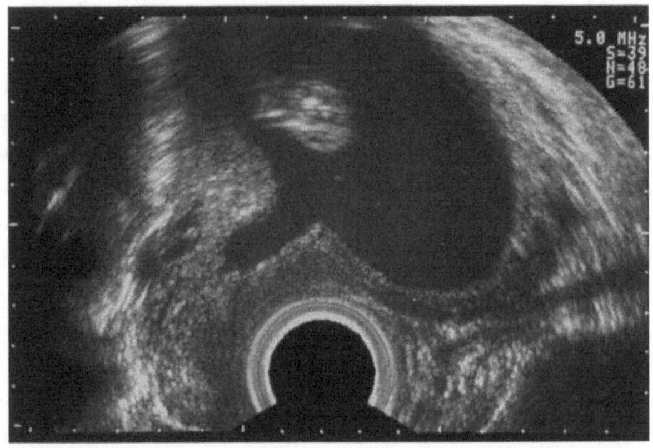

Fig. 8. Longitudinal vaginal scan through the cervix. Concurrent pressure on the abdomen with the free hand provokes dilatation of the internal os in this case of "dynamic" cervical incompetence

cervix and lower uterine segment in a larger percentage of patients (83%) than the transabdominal technique (76%). Vaginosonographic exploration immediately following digital examination of the cervix should be able to increase this percentage further.

We feel that abdominal sonography is superior to vaginosonography following membrane prolapse. Abdominal scanning is also best for visualizing the cervix following membrane rupture, as it reduces the risk of infection. However, vaginal scanning under aseptic conditions could still be helpful in these cases of transabdominal findings are equivocal.

In summary, evaluation of the cervix by vaginosonography offers the following advantages:

1. Optimum visualization of the cervix allows an accurate measurement of cervical length.
2. There is no compression of the lower uterine segment by a distended bladder, so dilatation of the internal os is more easily recognized and interpreted.
3. The early, reliable visualization of anatomic changes in the lower uterine segment by vaginosonography supports its use for the detection of impending preterm delivery, at least in high-risk populations.

References

Brown JE, Thieme GA, Shah DM, Fleischer AC, Boehm FH (1986) Transabdominal and transvaginal endosonography: evaluation of the cervix and lower uterine segment in pregnancy. Am J Obstet Gynecol 155: 721–726

Christopher CR, Spinelli A, Collins ML (1979) Ultrasound detection of hourglass membranes with funic presentation. Obstet Gynecol 54: 130-132

Hendricks CH, Brenner WE, Kraus G (1970) Normal cervical dilatation pattern in late pregnancy and labor. Am J Obstet Gynecol 106: 1066

Mahran M (1980) The role of ultrasound in the diagnosis and management of the incompetent cervix. In: Kurjak A (ed) Recent advances in ultrasound diagnosis, vol 2. Excerpta Medica, Amsterdam, p 505

Michaels WH, Montgomery C, Karo J, Temple J, Ager J, Olson J (1986) Ultrasound differentiation of the competent from the incompetent cervix: prevention of preterm delivery. Am J Obstet Gynecol 154: 537-546

Parikh MN, Mehta AC (1961) Internal cervical os during the second half of pregnancy. J Obstet Gynaecol Br Commonw 68: 818-821

Parulekar SG, Kiwi R (1988) Dynamic incompetent cervix uteri. J Ultrasound Med 7: 481-485

Pfersmann C, Deutinger J, Bernaschek G (1986) Die Zervixlänge gegen Ende der Schwangerschaft - eine sonographische Studie. Geburtshilfe Frauenheilkd 46: 213-214

Sarti DA, Sample WF, Hobel CJ, Staisch KJ (1979) Ultrasonic visualisation of a dilated cervix during pregnancy. Radiology 130: 417-420

Varma TR, Patel RH, Pillai U (1987) Ultrasonic assessment of cervix in "at risk" patients. Int J Gyneaecol Obstet 25: 25-34

Zemlyn S (1981) The length of the uterine cervix and its significance. J Clin Ultrasound 9: 267-269

Placenta Previa

1 General

Localization of the placenta has always been one of the major applications of diagnostic ultrasound (Gottesfeld et al. 1966; Kratochwil 1967). From its inception, sonography consistently proved superior to X-ray placentography, angiography, thermography, and isotope scanning for evaluations of the placenta. With the advent of grayscale imaging, ultrasound was able to resolve structural differences in placentas and relate them to pathoanatomic changes (maturity grades, etc.). Sharper resolution also enabled early identification of the villous chorion frondosum and its differentiation from the chorion laeve.

Vaginosonography has made it possible to identify the chorion frondosum earlier and with absolute confidence (see chapter "Normal Early Pregnancy"), a circumstance that is used to good advantage in the conduct of ultrasound-guided chorionic biopsies. A particular advantage of vaginosonography is that it eliminates many drawbacks of abdominal ultrasound in detecting an abnormal placental location and thus contributes greatly to the diagnosis of placenta previa. Reports on the detection of placenta previa in the second trimester with abdominal ultrasound indicate an overall detection rate of approximately 5% (Zanke 1985, 4.0%; Schmidt et al. 1986, 4.2%; Wexler and Gottesfeld 1979, 5.6%; Comeau et al. 1983, 5.6%; Rizos et al. 1979, 5.3%; Gillieson et al. 1982, 6.0%).

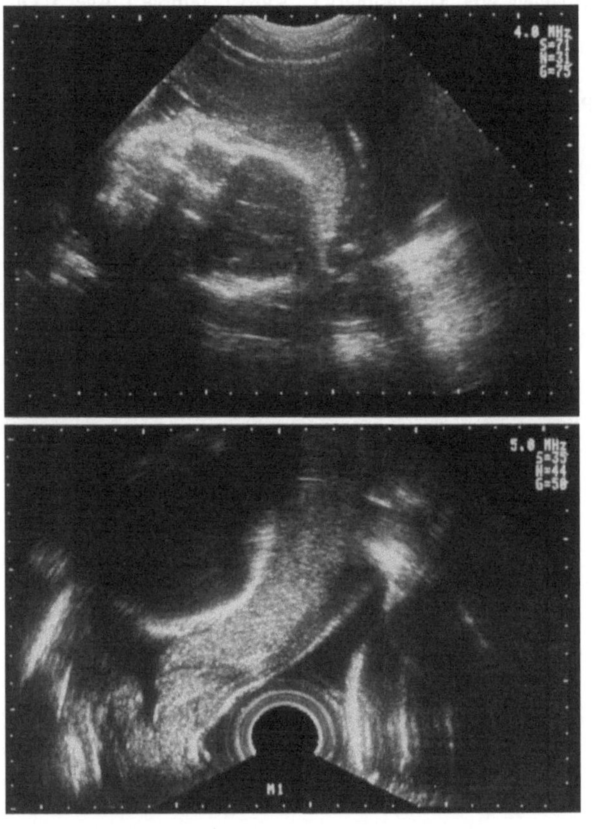

Fig. 1. Marginal placenta previa implanted on the anterior wall (27th week of pregnancy). *Top:* transabdominal scan; *bottom:* vaginal scan

2 Vaginosonography

Although vaginosonography is applied mostly in early pregnancy, it has selected indications in the second and third trimester, most notably in the diagnosis of low-lying placenta (Bernaschek et al. 1988). While the purely visual distinction between the transabdominal and transvaginal sonographic appearances of low-lying placenta can be impressive (Fig. 1), it is difficult to quantitate this distinction in objective terms. Our own study represents an attempt to provide data to substantiate the visual impression.

A total of 1610 women between the 16th and 24th weeks of pregnancy were examined by transabdominal sonography. An abnormal placental location was suspected in 66 (4.1%). A low-lying placenta was diagnosed in 54 women, 38 of whom underwent immediate vaginal scanning. Vaginosonography disclosed 15 cases of marginal placenta previa (Fig. 2) and 23 cases in which the lower edge of the placenta was 1–5 cm from the internal os (Fig. 3).

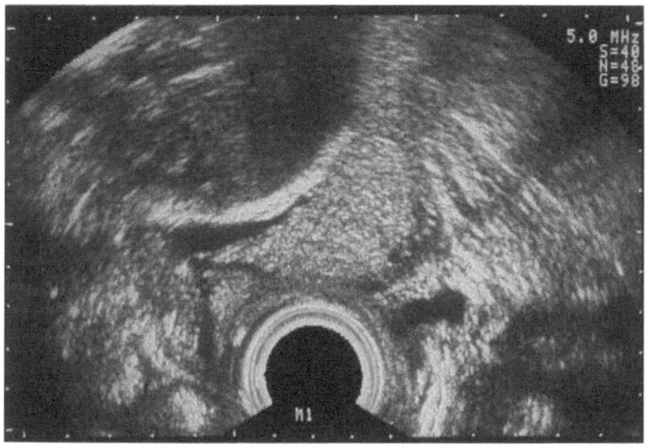

Fig. 2. Longitudinal scan showing extension of the anteriorly implanted placenta to the internal os

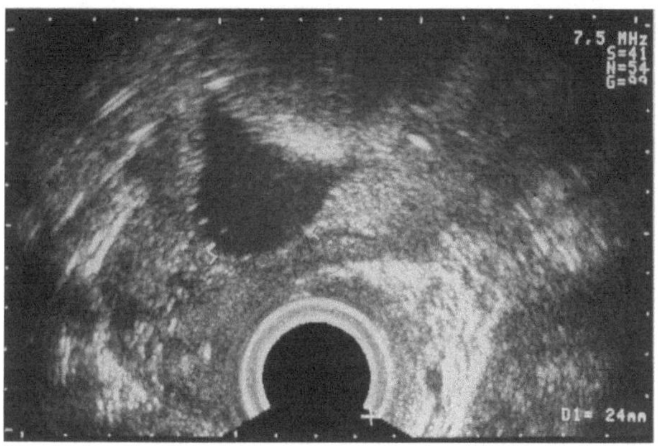

Fig. 3. The distance from the lower edge of the anteriorly implanted placenta to the internal os is 24 mm

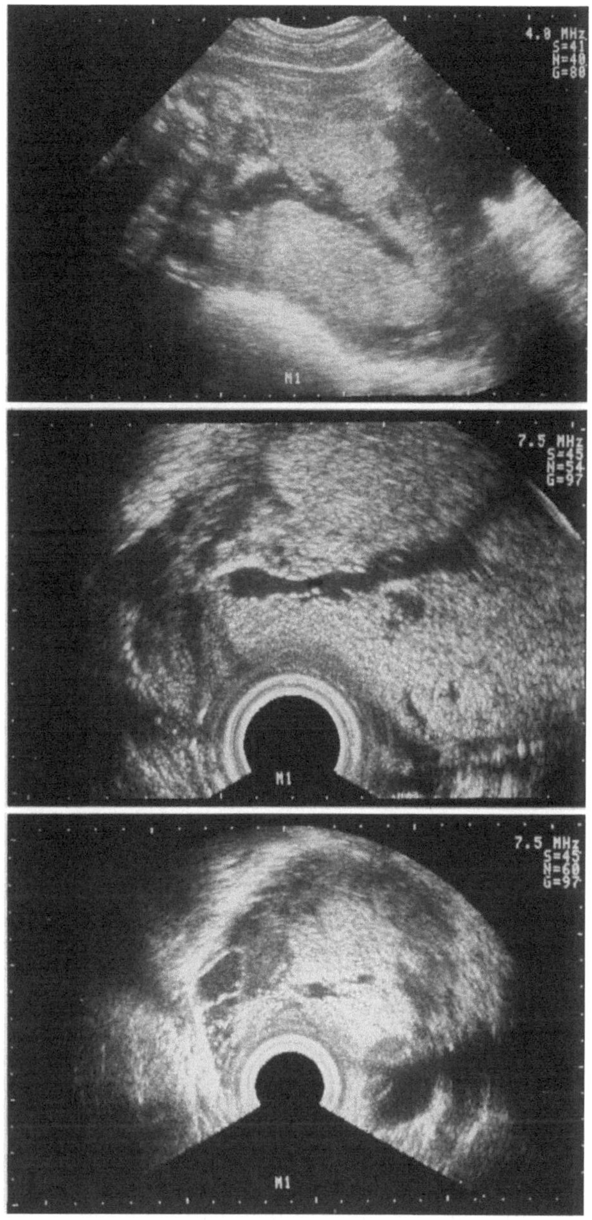

Fig. 4. Total placenta previa (35th week of pregnancy). *Top:* ▶ transabdominal scan; *center:* sagittal vaginal scan; *bottom:* horizontal vaginal scan

Total placenta previa had been diagnosed transabdominally in 12 of the women, but in only six cases was the diagnosis confirmed by vaginosonography (Fig. 4). Five of the placenta previas were classified as marginal, and in one case the placental edge was 2 cm from the internal os (Fig. 5). We disregarded "partial" placenta previa as a diagnostic entity, considering that distinction to be questionable on methodologic grounds and even by definition, since it presupposes at least some dilatation of the internal os.

In all cases the probe was inserted carefully into the vagina under ultrasound guidance so as not to provoke or exacerbate hemorrhage. Longitudinal and horizontal (coronal) scans were taken from the anterior vaginal fornix without exerting pressure on the cervix. None of the vaginal scans provoked hemorrhage or exacerbated a preexisting bleed. Probes with a large view angle (panoramic scanners) are best for diagnosing an abnormal placental location. Only a large field of vision can demonstrate the whole cervical canal, the internal os, and the lower placental margin on one image.

Comparing transabdominal sonography with vaginosonography in our series, we find that vaginosonography is superior in its ability to demon-

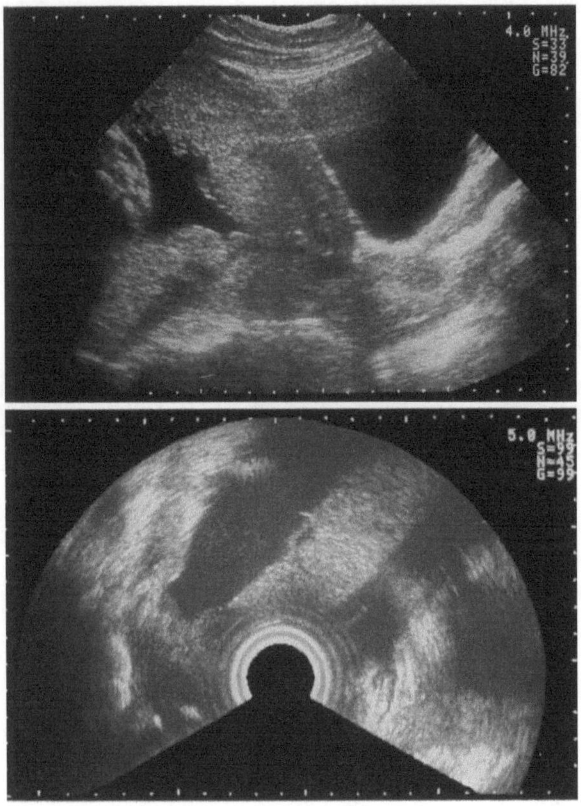

Fig. 5. *Top:* compression of the lower uterine segment on the transabdominal scan mimics a total placenta previa. *Bottom:* vaginosonography shows that the placenta does not cover the internal os

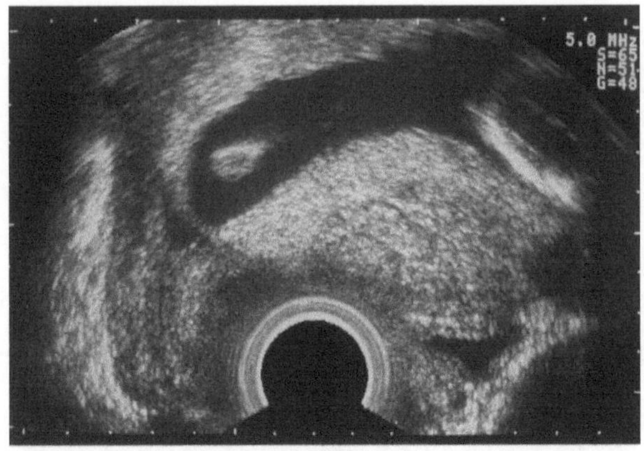

Fig. 6. Total placenta previa with extreme thinning of the placenta above the internal os

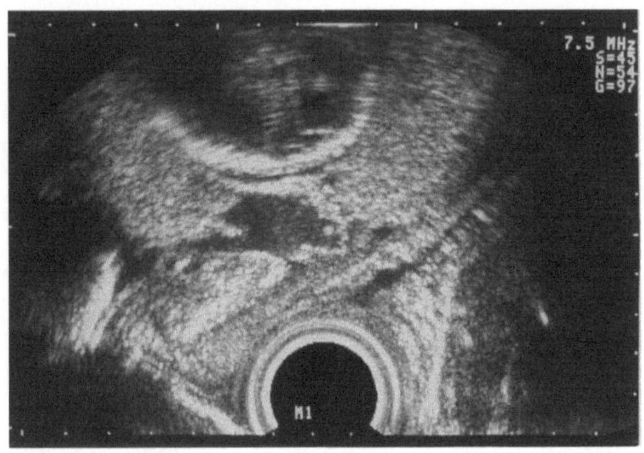

Fig. 7. Total placenta previa with degeneration of placental tissue above the internal os

strate accurately and consistently the entire cervical canal and especially the internal os. In addition, vaginosonography is free of the problems that result from varying states of bladder distention in abdominal sonography. This eliminates the dilemma of having to choose between a full bladder, which gives optimum delineation of the cervix but compresses the lower uterine segment, and a moderately filled bladder that causes less compression but compromises the image quality. A full bladder in transabdominal scanning, while necessary for accurate evulation of the cervix, can lead to overestimation of the degree of placenta previa or even to a false-positive diagnosis due to compression of the lower uterine segment mimicking an abnormal placental location.

Owing to the lack of compression of the lower uterine segment in vaginosonography, it is generally possible to demonstrate a marked thinning of the placenta directly above the internal os in patients with a total placenta previa (Figs. 6 and 7). This thinning, which is sometimes extreme, apparently relates to a deficiency of cell proliferation in that area.

Vaginosonography is especially valuable in demonstrating placentas implanted on the posterior uterine wall (Fig. 8). The view angle of the vaginal probe is such that the placenta is not shadowed by overlying fetal parts. This avoids the need for transabdominal manual elevation of the fetus or examination in the Trendelenburg position that is sometimes necessary for the abdominal scanning of a posterior placenta.

Cranial "migration" of the placental edge in relation to the internal os was first described sonographically by King (1973), although there is no proof that this represents an active migratory phase of placentation. It is more likely that "placental mi-

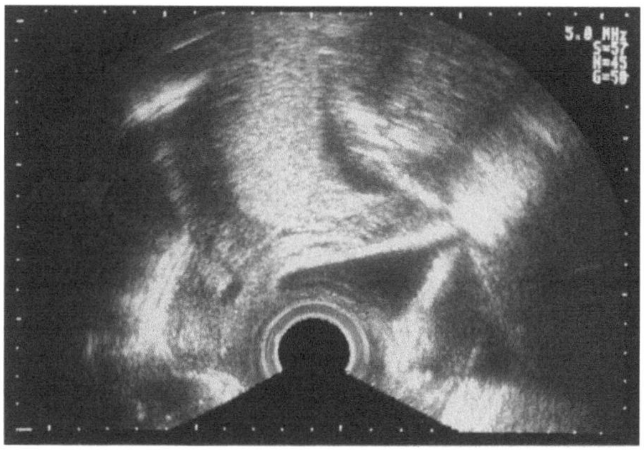

Fig. 8. Posteriorly implanted total placenta previa

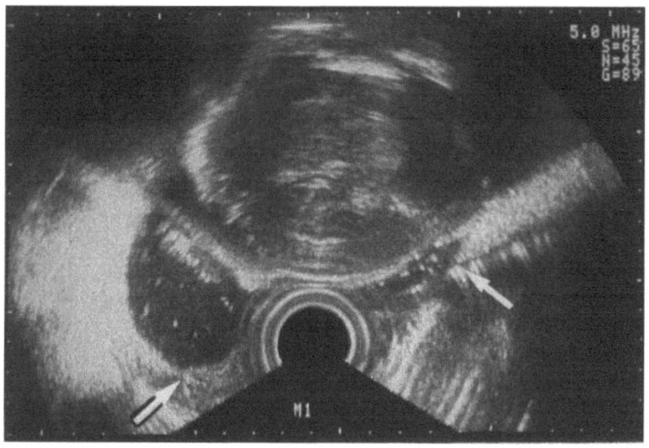

Fig. 9. Same case as in Fig. 1, 38th week. Note the increased distance between the internal os and the placental edge *(arrows)*

gration" is based on longitudinal growth of the uterus and especially on passive stretching of the lower uterine segment. While we did not see any migration in our cases of total placenta previa confirmed by vaginosonography, we did observe a sometimes marked increase in the distance of the placental edge from the internal os in all cases of marginal placenta previa diagnosed by vaginosonography. This occurred between the 25th and 35th weeks of pregnancy (Fig. 9). Accordingly, there may be no need to monitor the placental location by follow-up scans in cases where the vaginosonographically measured distance of the placental edge from the internal os is 1 cm or more in the second trimester or early third trimester.

False-positive transabdominal findings in the second trimester should cause few problems because they are likely to be resolved by subsequent scans. However, false-positive findings in late pregnancy, avoidable by vaginosonography, can be disconcerting to both the patient and the obstetrician. Even vaginosonography, though, cannot furnish a more accurate prognosis regarding the further clinical course and especially the course of the delivery. However, it can be very helpful in deciding whether it is reasonable to await spontaneous delivery or make an immediate recommendation for cesarean section.

Although the case numbers in our study are too small to be conclusive, the visual results are impressive when transabdominal and vaginal findings are compared in the diagnosis of placenta previa. They justify the use of vaginosonography as an aid to decision making in all cases where there is presumptive evidence of an abnormal placental location, especially in the third trimester.

3 Summary

1. Vaginosonography significantly improves the delineation of the cervical canal and internal os compared with transabdominal sonography, enabling a more exact diagnosis and classification of low-lying placenta.
2. The absence of a distended bladder in vaginosonography eliminates compression effects and provides an undistorted picture of topographic relations in the lower uterine segment.
3. Vaginosonography is particularly advantageous for evaluating low-lying placentas implanted posteriorly.

References

Bernaschek G, Rudelstorfer R, Kölbl H (1988) The contribution of vaginal endosonography in the diagnosis of bleeding in pregnancy. J. E. M. U. 9: 204-206

Comeau J, Shaw L, Marcell KCC, Lavery PL (1983) Early placenta previa and delivery outcome. Obstet Gynecol 61: 577-580

Gilieson MS, Winer-Muram HT, Muram D (1982) Low-lying placenta. Radiology 144: 577-580

Gottesfeld KR, Thompson HE, Holmes JH, Taylor ES (1966) Ultrasonic placentography - a new method for placental localisation. Am J Obstet Gynecol 96: 538-547

King DL (1973) Placental migration demonstrated by ultrasonography. A hypothesis of dynamic placentation. Radiology 109: 167-170

Kratochwil A (1967) Möglichkeiten der Ultraschalldiagnostik in Geburtshilfe und Gynäkologie. Wien Klin Wochenschr 79: 399-402

Rizos N, Deran TA, Miskin M, Benzie RJ, Ford JA (1979) Natural history of placenta praevia ascertained by diagnostic ultrasound. Am J Obstet Gynecol 133: 287-291

Schmidt W, Boos R, Hendrik HJ, Schmidt R (1986) Pathologischer Plazentasitz nach der 20. Schwangerschaftswoche - Bedeutung für den Schwangerschafts- und Geburtsverlauf. Geburtshilfe Frauenheilkd 46: 206-212

Wexler P, Gottesfeld KR (1979) Early diagnosis of placenta previa. Obstet Gynecol 54: 231-234

Zanke S (1985) Die ultrasonographische Frühdiagnose der Placenta praevia und ihr klinischer Stellenwert. Geburtshilfe Frauenheilkd 45: 710-713

Vaginosonographic Pelvimetry

1 General

Although skeletal deformities caused by malnutrition have become rare in developed countries, the reported incidence of cephalopelvic disproportion has risen in the last 20 years from 1.5% to 3.8% of all births. This is believed to be due to an average 100-g increase in birthweight and also to improved methods of perinatal monitoring (Silbar 1986).

Several factors must be considered in determining the feasibility of vaginal delivery: the size and shape of the bony pelvis, the size and moldability of the fetal head, the efficacy of labor, the fetal presentation, and the position of the fetal head. In the past, early detection of cephalopelvic disproportion has relied on both the clinical examination (pelvic exam, Zangenmeister's maneuver) and X-ray pelvimetry (Guthmann 1929). While the latter method permits the evaluation and measurement of critical pelvic dimensions, it should not be used routinely due to the hazards of fetal radiation exposure (Pritchard et al. 1985). X-ray pelvimetry can successfully measure not only the true conjugate, but also the inclination of the pelvic inlet plane and the transverse diameter of the pelvic inlet and the interspinous distance.

Before X-ray pelvimetry is elected, however, it must be asked whether the information provided will contribute significantly to subsequent management of the delivery (Pritchard et al. 1985). Recent studies cast doubt on the value of X-ray pelvimetry as a tool for selecting patients for cesarean delivery (Mandry et al. 1983; Alder et al. 1987). Also, even if cephalopelvic disproportion is suspected, there is no reason to expect that a poor fetal outcome is inevitable (Parsons and Spellacy 1985). Assuming that reasonably close supervision of the pregnancy is maintained, it does not appear that X-ray pelvimetry contributes significantly to obstetric decision-making in patients who have had a previous cesarean section (Mahmod and Grant 1987). The American College of Obstetricians and Gynecologists recommends that fetal radiation exposure be kept as low as possible, because there is still no evidence to refute the studies of Stewart indicating an increased risk of leukemia in children exposed to diagnostic radiation in utero (Stewart et al. 1956; ACOG 1979). Brent (1981) cites a 40% incidence of leukemia in children thus exposed. The dose from a single X-ray pelvimetric examination is in the range of 0.15–0.9 rad, which is close to the maximum dose allowable during pregnancy (Hochuli 1985). One recognized indication for radiographic pelvimetry in pregnancy is a breech presentation in a primipara (Gimovsky et al. 1985).

Following a protracted delivery with suspicion of a contracted pelvis, X-ray pelvimetry may be used rather freely in the puerperium to provide information on possible anomalies of the bony pelvis that might affect a future pregnancy.

Given the many objections to the use of radiography for prenatal pelvic measurements, it is only natural that attention was turned very early to ultrasonography as a possible alternative (Kratochwil and Zeibekis 1972; Kratochwil et al. 1973). In the past, compound scanners were used exclusively to determine critical pelvic dimensions. This method can provide a satisfactory measurement of the true conjugate, which usually is determined in late pregnancy (Kratochwil and Zeibekis 1972). Measurements can be performed on the two-dimensional image as well as on a simultaneously recorded A-mode trace using frequencies of 1–2 MHz. Studies in large series have confirmed the value of this method (Schlensker 1979). So far the newer real-time scanners have not proven satisfactory for ultrasound pelvimetry.

We have used a vaginal scanner to measure critical pelvic dimensions and have compared the results with those obtained using a compound scanner during pregnancy and X-ray pelvimetry after delivery (Deutinger and Bernaschek 1986 a, 1987).

2 Technique and Preliminary Results

Vaginal pelvimetry is performed with a 5-MHz vaginal probe. Sector scanners with a wide-angle view are the only suitable scanner types for this indication (Deutinger and Bernaschek 1986 b). The panoramic scanner is characterized by a 240° forward scan angle.

As for all vaginosonographic examinations, the lithotomy position has proven best for it allows the greatest scanner mobility. The true conjugate can usually be measured vaginally without difficulty, although in some cases the examination must be repeated after evacuation of the bowel because a full

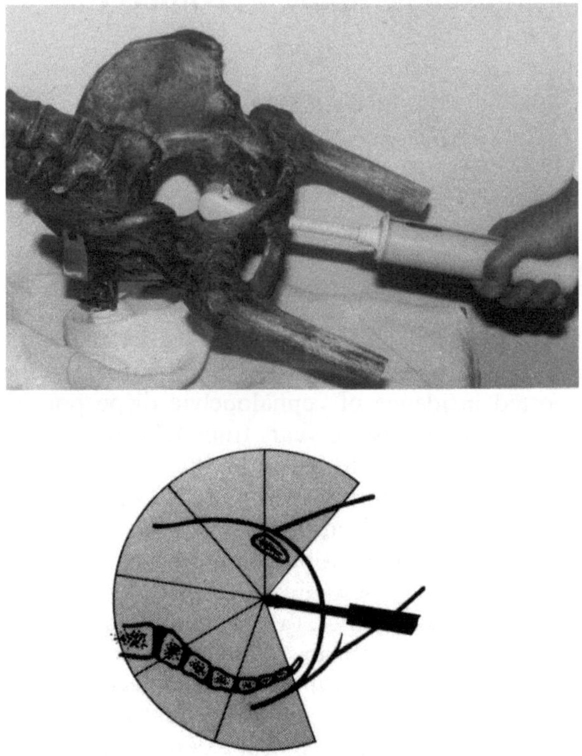

Fig. 1. *Top:* position of the vaginal scanner for measuring the ▶ true conjugate, illustrated in a pelvic model. *Bottom:* the symphysis and sacral promontory can be imaged concurrently with a panoramic scanner

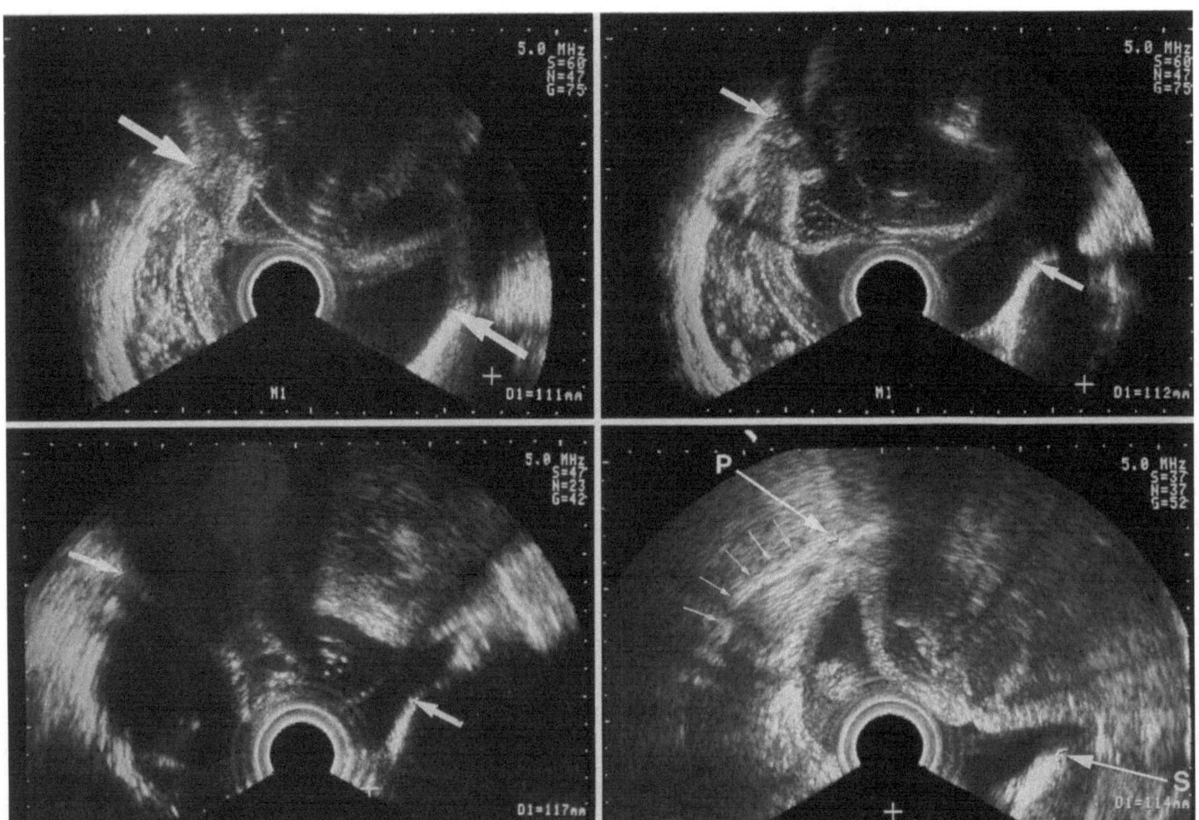

Fig. 2. *Top:* vaginosonographic measurement of the true conjugate in a vertex presentation. *Arrows* indicate the points of measurement. *Bottom:* vaginosonographic measurement of the true conjugate in a breech presentation. *Small arrows* in the right image mark the course of the promontory ▼

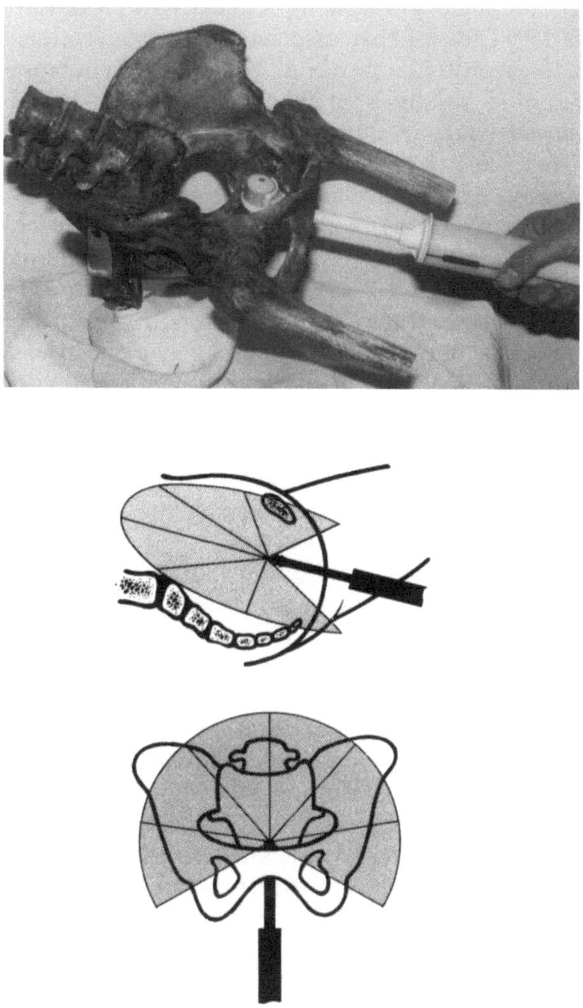

Fig. 3. *Top:* position of the vaginal scanner for measuring the transverse conjugate, illustrated in a pelvic model. *Center:* schematic diagram of the coverage of the panoramic scanner for measuring the transverse conjugate, viewed from the lateral aspect. *Bottom*: schematic diagram of the coverage of the panoramic scanner, viewed from the perspective of the examiner

rectum can cast an acoustic shadow on the anterior border of the sacral promontory.

The probe is first inserted into the vagina with the beam directed so that a sagittal scan is obtained. It is important to position the scan plane exactly on the midline so that the symphysis and the promontory are imaged concurrently (Figs. 1 and 2). For measurement of the true conjugate, the handle of the probe is lowered slightly so that the symphysis is depicted in its complete height. The distance from the posterior superior border of the symphysis to the promontory is measured. The end points of the true conjugate can be demonstrated regardless of the fetal lie.

For measurement of the transverse pelvic diameter, the probe is rotated 90°, and the handle is moved vertically until the linea terminalis on either side is demonstrated and the distance between them can be measured (Figs. 3 and 4).

We performed vaginal pelvimetry with a panoramic scanner in 74 patients, 22 of whom were examined shortly before term. Preterm deliveries were excluded from the evaluation. In 58 cases the true conjugate was measured with a compound scanner, and postpartum pelvic X-ray films were taken in 36 patients. Sixty-five of the pregnancies examined by us were vertex presentations. Eight of the 58 measurements performed with a compound scanner proved to be unrewarding; two of those cases were transverse lies and three were breech presentations. In one case the true conjugate could not be determined because of an oblique lie, and in two cases because of marked obesity.

The measurements of the true conjugate performed by vaginosonography correlated well with

Fig. 4. Vaginosonographic measurement of the transverse conjugate in a vertex presentation *(left)* and a breech presentation *(right)*. *Arrows* indicate the points of measurement ▼

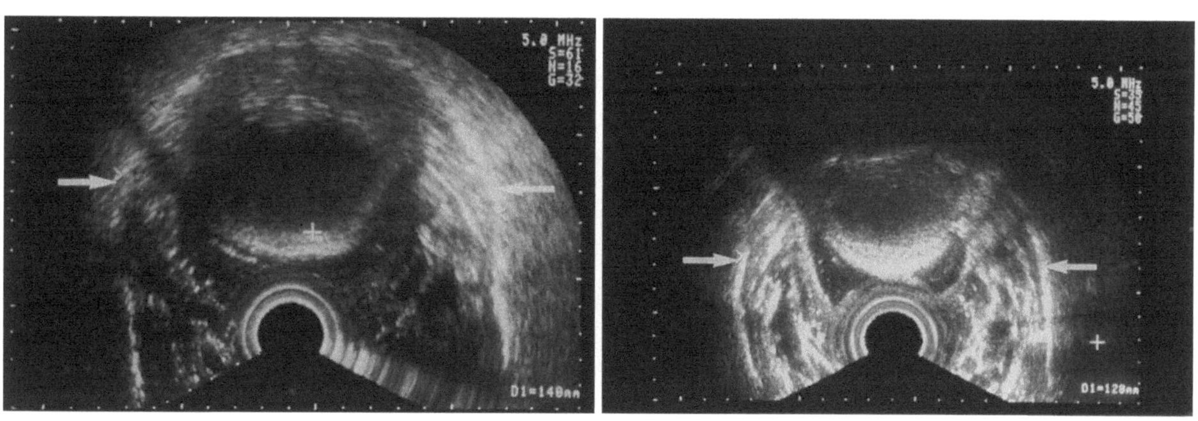

Table 1. True conjugate, fetal BPD, and birthweight in spontaneous and cesarean deliveries

	Spontaneous ($n=49$)	Cesarean ($n=11$)
True conjugate	11.8 cm (±0.4)*	10.9 cm (±0.3)
BPD	9.1 cm (±0.3)	9.3 cm (±0.2)
Birthweight	3340 g (±440)	3470 g (±450)

* $p<0.05$

measurements made by compound scanners. The discrepancy with respect to radiographic measurements was 5 mm (Deutinger 1988). Differences of 4 mm were noted in measurements of the transverse diameter of the pelvic inlet. After 37 weeks the largest transverse pelvic diameter could not be measured in seven primiparae because the fetal head was already too low in the pelvis. In the remaining patients this measurement could be performed without difficulty.

Because cephalopelvic disproportion represents a relation between the maternal pelvis and fetal size, we calculated the ratio of the true conjugate to the fetal biparietal diameter (BPD) in vertex presentations. We found that this ratio averaged 1.29 in spontaneous deliveries and 1.17 when cephalopelvic disproportion was present (Table 1). This difference was of high statistical significance. Interestingly, no value higher than 1.24 was found in any case of cephalopelvic disproportion.

The babies delivered spontaneously and by cesarean section showed no significant difference in terms of the BPD measured sonographically within 1 week of delivery or in terms of birthweight.

3 Summary

The major advantages of vaginosonographic pelvimetry are its relative simplicity and the absence of radiation exposure. Vaginosonography also makes it possible to determine the transverse diameter of the pelvic inlet, providing information useful for evaluating the pelvic configuration on the inlet plane. The examination can be performed at any stage of pregnancy and is unaffected by maternal obesity. Moreover, a full urinary bladder is not required. The result of the examination is not affected by the fetal position, although measurement of the transverse diameter can be difficult after 37 weeks in primiparae if the low station of the head limits transducer mobility or obscures the lateral pelvic boundaries.

Computed tomography reduces X-ray exposure to 15%-30% of that associated with conventional radiography (Claussen et al. 1985), and the interpretation of results is greatly facilitated by the enhanced image quality compared with conventional X-ray films (Kopelman et al. 1986). However, CT findings can be difficult to interpret in very obese patients (Claussen et al. 1985), whereas vaginosonographic pelvimetry is not affected by maternal obesity.

Of course, there are other factors that influence the course of the delivery besides pelvic dimensions and fetal size, such as the overall configuration of the pelvis, the inclination of the pelvic inlet plane, and the angle of the pubic ramus. However, the true conjugate and the transverse diameter of the pelvic inlet are the most critical measurements in terms of evaluating the bony birth canal.

We have compared the results of vaginosonographic pelvimetry with compound scans and postpartum X-rays, but it would also be desirable to compare vaginal measurements with the results of CT scanning and magnetic resonance imaging (MRI). MRI reportedly can provide very high quality images of the bony maternal pelvis from which accurate measurements can be made without risk of fetal injury (Hricak et al. 1983; Johnson et al. 1984; Powell et al. 1988). Measurements of the transverse pelvic diameter showed differences of up to 2 cm relative to X-ray pelvimetry (Powell et al. 1988). At present, however, the high costs of MRI would seem to preclude its routine use for pelvimetry in the foreseeable future.

Our investigations indicate that vaginosonographic pelvimetry is a well-tolerated, highly informative procedure that could provide an excellent routine obstetric examination, especially in primiparae with suspected cephalopelvic disproportion.

References

Alder C, Aebi S, Bernhard M (1987) Der Stellenwert der radiologischen Beckenmessung. Geburtshilfe Frauenheilkd 47: 483-486

American College of Obstetricians and Gynecologists (1979) X-ray pelvimetry in pregnancy? ACOG Bull 23: 10

Brent RL (1981) Irradiation in pregnancy. In: Sciarra J (ed) Gynecology and obstetrics. Harper and Row, Philadelphia

Claussen C, Köhler D, Christ F, Golde G, Lochner B (1985) Pelvimetry by digital radiography and its dosimetry. J Perinat Med 13: 287-292

Deutinger J (1988) Die vaginosonographische Beckenmessung. In: Dudenhausen JW, Saling E (eds) Perinatale Medizin. Thieme, Stuttgart, pp 267-268

Deutinger J, Bernaschek G (1986 a) Die vaginosonographische Pelvimetrie als neue Methode zur sonographischen Bestimmung der inneren Beckenmaße. Geburtshilfe Frauenheilkd 46: 345-347

Deutinger J, Bernaschek G (1986b) Die vaginosonographische Pelvimetrie - eine neue Methode zur sonographischen Bestimmung der inneren Beckenmaße. In: Otto RC, Schnaars P (eds) Ultraschalldiagnostik 85: Drei-Länder-Treffen Zürich. Thieme, Stuttgart, pp 2-3

Deutinger J, Bernaschek G (1987) Vaginosonographical determination of the true conjugate and the transverse diameter of the pelvic inlet. Arch Gynecol Obstet 240: 241-246

Gimovsky ML, Willard K, Neglio M, Howard T, Zerne S (1985) X-ray pelvimetry in a breech protocol: a comparison of digital radiography and conventional methods. Am J Obstet Gynecol 153: 887-888

Guthmann H (1929) Die röntgenologische Messung der Conjugata vera. Fortschr Röntgenstr 36: 257

Hochuli E (1985) Geburtshilfe, Gynäkologie und Grenzgebiete. Huber, Bern

Hricak H, Alpers C, Crooks LE, Sheldon PE (1983) Magnetic resonance imaging of the female pelvis - initial experience. Am J Radiol 141: 1119-1128

Johnson IR, Symonds EM, Worthington BS, Broughton Pipkin F, Hawkes RC, Gyngell M (1984) Imaging the pregnant human uterus with nuclear magnetic resonance. Am J Obstet Gynecol 148: 1136-1139

Kopelman JN, Duff P, Karl RT, Schipul AH, Read JA (1986) Computed tomographic pelvimetry in the evaluation of breech presentation. Obstet Gynecol 68: 455-458

Kratochwil A, Zeibekis N (1972) Ultrasonic pelvimetry. Acta Obstet Gynecol Scand 51: 357

Kratochwil A, Jantsch K, Brezina KF (1973) Ultraschallanatomie des weiblichen Beckens und ihre klinische Bedeutung. Arch Gynecol 214: 273-275

Kratochwil A (1980) Ultraschalluntersuchung in der Geburtshilfe. In: Diethelm L, Heuck F, Olsson O, Strnad F, Vieten H, Zuppinger A (eds) Handbuch der medizinischen Radiologie. Springer, Berlin Heidelberg New York, pp 349-401

Mahmood TA, Grant JM (1987) The role of radiological pelvimetry in the management of patients who have had a previous caesarean section. J Obstet Gynaecol 8: 24-28

Mandry J, Grandjean H, Reme JM, Pastor J, Levade C, Pontonnier G (1983) Assessment of the predictive value of X-ray pelvimetry and biparietal diameter in cephalopelvic disproportion. Eur J Obstet Gynec Reprod Biol 15: 173-179

Parsons MT, Spellacy WN (1985) Prospective randomized study of X-ray pelvimetry in the primigravida. Obstet Gynecol 66: 76-79

Powell MC, Worthington BS, Buckley JM, Symonds EM (1988) Magnetic resonance imaging (MRI) in obstetrics. I. Maternal anatomy. Br J Obstet Gynaecol 95: 31-37

Pritchard JA, MacDonald PC, Gant NF (1985) Williams obstetrics, 17th edn. Appleton-Century-Crofts, East Norwalk, pp 229-231

Schlensker KH (1979) Ultraschallmessungen der Conjugata vera obstetrica. Geburtshilfe Frauenheilkd 39: 333-337

Silbar EL (1986) Factors related to the increasing cesarean section rates for cephalopelvic disproportion. Am J Obstet Gynecol 154: 1095-1098

Stewart A, Webb J, Giles D, Hewitt D (1956) Malignant disease in childhood and diagnostic irradiation in utero. Lancet ii: 447-457

Endosonography of the Uterus

1 Normal Anatomy

1.1 Size

It is apparent on manual examination that nulliparous, uniparous, and multiparous women differ in the size of their uterus and especially in uterine length. The average length of the normal uterus is 85 mm, with a range from 60 to 110 mm (Bernaschek et al. 1981). The average uterine length in nulliparae is only 76 mm, which is significantly lower than the 80.5 mm average length in uniparae. Similarly, the average uterine length in multiparae, at 87 mm, is significantly greater. Sample et al. (1977) found similar values in their investigations. After menopause the uterine length decreases to the range of 35-65 mm, comparable to prepubertal values (Miller et al. 1977).

The anteroposterior diameter of the uterus also depends on parity and ranges from 30 to 40 mm (Bernaschek et al. 1981). The transverse uterine diameter is 24-50 mm in the childbearing years (Sample et al. 1977) and 12-18 mm after menopause (Miller et al. 1977). Vaginosonographic demonstration of the whole uterus in the long axis is possible only with the panoramic scanner; it follows that precise measurement of uterine length can be achieved only with this type of scanner.

1.2 Version and Flexion

Ultrasound demonstrates an anteverted uterus in approximately 90% of women, although transabdominal scans, at least, show marked angulation of the cervix relative to the corpus in only about 20% of cases (Bernaschek 1982). However, the distended urinary bladder in abdominal sonography tends to straighten the uterus, so it is likely that vaginosonography, performed with an empty bladder, would demonstrate uterine anteflexion in a substantially higher percentage of cases (Fig. 1).

Vaginosonography is especially useful for evaluating the retroverted uterus, which is almost always combined with retroflexion. When scanned from the posterior vaginal fornix, this type of uterus can be delineated with the same clarity as an anteverted uterus scanned from the anterior fornix (Fig. 2). Difficulties can arise when the fornices are obliterated. In these cases the uterus "sits atop" the probe with its long axis following the probe axis, and only the cervix can be adequately evaluated.

It is noteworthy that in a series of 152 ultrasound examinations, the longitudinal uterine axis was approximately on the midline in only about one-fourth of cases. In 60% of anteverted uteri the long axis was found to deviate toward the right side, while levoversion was noted in 80% of the retroverted uteri (Bernaschek 1982). The endometrial echo is

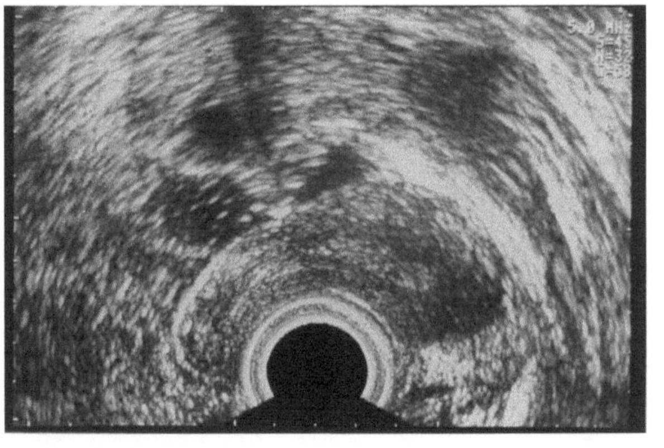

Fig. 1. Anteverted uterus

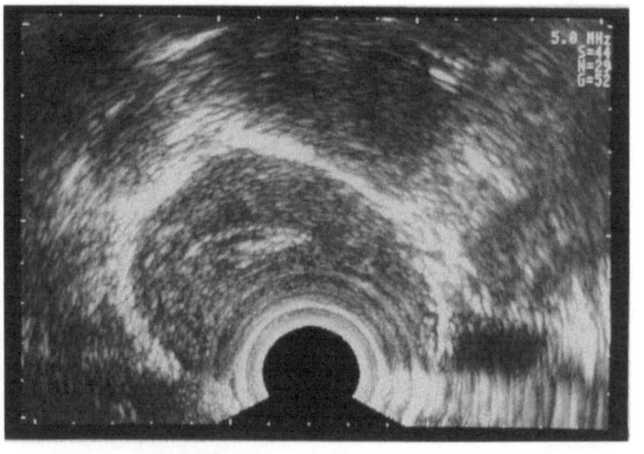

Fig. 2. Retroverted uterus

the best guide for obtaining a complete long-axis view of the uterus. Accordingly, the scanner shaft should be moved on the horizontal plane until the endometrial echo in the long uterine axis is completely visualized. The percentages stated above can help the examiner to locate the long-axis plane more quickly at the start of the examination.

2 Congenital Anomalies

The overall incidence of congenital uterine anomalies is between 0.1% and 2%. These anomalies frequently are associated with other malformations of the urogenital tract (Fried et al. 1978).

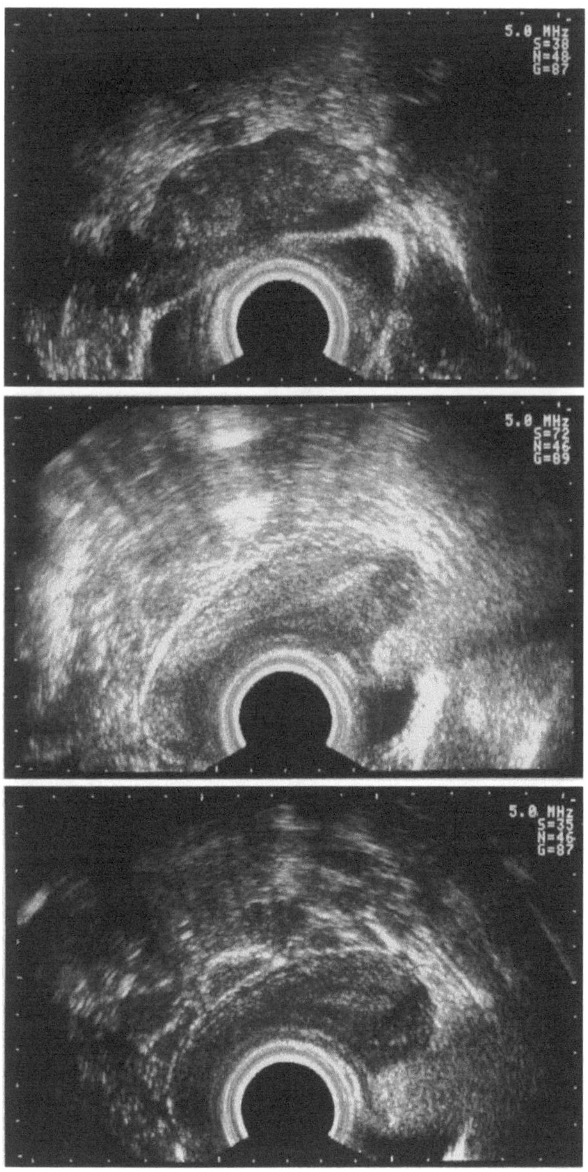

If the presence of a uterine anomaly is suspected, the first step is to evaluate the shape and size of the uterine body and cervix on transverse ultrasound scans. Next a search is made for the presence of an extra uterine cavity or a second cervical canal. Finally, septa may be identified.

The sonographic hallmark of a uterine anomaly is an exceptionally wide corpus on the transverse scan (Funk and Fendel 1988). This excludes anomalies, of course, in which two completely separate uterine bodies are present. If the transverse uterine diameter is disproportionately larger than the anteroposterior diameter, a detailed survey should be made of the endometrial echo (Fig. 3). It is unusual to find simply a wide corpus with a broad endometrial echo that is not doubled. This finding would be consistent with uterus arcuatus, but additional studies would be needed to confirm that diagnosis.

Transverse scans can demonstrate indentations in the anterior or posterior wall of the uterus or in the fundus. Differentiation between an uterine septum and a duplication is made by measuring the width of the partition, which is markedly wider in a duplication due to the presence of muscular elements. It should be noted that the appearance of double endometrial echoes at the tapering angles of the fundus is not in itself an anomalous sign, but should be considered a normal finding (Fig. 4). However, it is still prudent to evaluate both uterine horns separately on longitudinal scans to exclude uterine myomas and solid adnexal tumors.

◀ **Fig. 3.** *Top:* The uterine body appears disproportionately widened in the transverse scan. *Middle, bottom:* Longitudinal scans reveal two separate uterine cavities, recognized by their endometrial echoes

Fig. 4. Transverse scan through the fundus. The wide endometrial echo is a normal finding in this area
▼

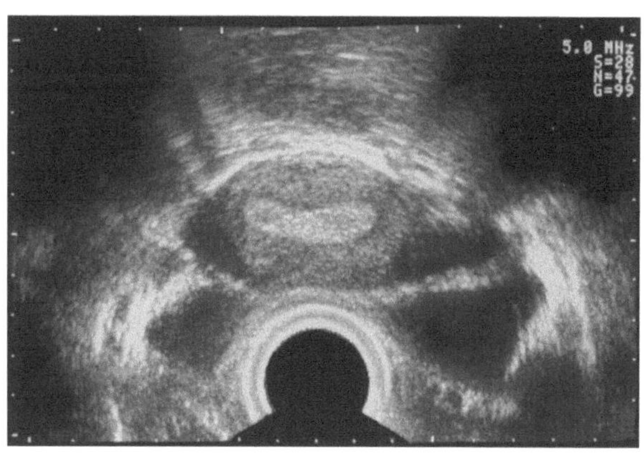

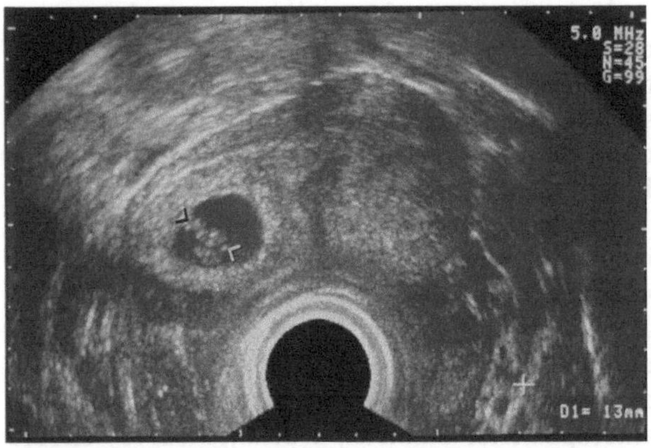

Fig. 5. Transverse scan through a bicornuate uterus. The right horn contains an intact pregnancy in the 7th week *(arrowheads)*

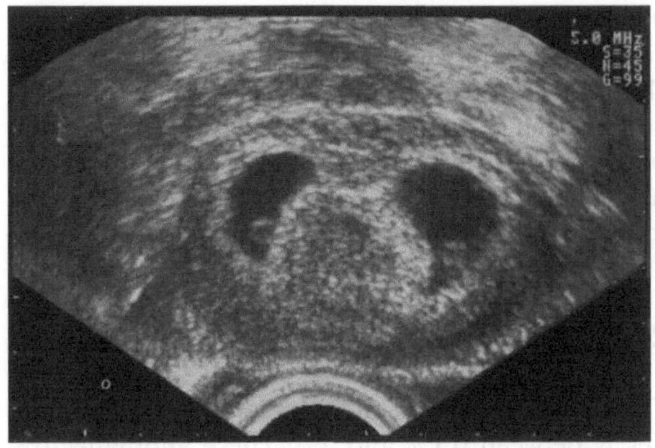

Fig. 6. Transverse scan through a bicornuate uterus. Both horns contain yolk sacs corresponding to the 6th week of pregnancy

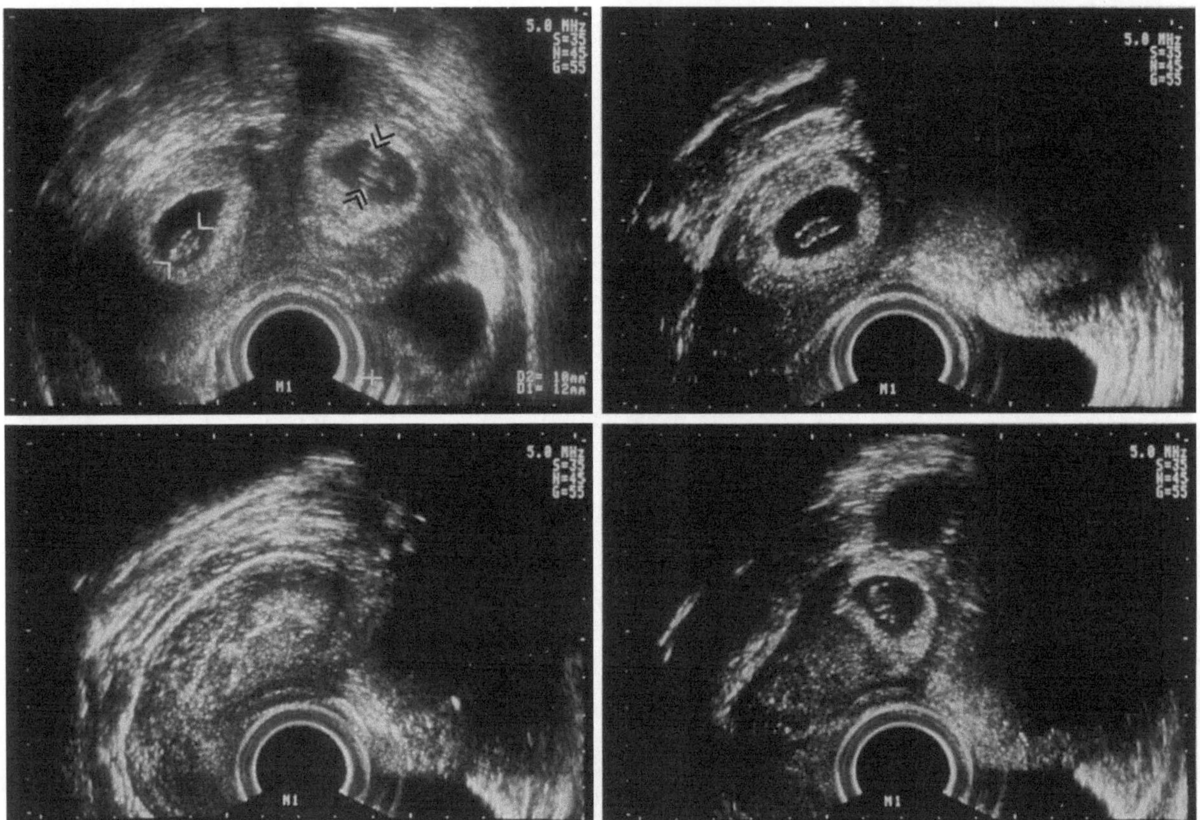

Fig. 7. *Top left:* Transverse scan through a bicornuate uterus showing two independent, intact gestational sacs *(arrowheads)* corresponding to the 7th week of pregnancy. *Right:* Longitudinal scans through both gestational sacs. *Bottom left:* Longitudinal scan of the uterine septum

In the cervical region, transverse scans are able to demonstrate the lumen of the cervical canal and reliably differentiate between a single and double cervix. Regarding duplication anomalies of the uterus, the superior resolution of vaginosonography can define septa with greater clarity than abdominal ultrasound, and it can delineate the endometrium with greater accuracy in both uterine horns. The latter capability is important, as it is the basis for differentiating uterine myomas from solid adnexal masses.

Vaginosonography usually cannot establish the exact degree of a duplication anomaly in the non-gravid uterus. This is a case where hysterosalpingography and other studies cannot be satisfactorily replaced by endosonography.

But when there is a coexisting pregnancy in the first trimester, vaginosonography is the method of choice for evaluating the bicornuate uterus (Fig. 5). During this period both uterine horns may show decidual reaction although normally only one horn contains a true gestational sac (Bernaschek and Kratochwil 1981). However, a saclike structure may develop in the opposite, nongravid horn and create the impression of a twin pregnancy. Vaginosonography can clarify the situation as early as the 6th week of gestation by the early demonstration of one intact pregnancy (Fig. 6), or in rare cases it may confirm two separate gestational sacs, one in each horn (Fig. 7). As gestation progresses, it becomes increasingly difficult to demonstrate a uterine duplication anomaly with ultrasound.

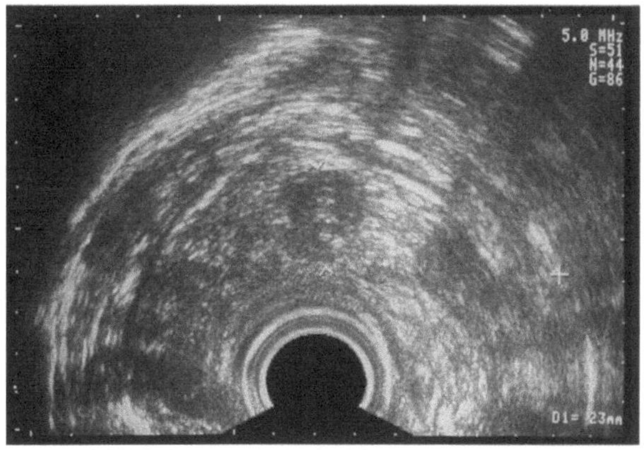

Fig. 8. Longitudinal scan through a uterus with a small intramural myoma in the fundal region *(arrowheads)*

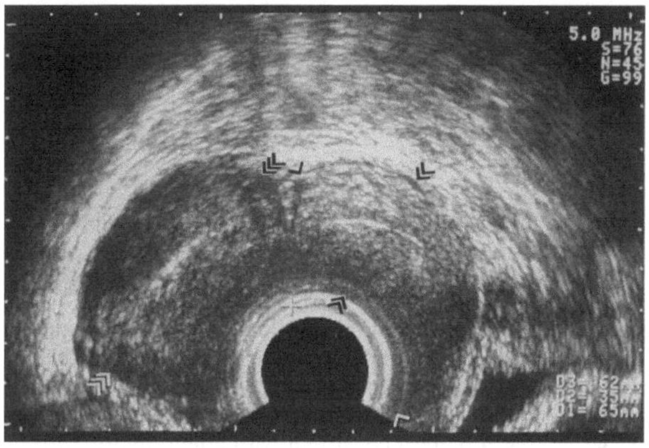

Fig. 9. Large myoma *(triple arrowheads)* in the uterine fundus that could not be isolated from the retroverted uterus by palpation. *Single and double arrowheads,* longitudinal and anteroposterior axes of uterus

3 Diagnosis of Myomas

Vaginosonography can provide highly detailed views of the uterus and its boundaries. In women of childbearing age the uterine cavity appears as a central linear echo that aids in the evaluation of pathologic changes. Vaginosonography is especially helpful when clinical findings are equivocal, e.g., when small subserous and submucous changes are presumed but are poorly demonstrated by abdominal scanning. The latter examination is especially likely to miss small, subserous myomas of the uterine fundus that are obscured by gas-containing loops of bowel (Fig. 8).

Evaluation of the central cavity echo is very helpful in the diagnosis of small submucous myomas, which alter or disrupt the course of the echo (Timor-Tritsch et al. 1988). Larger, pedunculated, subserous myomas may simulate a solid ovarian tumor. Differentiation from ovarian masses, which can have the same sonographic appearance, is accomplished by watching for concomitant motion of the tumor when the associated organ is moved. Vaginosonography can provide a very clear view of myomas in the retroverted uterus, which may be poorly accessible to digital palpation (Fig. 9).

In large myomas, vaginosonography can establish the presence and degree of degenerative changes. These foci usually appear as small, ring-shaped sonolucent areas within the myomatous nodule. Some myomas may contain strongly echogenic areas with adjacent acoustic shadowing, representing foci of calcification.

Myomas, like the uterus itself, diminish in size after menopause. At this time there may be little evidence of a central cavity echo.

Uterine adenomyosis is characterized by the presence of endometrial tissue in the uterine muscular layer. Often these changes are not well demonstrated by abdominal sonography but are shown more clearly by vaginosonography. Scans show a collection of ringlike structures permeating the affected area (Fig. 10), the extent of changes depending on whether the diffuse or circumscribed form of the disease is present.

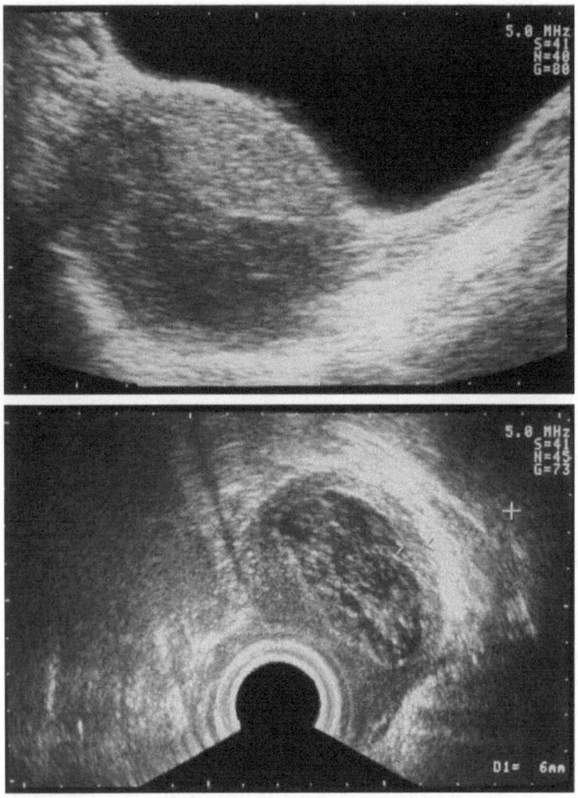

Fig. 10. *Top:* Longitudinal transabdominal scan through the uterus and an adjacent, indeterminate mass. *Bottom:* The superior resolution of vaginosonography localizes the suspicious area to the uterus and shows it to be permeated by ring-like structures, a pattern consistent with adenomyosis *(arrowheads)*

References

Bernaschek G (1982) Die Lageverhältnisse des Uterus in vivo. Wien Klin Wochenschr 94: 124-127

Bernaschek G, Kratochwil A (1981a) Schwangerschaften im Uterus bicornis. Sonographische Diagnose und Verlaufskontrolle. Z Geburtshilfe Perinatol 185: 243-247

Bernaschek G, Endler M, Beck A (1981b) Zur Lagekontrolle von Intrauterinpessaren. Geburtshilfe Frauenheilkd 41: 566-569

Fried AM, Oliff M, Wilson EA, Whisnant J (1978) Uterine anomalies associated with renal agenesis: role of gray-scale ultrasonography. Am J Radiol 131: 973-975

Funk A, Fendel H (1988) Beitrag zur Diagnostik kongenitaler Uterusanomalien. Z Geburtshilfe Perinatol 192: 77-82

Miller EI, Thomas RH, Lines P (1977) The atrophic postmenopausal uterus. J Clin Ultrasound 4: 261-263

Sample WF, Lippe BM, Gyepes MT (1977) Gray-scale ultrasonography of the normal female pelvis. Radiology 125: 477-483

Timor-Tritsch IE, Rottem S, Thaler I (1988) Review of vaginal ultrasonography. A description with clinical application. Ultrasound Q 1988 (6): 1-34

Endosonography of the Ovaries

1 The Normal Ovary

Today the main indication for ultrasound evaluation of the ovaries is the monitoring of follicular maturation. Vaginosonography also has a role in the evaluation of equivocal adnexal findings and in the differential diagnosis of ovarian masses. It must be emphasized, however, that vaginosonography is a macroscopic examination technique, and that one should not be tempted to make a histologic diagnosis on the basis of sonographic findings. It should also be remembered that a sonographic diagnosis made by a gynecologist is influenced by the experience of the examiner and his knowledge of clinical findings and all other available studies. This is by no means unfortunate, because the gynecologic specialist should be able to make a better evaluation in these cases than the nonspecialist.

The normal almond-shaped ovary, situated on the pelvic wall, is easily delineated with vaginal ultrasound. It has echogenic outer borders and typically contains low-level internal echoes. In transabdominal sonography, it is common practice to locate the ovary by following vascular landmarks in the infundibulopelvic ligament. In vaginosonography the ovary is most easily located by tracing the uterine artery laterally to its origin from the iliac artery (Rodriguez et al. 1988). In vaginosonography as in abdominal sonography, delineation of the left ovary may be hampered or prevented by overlying gas-containing sigmoid colon, and the examination may have to be repeated after a cleansing enema. After menopause the ovarian volume gradually diminishes until the organ is no longer palpable and appears no larger than about 1.2 cm on sonograms (Rodriguez et al. 1988).

2 Ovarian Cysts

2.1 Corpus Luteum Cysts

Corpus luteum cyst is one of the most frequently observed uterine changes. Sonographically it presents the rounded shape and smooth borders that are typical of cysts (Fig. 1). Its appearance on vaginal scans differs from that on abdominal scans in that the predominantly cystic area often contains small, low-level echoes caused by internal hemorrhage within the cyst.

Similar changes can be recognized in functional ovarian cysts. As long as these cysts do not exceed 4 cm in diameter, operative treatment is not justified unless a pedunculated cyst has undergone torsion. It is sufficient to repeat the examination in

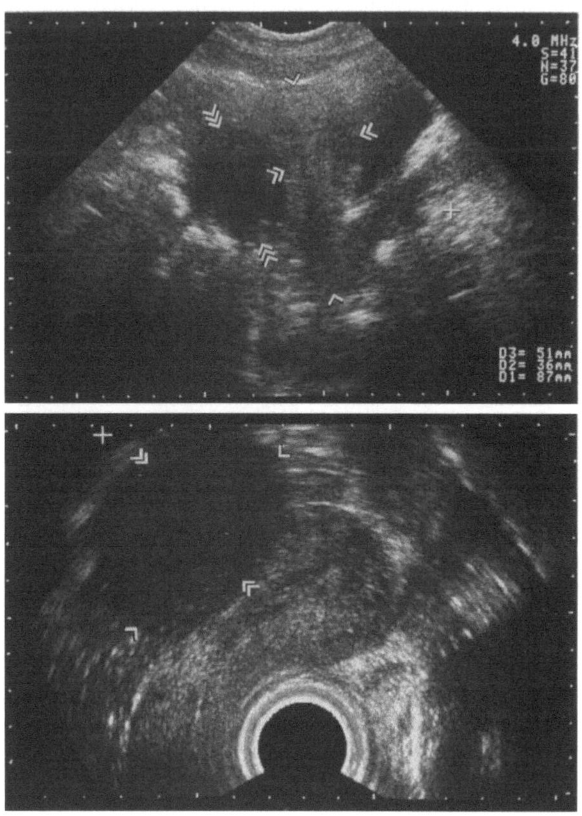

Fig. 1. *Top:* Transabdominal sonography showing a simple ▶ cyst *(triple arrowheads)* behind the uterus (*single and double arrowheads*, longitudinal and anteroposterior axes). *Bottom:* Vaginosonography showing a spherical mass consistent with corpus luteum cyst

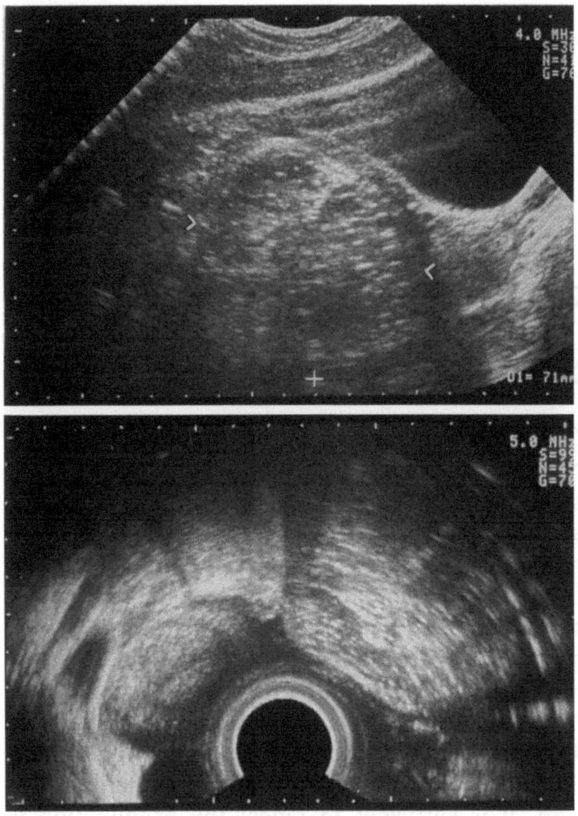

Fig. 2. *Top:* Longitudinal abdominal scan depicts an echogenic mass with peripheral shadowing *(arrowheads)* behind the bladder. *Bottom:* The mass, a dermoid cyst, is also poorly delineated on the vaginal scan

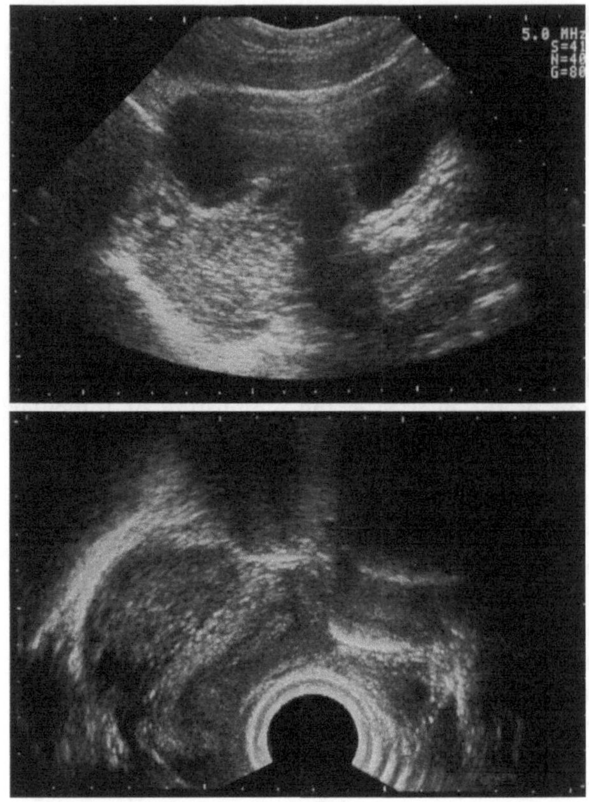

Fig. 3. *Top:* Longitudinal transabdominal scan shows a mass similar to that in Fig. 2. *Bottom:* On the vaginal scan the tumor is plainly demarcated from the uterus. Dermoid cyst

6–8 weeks. Usually the cyst can no longer be demonstrated by that time, especially if concurrent hormonal therapy is administered. This conservative approach can often spare the patient the risks of an unnecessary operation.

2.2 Dermoid Cysts

Dermoid cysts, though easily palpated during the gynecologic examination, are often poorly demonstrated by vaginal as well as transabdominal sonography (Figs. 2, 3). This has to do with their contents, which may cause a high degree of sound attenuation. If the cyst can be visualized, it appears very echogenic and may be indistinguishable from bowel on transabdominal scans. If the cyst contains hair or teeth or if large calcifications are present, acoustic shadowing may be noted adjacent to the mass (Fig. 4). If the patient remains in one position for some time, sedimentation effects with fluid levels may be depicted in the interior of the cyst. The main advantage of vaginosonography is that palpation can be performed almost concurrently with the scanning; indeed, the palpable mass can even be moved with the free hand to allow better evaluation of its shape and size, which otherwise may be poorly demonstrated by the scan.

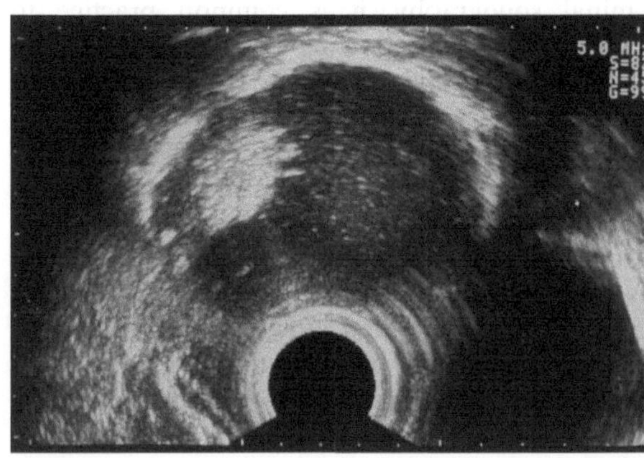

Fig. 4. Transverse vaginal scan of a dermoid cyst with calcifications

2.3 Endometriotic Cysts

The cystic lesions of endometriosis resemble dermoid cysts in their sonographic appearance. However, they cause less sound attenuation, so they are more easily penetrated by the ultrasound beam (Fig. 5).

Often the cyst wall is thickened and has a serrated appearance caused by intramural hemorrhage. Typically the echo pattern of endometriotic cysts is somewhat more homogeneous than in dermoid cysts. Internal echoes (chocolate cysts) are demonstrated more clearly by vaginosonography. Generally, however, endometriotic cysts, dermoid cysts, and corpus luteum cysts cannot be differentiated on the basis of sonographic findings, or at best this differentiation can be made only when typical sonographic features are present.

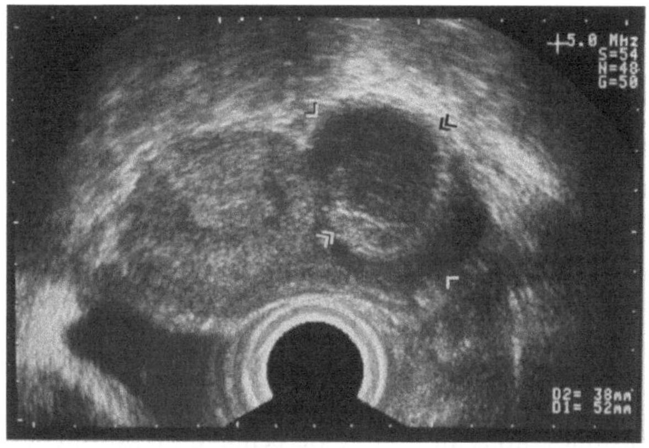

Fig. 5. Transverse vaginal scan showing a well-demarcated, relatively echogenic mass with sites of wall thickening to the left of the uterus *(arrowheads)*. Endometriosis

2.4 Multiloculated Cysts

These cysts are easily recognized on vaginal scans. As on abdominal scans, the septa should be evaluated for their thickness and any deposits that may be present. This applies to papillary cysts as well. Although, as noted, it is not appropriate to use ultrasound to make a histologic diagnosis, thickening and irregularity of the wall may be suggestive of malignancy. This suspicion is heightened by the presence of free fluid in the cul-de-sac (Fig. 6). For this reason, evaluation of the ovaries by vaginosonography is recommended in all patients with ascites of undetermined origin to exclude possible ovarian carcinoma (Timor-Tritsch et al. 1988). Usually, large cysts can be only partly delineated due to the limited penetration of the ultrasound beam.

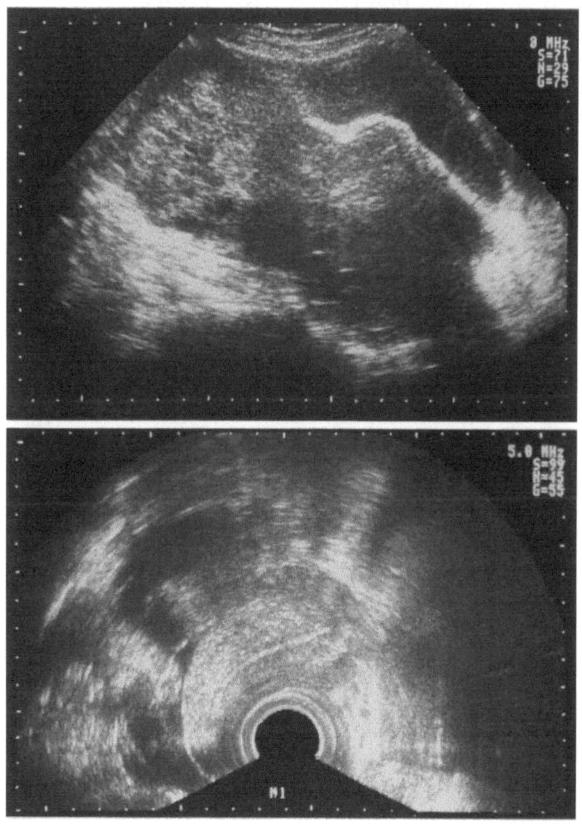

Fig. 6. *Top:* Longitudinal abdominal scan showing a large, solid mass that cannot be separated from the uterus. *Bottom:* Vaginal scan plainly delineates the mass with respect to the uterus

2.5 Ovarian Carcinoma

Most cases of ovarian carcinoma are already far advanced when the patient comes to operation. Earlier detection would be an important key to improving survival rates, and much hope is being placed in vaginosonography as a means toward that end. Campbell et al. (1982) examined 1041 women in the highest-risk age group for ovarian cancer using transabdominal ultrasound. Scans revealed abnormal findings in 15 of the women. Subsequent operation disclosed one true carcinoma and numerous smaller cysts judged to be potentially malignant. Similar results are reported by Rodriguez et al. (1988), who noted nine suspicious findings in 104 postmenopausal women examined by vaginosonography. None of the suspicious findings was a carcinoma.

These studies suggest that mass screening, which is costly and labor-intensive, would hold little promise of success. Added to this is the presumed lack of motivation among older women to undergo screening tests for ovarian cancer, since most women do not even avail themselves of the well-established Pap smear screening for cervical carcinoma.

Even if motivation were not a problem, it might be difficult to impossible to perform successful vaginosonography in a large percentage of postmenopausal women due to involutional changes in the introitus and posterior third of the vagina as a result of estrogen deficiency. Obliteration of the fornices would pose an additional problem. The early diagnosis of ovarian carcinoma continues to be a goal worth striving for, but it is unlikely that dramatic progress will be made in the near future.

3 Inflammatory Adnexal Changes

Sacks et al. (1988) describe the appearance of the normal fallopian tube on vaginosonograms. The tube first runs parallel to the round ligament before passing laterally and posteriorly toward the pelvic wall. Delineation of the tube is enhanced by fluid in the cul-de-sac. Our experience has been that these anatomic structures, which vary considerably in their position, cannot always be clearly identified with a 7-MHz vaginal probe.

In the acute stage of pelvic inflammatory disease, the tube cannot be delineated. Pain can be elicited by pressing gently with the probe during the examination. The poor delineation of the tube and of other genital organs involved by inflammation is easily understood (Fig. 7). The inflammation incites hyperemia in the pelvic viscera, which leads to enhanced sound penetration and a consequent obscuring of organ boundaries (Kratochwil et al.

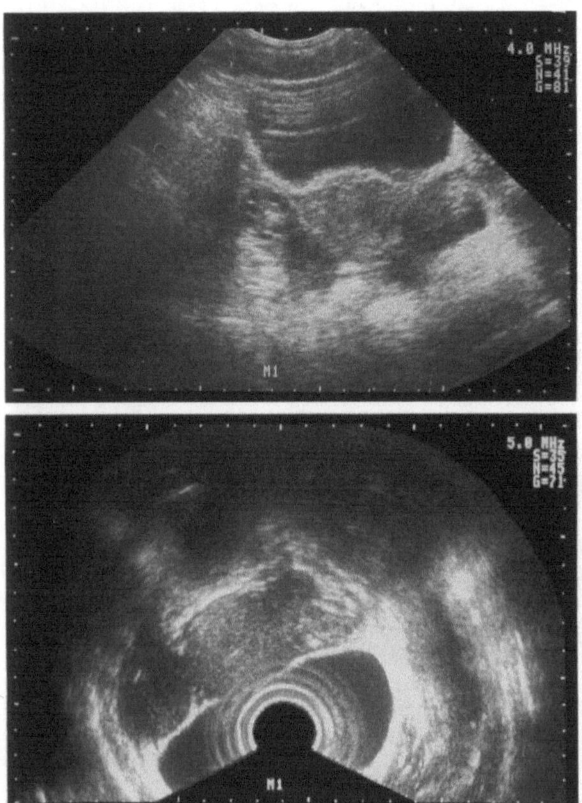

Fig. 7. *Top:* Transverse abdominal scan shows inflammatory adnexal masses flanking the uterus. *Bottom:* Vaginal scan of the same case

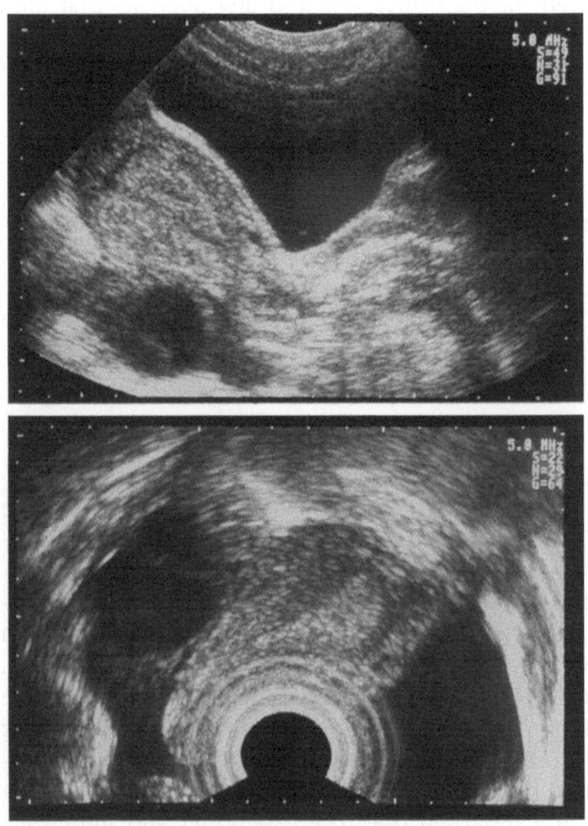

Fig. 8. *Top:* Transabdominal scan showing a cystic mass located behind the uterus. *Bottom:* Vaginal scan of the same mass: hydrosalpinx

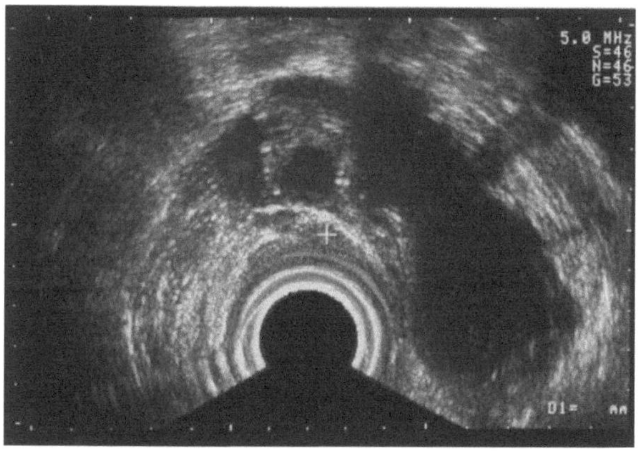

Fig. 9. Vaginal scan of an extensive hydrosalpinx on the left side

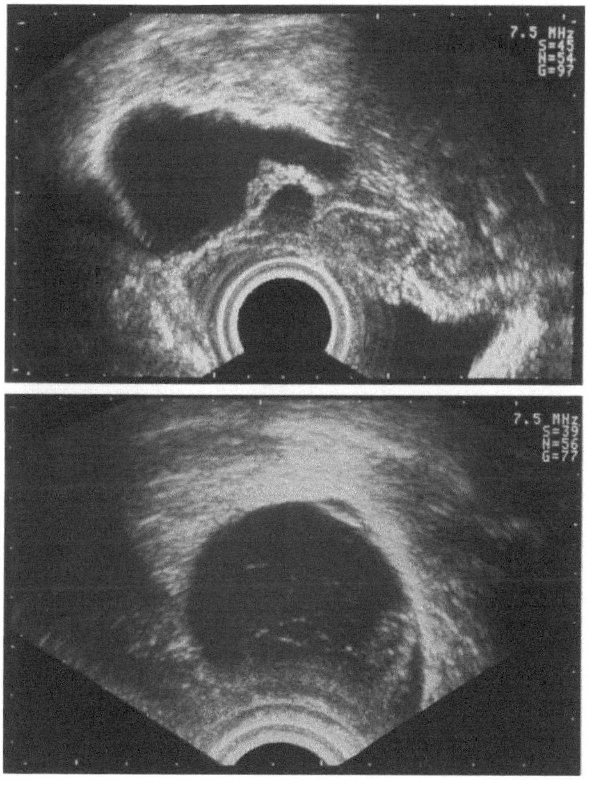

Fig. 10. *Top:* Old hydrosalpinx. *Bottom:* Pyosalpinx with lowlevel internal echoes

1978). In the subacute and chronic stage of the inflammation, fluid collects within the obstructed tubes, and hyperemia decreases. The tubes are now depicted as fluid-filled structures (Fig. 8, 9).

With tubo-ovarian abscess, the ovary is involved in the disease process and can no longer be demonstrated as a separate organ. In the acute stage of pyosalpinx, low-level echoes are seen within the purulent fluid (Fig. 10). Serial examinations during the course of the disease are helpful in evaluating response to therapy.

References

Campbell S, Goessens L, Goswamy R, Whitehead DM (1982) Real time ultrasonography for determination of ovarian morphology and volume. A possible early screening test for ovarian cancer. Lancet i: 425-426

Kratochwil A, Altmann G, Wollmann G (1978) Ultraschalldiagnostik von Adnextumoren. Wien Klin Wochenschr 90: 569-575

Rodriguez H, Platt LD, Medearis AL, Lacarra M, Lobo RA (1988) The use of vaginal sonography for evaluation of postmenopausal ovarian size and morphology. Am J Obstet Gynecol 195: 810-814

Sacks GA, Fleischer A, Kepple D (1988) Clinical applications of vaginal sonography in obstetrics and gynecology. Toshiba Med Rev 4: 15-24

Timor-Tritsch IE, Bar-Yam Y, Elgali S, Rottem S (1988) The technique of vaginal sonography with the use of a 6 MHz probe. Am J Obstet Gynecol 158: 1019-1024

Postoperative Endosonography

The investigation and localization of intra-abdominal complications following operative procedures can be very challenging. The value of transabdominal ultrasound may be compromised by residual intraabdominal air or the fresh surgical wound. But if the source of the complications is in the lesser pelvis, endosonography can contribute greatly to evaluation of the postsurgical patient.

Fresh intra-abdominal hemorrhages present as free fluid collections of variable echogenicity, depending on their degree of organization. Vaginosonography can demonstrate even small amounts of free fluid in the cul-de-sac. Hematomas in the parametrial region appear sonographically as rounded masses with ill-defined margins whose echogenicity varies with their stage of organization (Fig. 1).

Abscesses and hematomas have very similar sonographic appearances, which is not surprising when one considers the frequency with which abscesses develop from infected hematomas. However, abscesses, especially when of long duration, tend to have a thicker, less regular wall that has a somewhat more mottled appearance as a result of the inflammatory process. Often their internal echoes are more diffuse and of higher intensity than in hematomas.

Lymph cysts can form as a sequela to operations for uterine carcinoma that involve a radical lymph node dissection. Early lymphomas require differentiation from postoperative hematomas, and late lymphomas from tumor recurrence on the pelvic wall (Bernaschek and Kratochwil 1980). Cysts developing in the blind vaginal pouch following hysterectomy may be merely functional or may result from the ligation of vessels in the ovarian pedicle (Fig. 2). Pseudocysts may develop postoperatively or as a sequel to inflammatory adnexal disease. They take the form of fluid collections between preexisting adhesions. Pseudocysts, like ovarian cysts, appear predominantly as echo-free areas, usually with smooth margins, but they are easily deformed due to a lack of internal pressure. This phenomenon can be plainly demonstrated by vaginosonography (Fig. 3). Sometimes these masses are radically deformed by gentle pressure from the probe.

Cysts occurring in a blind vaginal pouch following surgery can be aspirated with a needle under light sedation, obviating the need for an open operation (Fig. 4). Sclerotherapy of the type used for renal cysts may be attempted in some patients (Schramek et al. 1988). In addition, abscesses or inflam-

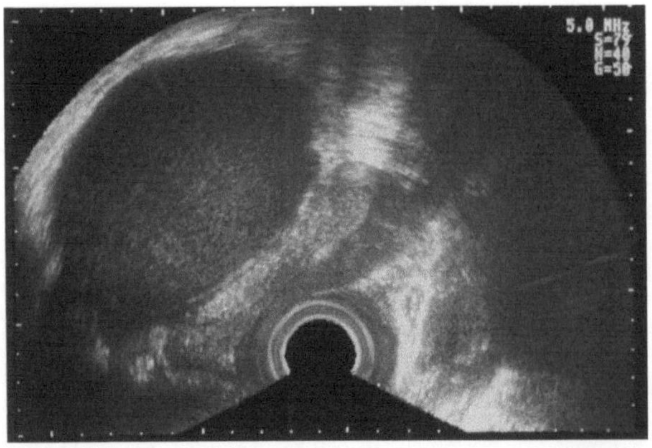

Fig. 1. Vaginosonography demonstrates a giant retrouterine hematoma that already contains high-level internal echoes, signifying that some organization has occurred

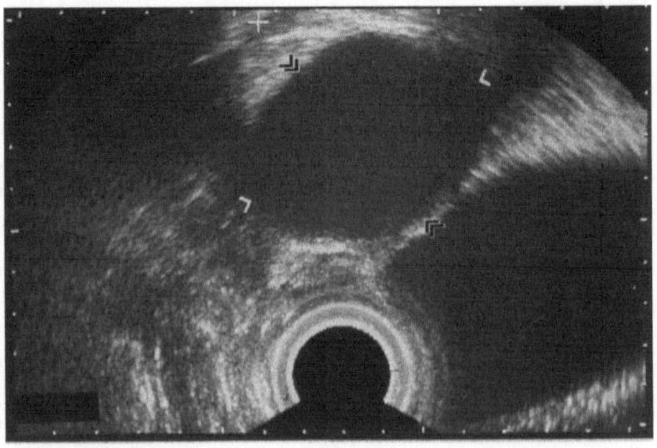

Fig. 2. Vaginal scan showing a 64 × 48 mm cyst *(arrowheads)* on a blind vaginal pouch

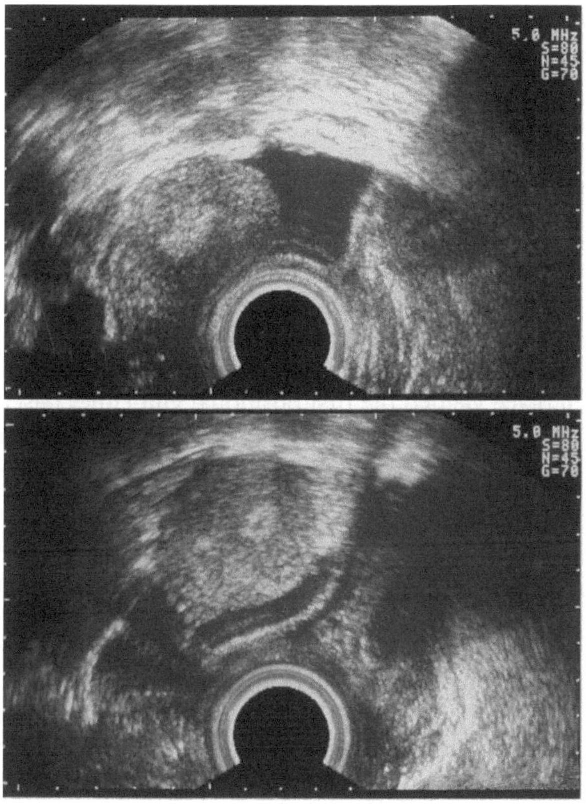

Fig. 3. *Top:* This pseudocyst appears as an echo-free area behind the uterus on a vaginal scan. *Bottom:* Same case, showing marked deformation of the cyst by gentle pressure from the vaginal probe

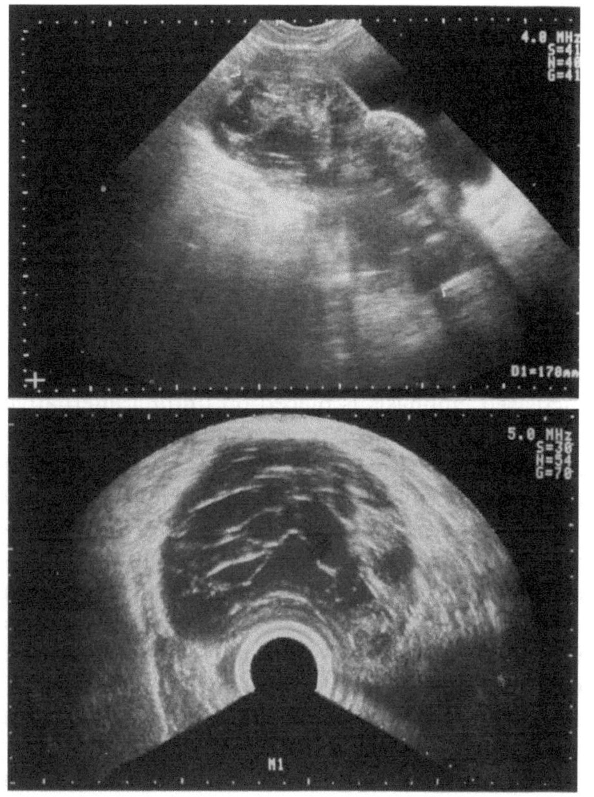

Fig. 4. *Top:* Transabdominal sonography: longitudinal scan shows a multicystic mass *(arrowheads)* that is poorly demarcated from the uterus. *Bottom:* Vaginosonography: transverse scan demonstrates the mass as a multiloculated, cystic structure

matory lesions may be aspirated to recover material for culturing and sensitivity testing. Further details may be found in the chapter "Endosonographically Guided Punctures."

References

Bernaschek G, Kratochwil A (1980) Echographische Diagnose von Komplikationen nach gynäkologischen Operationen. Ultraschall 1: 278-283

Schramek P, Hübner W, Dünser E, Umek H, Porpaczy P (1988) Ultraschall gezielte Nierenzystenpunktion. Bedeutung für Diagnostik und Therapie. Ultraschall Med 9: 169-171

Intrauterine Contraceptive Devices

1 General

In times of relative disenchantment with "the pill," increased emphasis is placed on finding alternative methods of pregnancy prevention. Intrauterine contraceptive devices (IUDs) have always occupied a dominant position in this regard. New IUDs are continually being designed in an effort to approximate the success rate of oral contraceptives. However, the efficacy of an IUD rests not on its design alone but on its position within the uterus and on individual anatomy. Accordingly, there has been much interest in finding objectively verifiable parameters for IUD position that would be useful in predicting the effectiveness of the device.

The sonographically measured distance between the upper end of the IUD and the uterine fundus has been suggested as a possible indicator of contraceptive efficacy (Meyenburg 1978; Schmidt et al. 1979). However, measurements of this distance so far have not taken into account individual differences in the thickness of the uterine wall, leading to uncertainties regarding the correct position of the IUD in the uterine cavity.

Echographic studies in 169 women with an IUD in place yielded the first positional guidelines based on the fundal distance (the distance between the upper end of the IUD and the uterine fundus) and the thickness of the anterior and posterior uterine walls (Bernaschek et al. 1981 a) (Fig. 1). Thus, the IUD is considered to be correctly positioned only if the fundal distance measured on a longitudinal scan through the uterus (Fig. 2) is no more than one-third greater than the thickness of the anterior or posterior uterine wall (Fig. 3). In patients with an improperly positioned IUD, our ultrasound studies indicated a significantly shorter uterine length ($p < 0.01$) as well as significantly more frequent complaints of IUD-related pain.

The connection between correct IUD placement and contraceptive efficacy is further demonstrated by the fact that, in patients who become pregnant while wearing an IUD, the device is found to be very low in the uterus or within the cervix in 80%

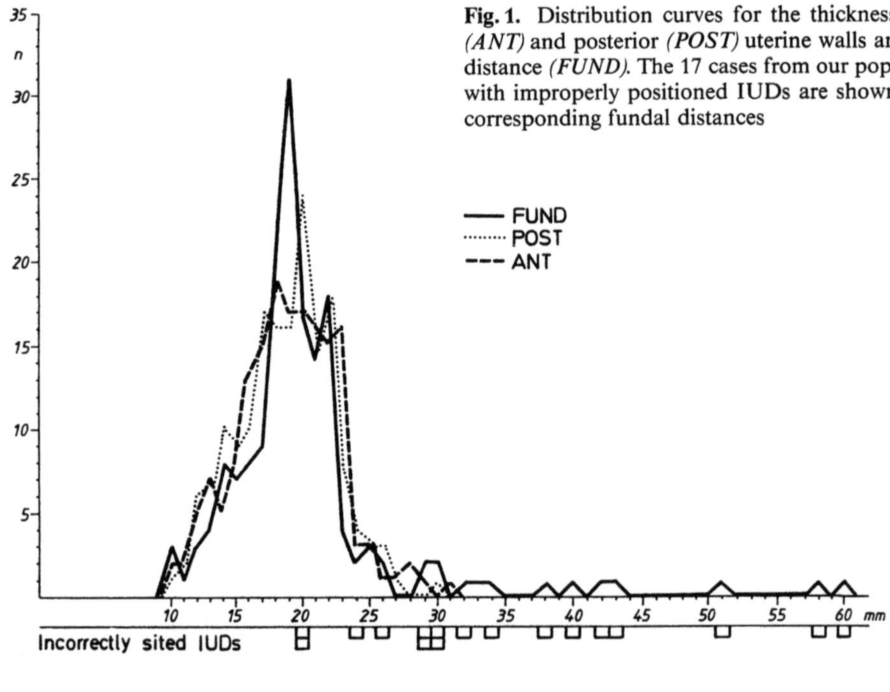

Fig. 1. Distribution curves for the thickness of the anterior *(ANT)* and posterior *(POST)* uterine walls and for the fundal distance *(FUND)*. The 17 cases from our population ($n = 169$) with improperly positioned IUDs are shown adjacent to the corresponding fundal distances

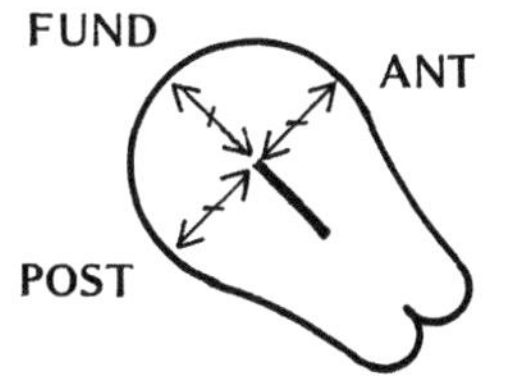

Fig. 2. Diagram showing the critical measurements (*FUND*, fundal distance; *ANT*, anterior wall thickness; *POST*, posterior wall thickness) performed on a longitudinal scan through the uterus to check correct placement of the IUD

(Bernaschek et al. 1981 b), while only 20% have a correctly positioned IUD adjacent to the gestational sac of early pregnancy. In cases where the patient wishes to continue the pregnancy, sonographic localization of the IUD relative to the gestational sac is crucial for further management. Given the high spontaneous abortion rate in "IUD pregnancies" (Tatum et al. 1976), an IUD located in the lower uterine segment or within the cervix below an intact gestational sac (as is usually the case) should be removed. Usually this can be accomplished without harming the pregnancy. In the rare cases where the IUD is located above or alongside the gestational sac, the device should not be withdrawn even if the marker thread is visible due to the high risk of provoking abortion. There is no evidence that leaving an IUD in place during pregnancy increases the incidence of congenital anomalies.

Disappearance of the marker thread may result from an unnoticed spontaneous expulsion of the IUD or simply from breakage of the eyelet to which the thread is attached. Another possibility is intrauterine displacement of the IUD, drawing the thread out of sight. A final possibility is an asymptomatic partial or complete uterine perforation by the device. Usually a perforating IUD can be detected sonographically only if at least a portion of the device is still visible within the uterus (Kratochwil 1976; Abramovici et al. 1985). In itself, a sonographically empty uterus does not prove expulsion of the IUD, and an X-ray film should be taken to exclude asymptomatic perforation by a device that has become lodged between bowel loops where it cannot be visualized with ultrasound.

Sonography can distinguish among the principal morphologic types of IUD currently in use. With the widely used T-shaped and Copper-7 devices, longitudinal scans demonstrate a very bright echo from the transverse limb of the IUD. The Lippes Loop, on the other hand, produces a characteristic pattern of interrupted echoes spaced at equal intervals, consistent with its spiral shape (Winters 1966; Ianniruberto and Mastroberadino 1972; Callen et al. 1980). Najarian and Kurtz (1986) described features that distinguish plastic IUDs with their "entrance-exit reflections" from copper-wrapped IUDs, where reverberations are visible behind the device.

Fig. 3. Correlation between the fundal distance and the anterior and posterior wall thickness for each case according to the formula $\frac{FUND \times 2 \cdot 100}{\text{ANT}+\text{POST}}$

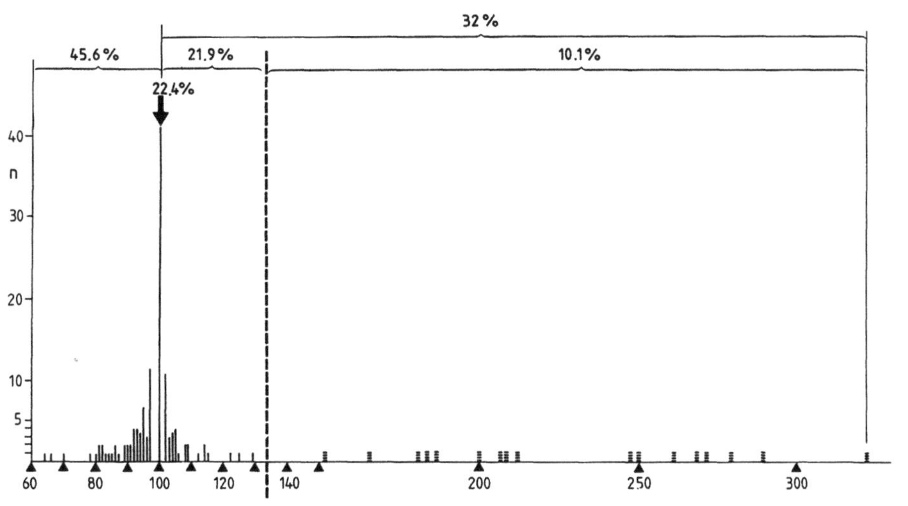

2 Vaginosonography

The main advantages of vaginosonography over traditional abdominal ultrasound in the detection and localization of IUDs are the relative brevity and simplicity of the examination. With vaginosonography it is possible to measure the uterine length and determine the appropriate IUD size immediately after performing the vaginal examination, while the bladder is still empty. This should be done with a vaginal probe having a large view angle. Probes with a view angle of only 90° or so usually cannot delineate both the uterine body and cervix on one longitudinal scan, making it more difficult to accurately measure the length of the uterine cavity (Fig. 4).

Since there is no need to fill the bladder, vaginosonography can also be used to check the placement of the IUD immediately following insertion. The fundal distance measured on the longitudinal scan is the critical parameter in this regard (Fig. 5). Transverse scans through the uterus may also be taken to verify that the arms of the IUD are properly extended (Fig. 6).

When transabdominal ultrasound is used to demonstrate a coil-shaped IUD in the retroverted uterus, problems can arise similar to those encountered in early pregnancy detection. Vaginosonography is decidedly superior in these cases owing to its ability to delineate retroverted and anteverted uteri with equal clarity.

Vaginosonography is also advantageous for localizing IUDs in early pregnancy (Fig. 7). The superior localizing ability of vaginosonography can provide more reliable information on the position of the IUD and the risk of its removal. Given the increased risk of ectopic pregnancy in IUD users, vaginosonography is useful for exploration of the adnexa, at least in patients who are being evaluated for amenorrhea.

Vaginosonography is also gaining a role in the detection of "lost" IUDs. Whereas abdominal sonography can demonstrate a perforated IUD only if at least a portion of the device is retained in the uterus (i. e., a partial perforation), endosonography can additionally cover the region of the cul-de-sac (Fig. 8). However, an IUD that has perforated between loops of bowel is inaccessible to vaginal ultrasound.

A significant advantage of abdominal sonography over vaginosonography is that it can be used to guide the removal of IUDs with detached threads.

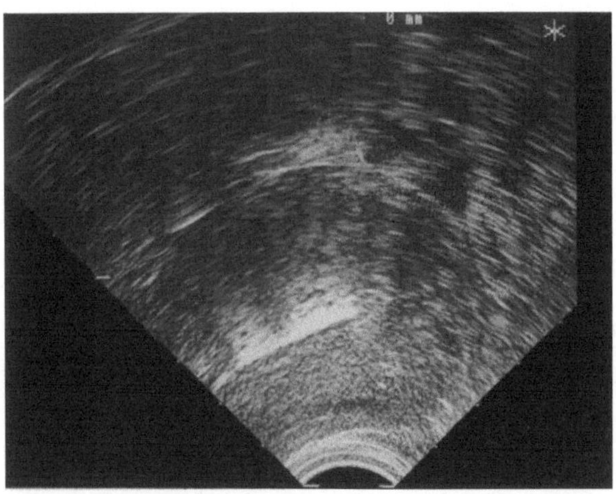

Fig. 4. Longitudinal scan through the uterine body with an improperly positioned IUD. The field of view of the transducer is inadequate to demonstrate the cervical region on the same scan

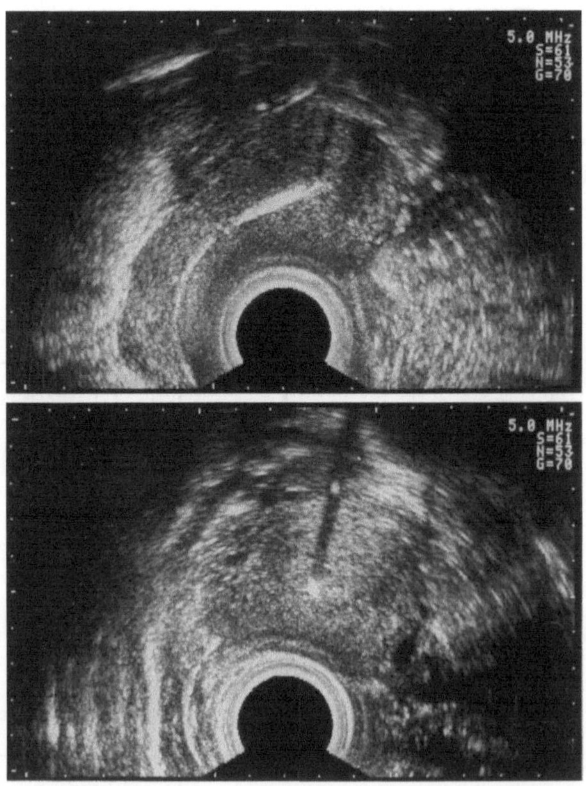

Fig. 5. *Top:* longitudinal scan through the uterus with a correctly positioned copper IUD (panoramic sector scanner). *Bottom:* corresponding transverse scan through the vertical limb of the IUD. Note the acoustic attenuation and reverberations behind the IUD on both images

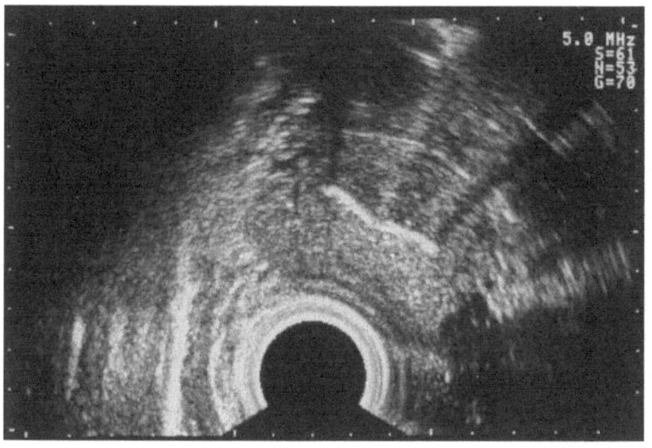

Fig. 6. Transverse scan through the uterus showing the lateral arms of a T-shaped IUD

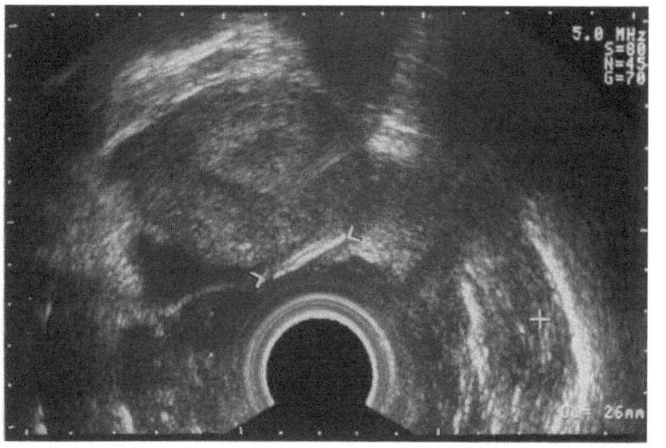

Fig. 8. Transverse scan through a retroverted uterus, showing the long limb *(arrowheads)* of a completely perforated IUD in the cul-de-sac

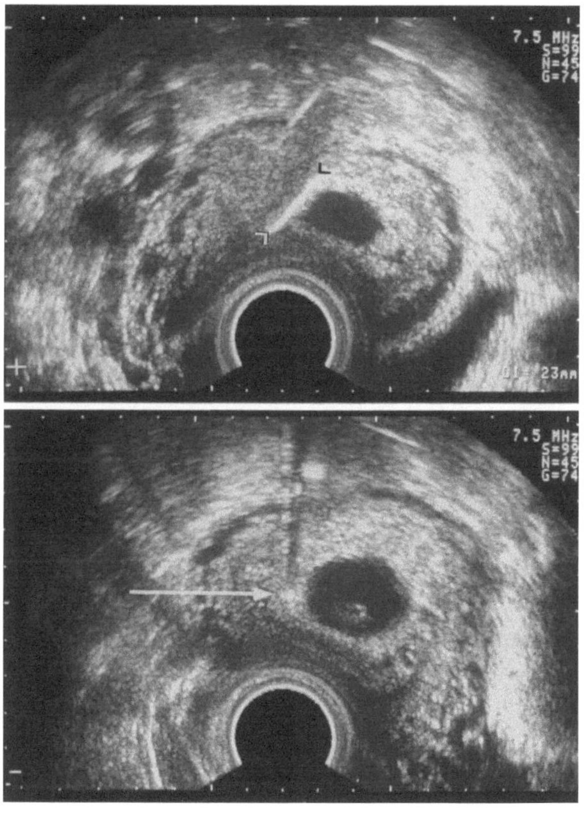

Fig. 7. Pregnancy coexisting with a correctly positioned copper IUD. *Top:* scan cutting the long axis of the IUD *(arrowheads;* note the reverberations). *Bottom:* transverse scan through the IUD, which is directly adjacent to the gestational sac *(arrow)*

The abdominal scan can direct insertion of the forceps or hook through the vagina to the site of the IUD.

In summary, the following ultrasound studies are recommended in conjunction with the use of IUDs:

1. Sonographic exploration of uterine anatomy prior to IUD insertion to detect an unacceptably short uterus or disclose the need for a smaller IUD.
2. Checking the IUD position after insertion to allow the immediate detection of improper placement. Removal may be deferred in these cases for up to 1 month to await possible spontaneous correction while an interim contraceptive method is used (note that the arms of an IUD may take time to extend, especially if they are directed upward during insertion).
3. Follow-up examinations at 3- to 6-month intervals. Most displacements and expulsions of IUDs are observed in the initial months after insertion.
4. Sonographic evaluation of IUD-related complaints and amenorrhea.
5. The main advantages of vaginosonography are the ability to perform the examination with an empty bladder and the improved image detail, especially in the cul-de-sac and the retroverted uterus.

References

Abramovici H, Sorokin Y, Bornstein J, Auslander R (1985) A partial uterine perforation (type 2) by a Copper-T IUD: sonographic diagnosis and management. J Ultrasound Med 4: 381-383

Bernaschek G, Endler M, Beck A (1981 a) Zur Lagekontrolle von Intrauterinpessaren. Geburtshilfe Frauenheilkd 41: 566-569

Bernaschek G, Spernol R, Beck A (1981 b) IUD-Lage bei intrauterinen Schwangerschaften. Geburtshilfe Frauenheilkd 41: 645-647

Callen PW, Filly RA, Munyer TP (1980) Intrauterine contraceptive devices: evaluation by sonography. Am J Radiol 135: 797-801

Ianniruberto A, Mastroberadino A (1972) Ultrasound localization of the Lippes Loop. Am J Obstet Gynecol 114: 78-82

Kratochwil A (1976) Ultraschalldiagnostik in der Gynäkologie. Gynäkologe 9: 166-180

Meyenburg H (1978) Anwendung der Ultraschall-Schnittbildtechnik zur Darstellung von Intrauterinpessaren. Geburtshilfe Frauenheilkd 38: 950-957

Najarian KE, Kurtz AB (1986) New observations in the sonographic evaluation of intrauterine contraceptive devices. J Ultrasound Med 5: 205-210

Schmidt EH, Wagner H, Quakernack K, Beller FK (1979) Ergebnisse der Lageüberwachung von Intrauterinpessaren durch Ultraschall. Geburtshilfe Frauenheilkd 39: 138-143

Tatum HJ, Schmidt FH, Jain AK (1976) Management and outcome of pregnancies associated with the Copper T intrauterine contraceptive device. Am J Obstet Gynecol 126: 869-879

Winters HS (1966) Ultrasound detection of intrauterine contraceptive devices. Am J Obstet Gynecol 95: 880-882

Endosonographic Diagnosis of Carcinoma

1 Cervical Carcinoma

1.1 General

Approximately 30% of all carcinomas in women are genital cancers, and approximately half of those are carcinomas of the cervix. However, successful early detection by screening examinations has led in some countries to a relative decline in the incidence of cervical carcinoma compared with corpus carcinoma in recent years, and it has become possible to detect about one-third of all cervical carcinomas in stage I and roughly another one-third in stage II (Michalica 1981; Bernoth et al. 1984).

The treatment of cervical carcinoma and the selection of patients for a planned treatment protocol are determined largely by the extent of disease. But a clear definition and classification of disease based on precise, objectifiable documentation is equally important for evaluating the efficacy of different modes of treatment. These considerations led Heymann, in the 1920s, to propose the first staging system for cervical carcinoma. As diagnostic technique improved and weaknesses in this system became apparent, constant efforts were made to find a better classification system and more precise definitions. The currently accepted staging system of the International Federation of Gynecology and Obstetrics (FIGO) is based on a clinical assessment of the extent of disease (Kottmeier 1982). Especially in stage II B, presently considered to represent the limit of operability, staging has traditionally relied on the subjective assessment of parametrial involvement by palpation. The greatest source of error in this method is the inability to distinguish consistently between neoplastic and inflammatory parametrial change. Moreover, different examiners may differ markedly in their interpretation of findings, especially with regard to the presence or absence of parametrial involvement and the extension of disease onto the pelvic wall.

As a result, clinical staging based on the findings of rectal palpation agrees with the histologic diagnosis in only a relatively small percentage of cases. Zander et al. (1981) found that clinical staging for stage I agreed with histologic findings in about 68% of cases, while for stage II A it was correct in less than 50%, and for stage II B in only 25%. This is consistent with our own findings (Bernaschek and Schaller 1983) that 65% of clinical stage II B patients were incorrectly evaluated. These large discrepancies demonstrate the difficulty of selecting the most appropriate therapeutic procedure, as they would cause Wertheim's radical hysterectomy to be incorrectly withheld from a large percentage of cervical carcinoma patients. Moreover, the frequent inaccuracy of clinical staging casts serious doubt on whether the results at surgical and radiotherapeutic centers can be validly compared.

In the search for a solution to the staging problem, a number of authors (Nelson et al. 1974; Buchsbaum 1972; Wharton et al. 1977; Piver and Barlow 1977; Lagasse et al. 1979; Avarette et al. 1972, 1981) have recommended pretherapeutic operative evaluation (staging laparotomy) in cervical carcinoma. They found that the discrepancies between clinical and operative staging are considerable not just in stages I B, II A, and II B (19%, 52%, and 63%, respectively) but especially in stages III A and III B (67% and 79%, respectively), which ordinarily are not managed by radical surgery. However, Kademian and Bosch (1977) point out that the information gained by surgical staging does not improve 5-year survival rates, and that the procedure itself is associated with significant added morbidity and mortality. What is more, treatment in approximately 85% of cases is not modified as a result of surgical findings.

In the past it has not been possible to assess preoperatively the depth of invasion of carcinoma in the cervical wall, so the extension of carcinoma within stage I B is disregarded in the FIGO classification. However, this knowledge would be of some importance prior to therapy, since clinical stage I B includes cases that vary considerably in their prognosis and their responsiveness to different modes of treatment (surgery or irradiation). Even within stage I B, where the cancer is confined to the cervix, the size of the tumor strongly affects the frequency of metastasis and thus the 5-year survival rate. Tumor volume is a significant factor in this regard, but even more significant is the maximum depth of invasion. Lange (1960) found that the incidence of lymph node metastasis was 11.1% when the tumor was smaller than 1 cm but rose to 26% when the tumor size was 1-3 cm. Piver and Chung (1975) found an 18% incidence of lymph node involvement in stage I B tumors smaller than 1 cm, 22% with tumors up to 3 cm, 33% up to 5 cm, and 50% over 6 cm. The special significance of the maximum depth of tumor invasion is demonstrated in a study by Baltzer and Goepcke (1979), who noted a 27% increase in micro- and macrometastases from cervical carcinoma when the depth of invasion was 11-15 mm, compared with a 43.7% increase when the depth of invasion was 16-20 mm.

1.2 Staging

1.2.1 Rectosonography Vs Vaginosonography

The staging of cervical carcinoma has traditionally relied mainly on the subjective interpretation of palpable findings in addition to inspection, cystoscopy, rectoscopy, and intravenous urography. CT and MRI (see Sect. 1.2.6), while costly, have been applied in an effort to improve staging, but the simpler procedure of endosonography has attracted particular interest as an alternative to traditional staging methods.

Of the endosonographic methods, rectosonography has attained the greatest importance in pretherapeutic staging of cervical carcinoma. Although the parametria and pelvic wall can be demonstrated by vaginosonography, which is better tolerated by most patients (Popp 1986), the rectal approach offers several advantages:

1. Optimum correlation with palpable rectal findings; suspicious areas can be imaged and compared immediately following the rectal examination.
2. The distended urinary bladder straightens the uterus so that the ultrasound beam strikes the parametria at a perpendicular angle, and the entire cervical region can be systematically explored and documented on transaxial planes.
3. Using a transaxial rotary scanner, both parametria can be compared on the same scan and traced as far as the pelvic wall, following the pattern of the digital examination. Vaginosonography is superior only for evaluating infiltration of the vagina itself. This study can be performed with the same probe used for rectosonography, and as in that examination a balloon is placed over the probe and filled with fluid once the probe has been inserted into the vagina. The transaxial scans cover the whole circumference of the vaginal wall and allow the wall thickness to be compared at specific distances from the introitus, which are read on the scanner shaft. In addition, the fluid-filled balloon makes the vaginal wall tense so that it is easier to evaluate. Longitudinal scans of the vagina are best obtained with a linear-array probe, which offers excellent near-field resolution as its principal advantage.

The technique of rectosonography was described in the chapter "Endosonographic procedures". For staging, however, it is important to determine the angle of inclination of the uterine long axis relative to the horizontal on the sagittal plane with the bladder distended so that correct transaxial scans can be obtained. This is best accomplished by abdominal sonography. Once the orientation of the uterus has been determined, the shaft of the rectal probe is angled correspondingly (e.g., lowered for an anteverted uterus) so that the long axis of the probe is approximately parallel to the long axis of the uterus.

Before the examination is begun, a small, liquid-filled rubber balloon may be placed in the posterior vaginal fornix and secured with a tampon to mark the junction of the cervix and vagina.

In the rare case where rectal scanning is painful, a topical anesthetic may be applied to the anus and also to the tip of the condom that is routinely placed over the rectal probe for hygienic purposes.

A transducer frequency of 7 MHz is recommended for evaluating intracervical structures and the portions of the parametria directly adjacent to the uterus. A frequency of 5 MHz is preferred for examinations of the entire pelvis. Because of the transaxial field of view of the scanner and its centralized position within the lesser pelvis, it is unnecessary to employ lower frequencies.

1.2.2 Intracervical Extent of Disease

The FIGO staging system for cervical carcinoma distinguishes between stage I A and I B lesions confined to the cervix. Stage I A denotes preclinical carcinoma in which microinvasion has already occurred. Accordingly, this is a purely histologic entity that cannot be diagnosed or evaluated by endosonography. The situation is different in stage I B, however, where rectosonography for the first time can demonstrate graduations in tumor size and

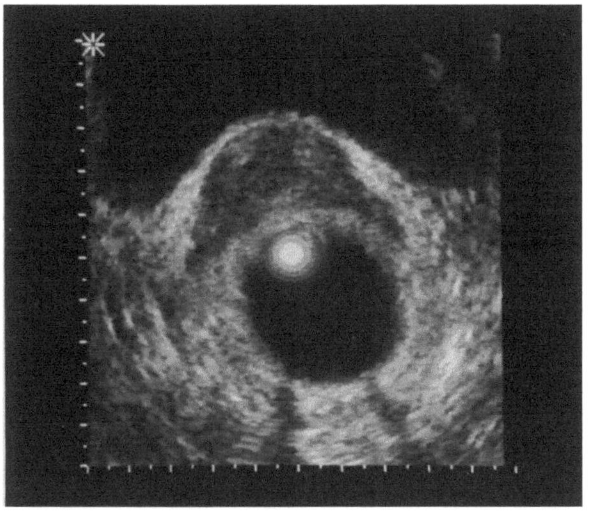

Fig. 1. Transverse scan through the cervix delineating a hypoechoic tumor area that extends to the outer third of the anterior cervical wall

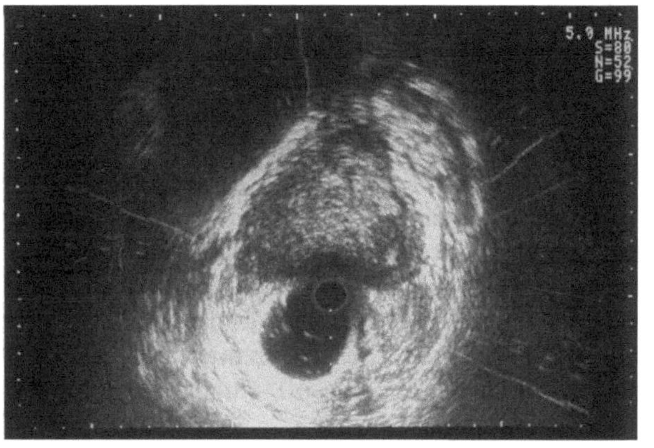

Fig. 2. Transverse scan through the cervix delineating a tumor area that is hyperechoic to surrounding cervical tissue (histology: undifferentiated small-cell carcinoma)

depth of cervical wall invasion. Most carcinomas are hypoechoic to the surrounding cervical tissue (Fig. 1), although some may appear hyperechoic (Fig. 2). Occasional isodense lesions are not detected with ultrasound and may lead to misinterpretation. It seems clear that the difference in the echogenicity of infiltrated tissue compared with healthy cervical tissue relates to the structure of the carcinoma. And while it is reasonable to assume that fast-growing or very cellular carcinomas will appear less echogenic than other types, larger and more directed studies are needed before final conclusions may be drawn.

The maximum depth of invasion on transaxial endosonograms can be graded on a scale of 1-3 according to whether the carcinoma is confined to the inner third of the cervix or has spread to the middle third (Fig. 3) or outer third (Fig. 4) of the cervical wall.

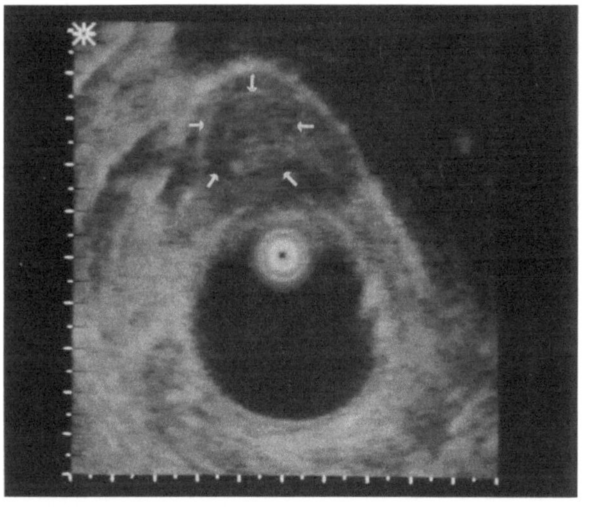

Fig. 3. Transverse scan through the cervix delineating a hypoechoic tumor area confined to the middle third of the cervix *(arrows)*

In a study of 41 patients who underwent radical hysterectomy for cervical carcinoma at the Second University of Vienna Women's Hospital, it was found that the rectosonographic prediction of depth of cervical invasion agreed with the histologic evaluation of the surgical specimen in 35 cases (85%). Based on the three-point scale, there were only four cases in which the extent of stage I carcinoma was overestimated by rectosonography, i.e., apparent extension of the carcinoma to the outer

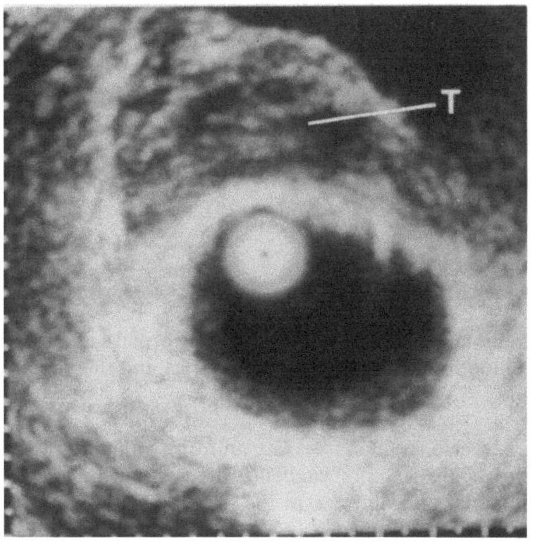

Fig. 4. Transverse scan through the cervix delineating a hypoechoic tumor area *(T)* that extends to the outer third of the cervix ▶

third of the cervix was not confirmed histologically. There were only two cases in which the histologically verified depth of invasion was greater than the sonographic impression.

These findings suggest that rectosonography could one day contribute significantly to the further pretherapeutic differentiation of stage I cervical carcinoma. Above all, it could provide a more accurate basis for comparing the results at surgical and radiotherapeutic centers, since at present cases with very different prognoses are treated and evaluated jointly under the heading of "clinical stage I B disease."

1.2.3 Vaginal Infiltration

When invasion and destruction of the cervix becomes advanced, the tumor may spread to involve the vagina, uterine corpus, and parametria. While extension to the corpus is disregarded in the FIGO classification, stage II A is defined as carinoma involving the vagina but sparing the lower third. The carcinomatous tissue may coat the vaginal vault and vaginal wall superficially, or the invasion may be far more extensive. Rectosonography (Fig. 5) and especially vaginosonography (Fig. 6) can local-

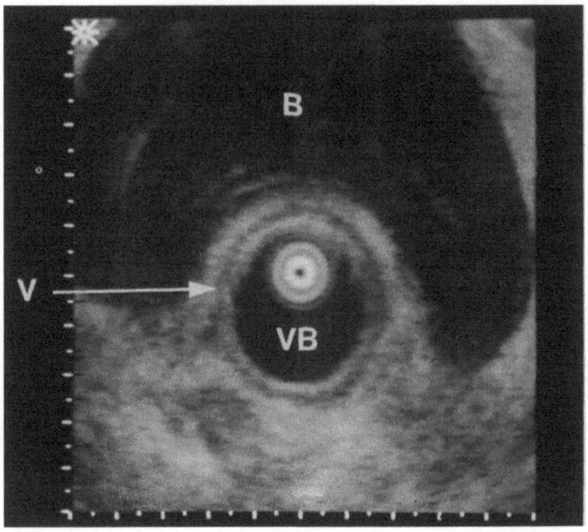

Fig. 6. Transverse vaginosonogram through the normal vagina *(V)* showing smooth, uniformly thick boundaries; urinary bladder *(B)*, vaginal balloon *(VB)*

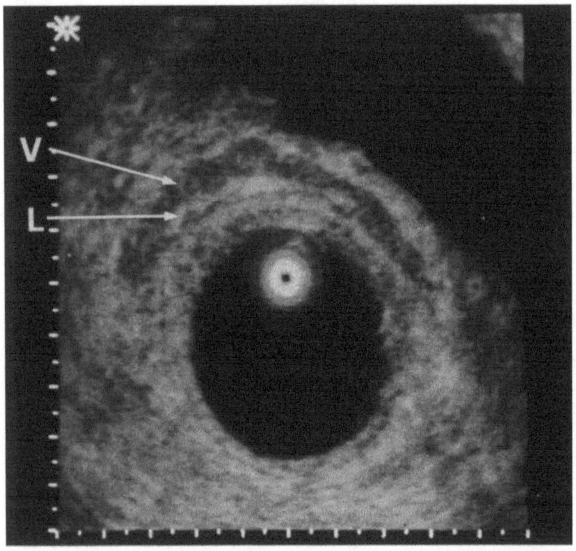

Fig. 5. Transverse rectal scan through the normal vagina *(V)* showing smooth, uniformly thick boundaries and the vaginal lumen *(L)*

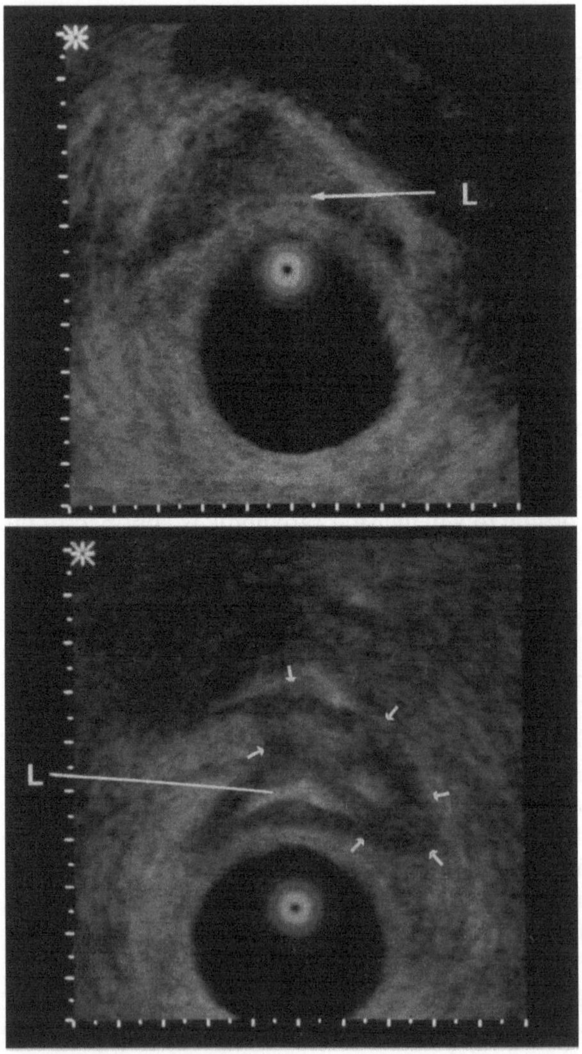

Fig. 7. Transverse rectal scans. *Top:* uniform widening of the upper third of the anterior vaginal wall; vaginal lumen *(L)*. *Bottom:* nodular area *(arrows)* in the left side of the anterior vaginal wall; vaginal lumen *(L)* ▶

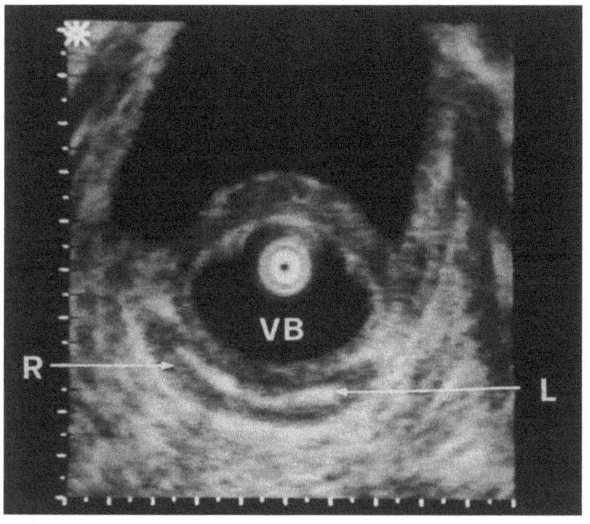

Fig. 8. Transverse vaginal scan showing infiltration of the lower third of the anterior vaginal wall. Note the difference in thickness between the infiltrated anterior vaginal wall and the uninvolved posterior wall; rectum *(R)* with lumen *(L)*, vaginal balloon *(VB)*

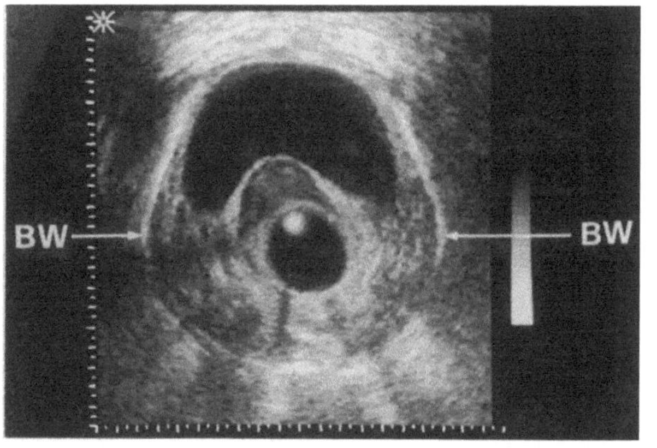

Fig. 9. Rectosonogram in a patient with histologically confirmed cervical carcinoma. Transverse scan through the cervix shows healthy parametria with tapered lateral extensions; pelvic wall (BW)

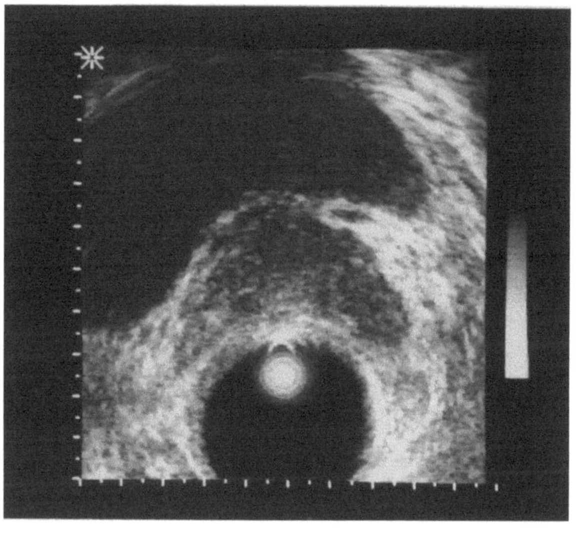

Fig. 10. Transverse rectal scan through the cervix showing lateral disparity in the parametria. The left parametrium appears widened and blunted

ize the involvement and determine the extent of the infiltration.

While the healthy vagina presents a uniform wall thickness over its whole circumference, carcinomatous infiltration may appear as a uniform thickening or may include nodular areas (Fig. 7).

Stage III A disease is characterized by invasion of the vagina that includes its lower third. In typical cases where there is involvement of the anterior vaginal wall, vaginosonography can supplement manual examination by documenting the extent of the infiltration (Fig. 8). Linear scanners are best suited for demonstrating the longitudinal extent of cancer infiltration. With rotary scanners, it is possible only to determine the level of infiltration by reading the depth of insertion on the scanner shaft. However, differentiation between inflammatory and carcinomatous change can be accomplished only by performing a sonographically guided needle biopsy.

1.2.4 Parametrial Infiltration

The spread of cervical carcinoma to the parametria is a crucial factor in terms of clinical staging as well as treatment planning. When primary irradiation is elected, parametrial involvement can be tracked as a means of monitoring therapeutic response. A contiguous series of transaxial rectal scans beginning at the junction between the vaginal and supravaginal parts of the cervix can accurately delineate the parametria, providing an exact description of their shape and margins and establishing the degree of cancer spread in relation to the pelvic wall. This technique can also demonstrate lateral disparities or unilateral shortening of the parametrium with an associated shift of the cervix. Normal parametria should appear equal on both sides, exhibit smooth margins, and taper off laterally in width (Fig. 9).

The value of rectosonography is particularly apparent in stage II B disease, where there is parametrial involvement but no extension of carcinoma onto the pelvic wall. This is an area in which rectosonography, by its objective documentation of the parametria, could contribute much to the accurate

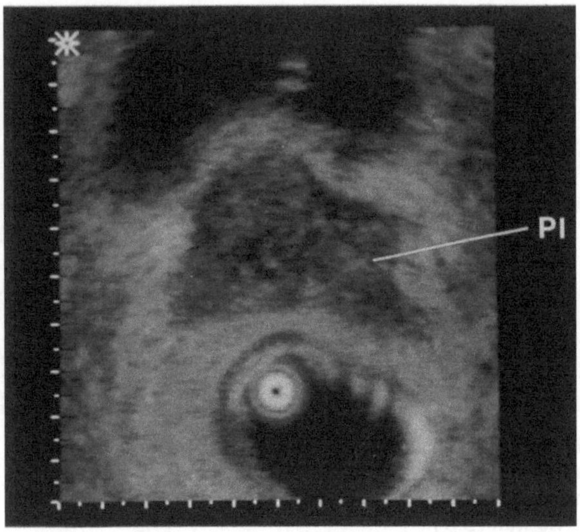

Fig. 11. Transverse rectal scan through the cervix showing marked lateral disparity in the parametrial echoes. The left parametrium *(Pl)* shows nodular expansion

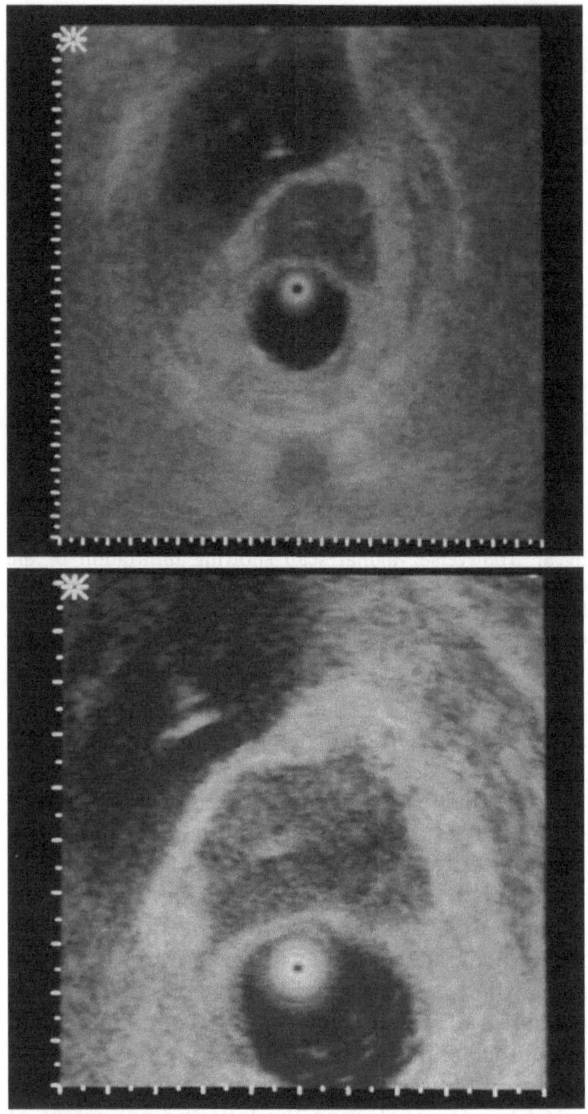

Fig. 12. Transverse rectal scans. *Top:* the left parametrium is ▶ widened and has an irregular contour, and the cervix is shifted toward the side of parametrial involvement. *Bottom:* enlarged view of the cervix and parametria

differentiation of stages I B, II B, and III B. The changes that are most easily documented sonographically are a lateral disparity with unilateral shortening or widening (Fig. 10) and especially the presence of irregular parametrial margins (Fig. 11). Additionally, a shift of the cervix toward the affected side should raise suspicion of extrauterine spread (Fig. 12). A major advantage of rectosonography over rectal palpation is that the former can demonstrate both the rectal and the vesical boundaries of the parametrium.

In stage III B disease the parametrial involvement has extended onto the pelvic wall, and operative treatment is no longer considered an option. In these cases, too, rectosonography supplements rectal palpation by providing a new, objectifiable technique that should assist accurate staging (Fig. 13). Scans usually show an obvious shift of the cervix toward the side of more severe parametrial involvement (Fig. 14). The pelvic wall appears as a highly echogenic structure which delimits the field of view through acoustic absorption. The typically hypoechoic parametrium appears widened and shows irregular borders. Usually it does not appear to reach the pelvic wall on rectal scans. However, minimally hypoechoic structures may be seen extending laterally and posteriorly from the parametrium, which in clinical stage III cases present as masses broadly fixed to the pelvic wall. The degree to

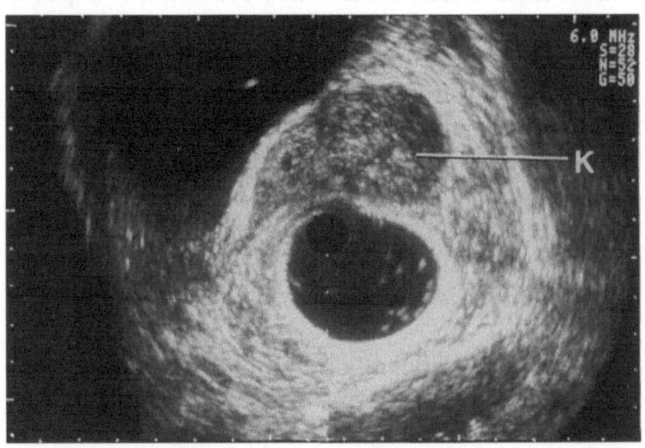

Fig. 13. Transverse rectal scan through the cervix. The normal right parametrium tapers laterally and presents smooth margins. The left parametrium is misshapen and widely infiltrated by histologically confirmed cervical carcinoma *(K)*

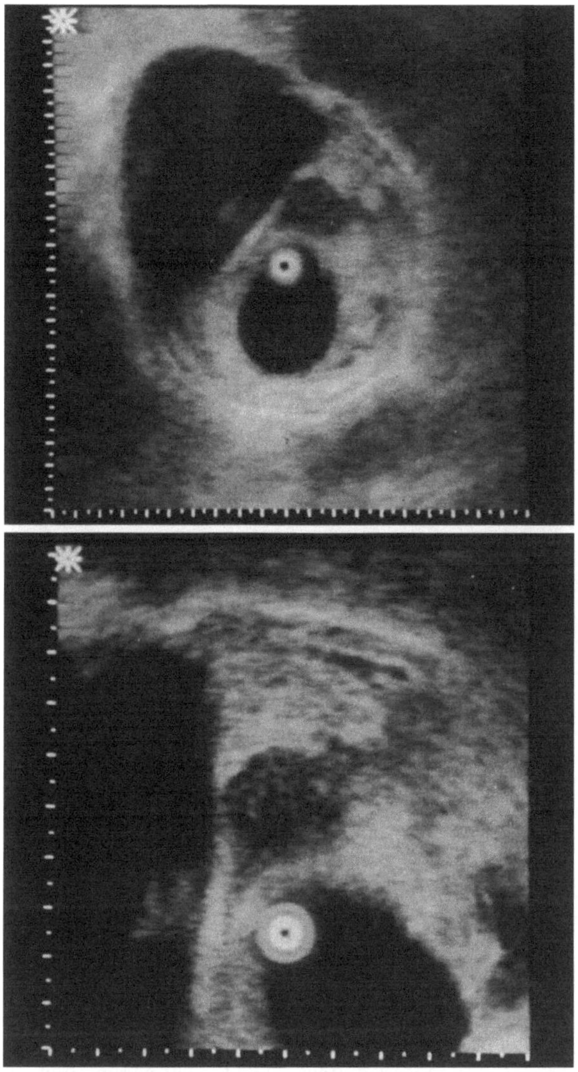

Fig. 14. Transverse rectal scan showing a shift of the cervix toward the side of parametrial involvement. The right parametrium is normal; the left shows nodular infiltration as far as the pelvic wall. *Top:* overall view; *bottom:* enlarged view

Table 1. Comparison of parametrial evaluation by rectosonography and palpation in 13 histologically positive parametria (infiltrated by carcinoma) and 36 histologically negative parametria in patients with cervical carcinoma

	Recto-sonog-raphy	Histo-logy	Palpation	
Parametrial invasion	12	**13**	6	Divergent assessments
			4	
No parametrial invasion	1		3	
Parametrial invasion	6		8	Divergent assessments
			4	
No parametrial invasion	30	**36**	24	

which these areas represent inflammatory infiltration remains unclear.

Although neither endosonography nor rectal palpation allows a primary differentiation between inflammatory and carcinomatous parametrial infiltration (Bernaschek and Janisch 1983), this differentiation can be made by means of ultrasound-guided needle biopsy and histologic evaluation, at least in positive cases. Thus if rectal palpation and rectosonography do not yield a definitive result such as normal-appearing parametria or an obvious lateral disparity with irregular parametrial margins (Bernaschek et al. 1986), a guided biopsy should be performed to ensure accurate pretherapeutic staging.

Our experience with pretherapeutic rectosonography was acquired in 102 cervical carcinoma patients. Fifty-three patients were referred for primary irradiation, but the 49 who underwent radical hysterectomy were of primary interest because ultrasound findings could be compared not just with rectal palpation but also with the histology of the surgical specimen. Histologic examination disclosed carcinomatous involvement of the parametria in 13 of the 49 operated cases; in 12 of those cases parametrial involvement had been suspected from rectosonographic findings. The one false-negative sonographic case was identified histologically as low-grade parametrial involvement close to the uterus. Whereas extensive parametrial invasion is marked sonographically by irregular margins, infiltration of only a few millimeters (eight patients) is marked by unilateral shortening and widening of the parametrial echo. The same sonographic pattern of lateral disparity was seen in six cases where histology did not demonstrate carcinomatous infiltration but did show some inflammatory infiltration of the parametria in the presence of stage I C or II A cervical carcinoma. In 30 cases the rectosonographic diagnosis of normal parametria was confirmed on histologic examination. Rectal palpation, which was performed by two surgically oriented gynecologists, tended to show less agreement with histology. In the 13 histologically confirmed cases of parametrial infiltration, both palpating gynecologists expressed suspicion of parametrial involvement in six cases, and only one on four cases. In the three cases where palpation findings were declared normal, histologic examination disclosed unilateral carcinomatous infiltration of the parametrium only a few millimeters thick. There was also a greater tendency to overstate parametrial changes in stages I B, I C, and II A on rectal palpation than with rectosonography (see Table 1).

Yuhara et al. (1987) demonstrated significant widening of the parametrial echo on rectal scans in 29 patients with histologically verified parametrial carcinomatous infiltration. The absolute value of parametrial thickness measured at the center of the parametrial echo (using a transaxial rotary scanner) was 1.06 ± 0.46 cm. This differed markedly from the value of 0.87 ± 035 cm measured in 121 cervical carcinoma patients without parametrial involvement. While the authors draw no case-related inferences in this study, they do state that a parametrial echo wider than 1 cm is highly suspicious for parametrial infiltration. Since the degree of parametrial thickening is small, the accuracy of the assessment depends strongly on the inclination of the probe axis, i. e., whether the parametrium is really struck at right angles by the beam. This also enables an accurate comparison with the opposite side, although contralateral parametria were not compared in the study cited.

Rectosonography, then, could one day contribute greatly to the preoperative differentiation of stage I C, II B, and III B disease. This should facilitate selection of the most appropriate mode of treatment, and it should also provide a more realistic basis for comparing results at surgical and radiotherapeutic centers. Although many experienced gynecologists may consider available case numbers too small to draw definitive conclusions, these initial results of rectosonographic parametrial staging are at least visually impressive and highly encouraging. Also, since infiltration of the parametria significantly worsens the prognosis of cervical carcinoma, the results of an endosonographically guided parametrial biopsy could contribute even further to therapeutic decision making.

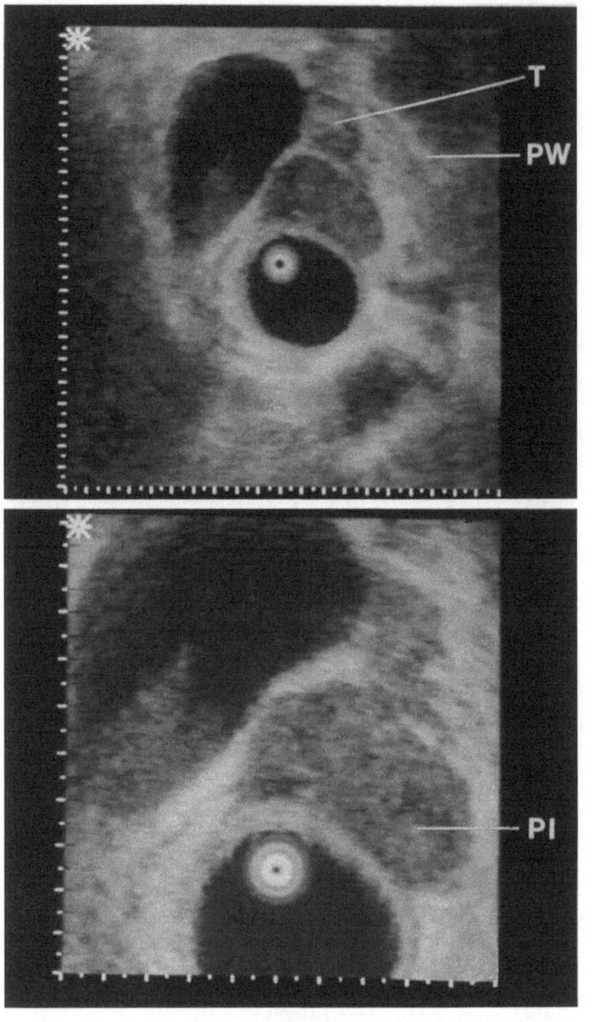

Fig. 15. Transverse rectal scan through the cervix. The right parametrium appears normal, while the left parametrium *(Pl)* is widely infiltrated. The tumor *(T)* has penetrated the anterior wall of the cervix and is invading the urinary bladder; the site of invasion is marked by a defect in the bladder wall contour; pelvic wall *(BW)*. *Top:* overall view; *bottom:* enlarged view

1.2.5 Infiltration of the Bladder Wall and Rectum

As cervical carcinoma advances to stage IV, it spreads to involve the urinary bladder and/or the rectum. Direct involvement of the bladder occurs in approximately 5% of cervical carcinomas (Bernoth et al. 1984). Infiltration of the outer layers of the bladder wall is manifested by a bullous edema of the bladder mucosa, which can be visualized and confirmed by cystoscopy. However, bullous edema alone does not justify assigning the case to stage IV. With further spread the carcinoma penetrates the bladder, and cystoscopy will disclose tumor tissue with a knobby, partially necrotic surface.

Penetration of the bladder creates a disruption of the bladder mucosa and typical bladder wall contour that is demonstrable by rectosonography (Fig. 15) and by cystosonography (Fig. 16). Bullous edema, by contrast, causes irregularity and widening of the mucosa (Fig. 17). Cystosonography is the method of choice, as it can demonstrate bladder wall changes with greater clarity than rectosonography while allowing a more accurate localization and size determination of pathologic changes.

Our own studies in 30 carcinoma patients with positive cystoscopic findings (mucosal edema or penetrating tumor) illustrated the advantages and disadvantages of cystosonography compared with the traditional staging method of cystoscopy. Whereas cystoscopy reveals only superficial

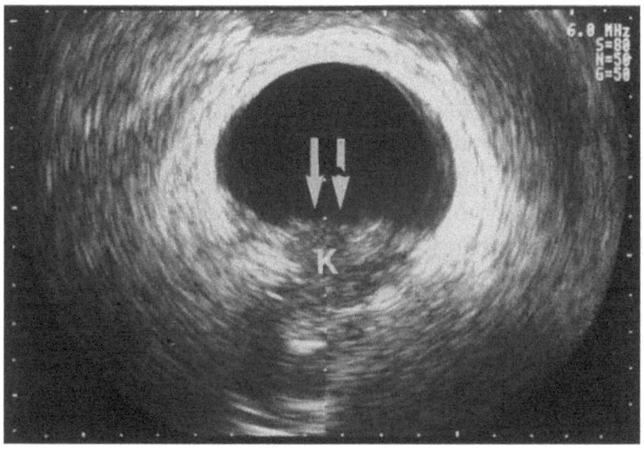

Fig. 16. Cystosonogram in a patient with stage IV cervical carcinoma. Transverse scan through the trigone region shows a small, central area of bladder penetration *(arrows)*, also detected by cystoscopy, and a wide, lateral edematous zone; carcinomatous area *(K)*

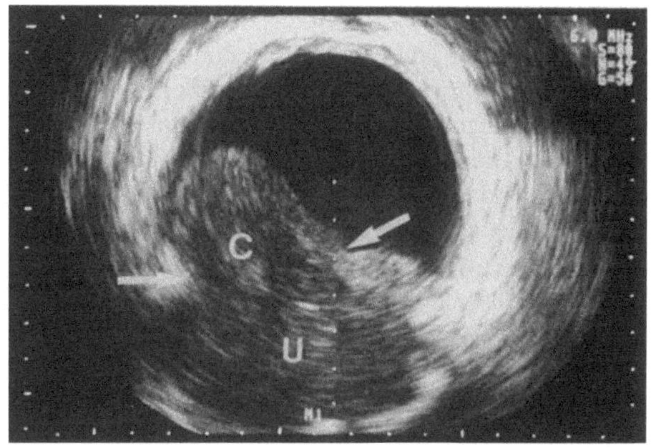

Fig. 17. Cystostonography. The histologically confirmed cervical carcinoma *(C)* has penetrated the anterior uterine wall and invaded the bladder, producing a typical contour defect in the bladder wall *(arrows)*. Cystoscopy reveals only bullous edema, which appears sonographically as an echogenic band stretched over the carcinomatous area

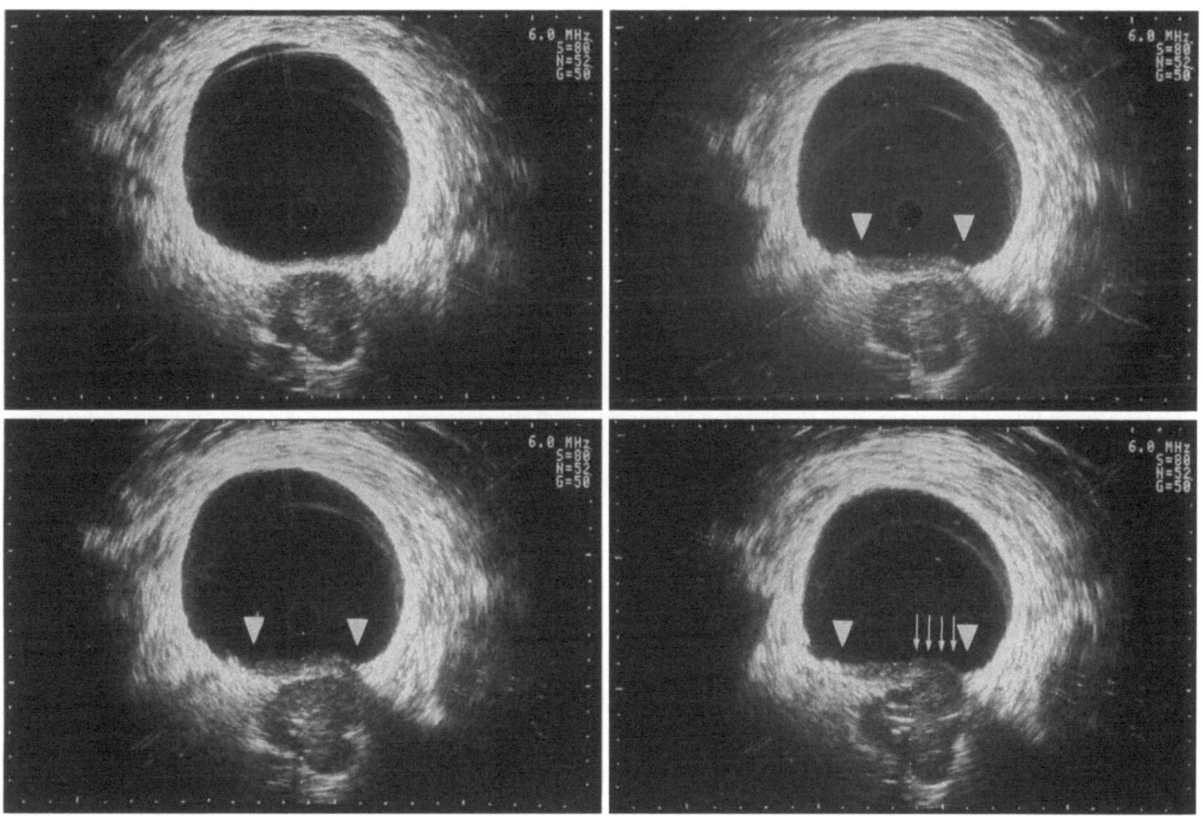

Fig. 18. Cystosonograms in a patient with stage IV cervical carcinoma. *Top left:* Transverse scan at the level of the uterine fundus shows an intact bladder wall. *Top right:* Transverse scan at a lower level shows widening of the mucosa consistent with bullous edema *(arrowheads)*; the bladder wall is still intact at this level. *Bottom left:* a more caudal scan shows incipient contour disruption of the bladder wall. *Bottom right:* a scan through the trigone region shows full-thickness infiltration of the bladder wall *(small arrows)*. Penetration at this site is confirmed by cystoscopy

changes in the bladder mucosa, cystosonography can demonstrate alterations in deeper layers of the bladder wall. Moreover, cystosonography permits a much more accurate assessment of tumor extent, especially in the perivesical region, than abdominal sonography owing to the proximity of the transducer to the affected area, thus enabling the use of higher frequencies (Kölbl and Bernaschek 1987). Our study also demonstrated the superiority of endosonography for evaluating the extent of tumor infiltration.

In all cases of cystoscopically confirmed bladder wall penetration, cystosonography revealed a contour defect in the bladder wall. However, in nearly all cases cystosonography demonstrated a greater extent of bladder wall infiltration than was appreciated by cystoscopy. Peripheral bullous edema, manifested sonographically by a widening of the mucosa, was usually present and limited the cystoscopic evaluation of the true carcinomatous involvement of the bladder wall (Fig. 18). In three cases where bullous edema was identified only by cystoscopy, cystosonography showed a defect in the typical bladder wall contour consistent with infiltration of the outer wall layers.

Besides allowing visual evaluation of the bladder mucosa and spaceoccupying lesions, cystoscopy provides a route of access for obtaining superficial biopsy specimens. Cystosonography is advantageous in this regard, as it allows selective sampling from malignant processes in deeper bladder wall layers that may not yet have reached the mucosa. Carcinomatous material biopsied from deeper bladder wall layers under endosonographic guidance is very helpful in assigning such a case to stage IV and avoiding intraoperative surprises caused by infiltration of the outer bladder wall that was not visible cystoscopically.

One disadvantage of cystosonography is its limited ability to evaluate the urethra. Neither can it evaluate vascular changes (e. g., varicosity) in the bladder mucosa. Therefore cystourethroscopy should be performed before endosonography for the reduction of bleeding.

Rectal infiltration is demonstrated to best advantage by rectosonography. Lesions disrupt the typical sonographic contour of the rectal wall and cause widening of the mucosa. There is associated shape distortion of the rectal balloon, especially in advanced cases. Occasionally the lumen of the rectum may be too constricted for passage of the transducer probe.

In summary, endosonography in patients with an established diagnosis of cervical carcinoma has made it possible to define the extent and boundaries of the tumor with a high degree of accuracy:

1. The extent of cervical carcinoma is particularly well demonstrated by rectosonography with a transaxial probe. Concurrent bladder involvement is demonstrated best by cystosonography.
2. Endosonography offers significant advantages in the objective documentation of parametrial invasion.
3. Endosonography can also establish the extent of the intracervical tumor area.
4. Infiltration of the vagina and paracolpium can be objectively documented.
5. Cystostonography permits evaluation of the bladder mucosa as well as deeper wall layers, especially in the presence of bullous edema.

Endosonography, then, should be an essential adjunct to the basic diagnostic workup and staging of cervical carcinoma. It is an important source of information for the surgeon and radiotherapist, and it provides a more accurate basis for comparing both treatment modalities.

1.2.6 Comparison of CT, MRI, and Endosonography

Modern imaging techniques have advanced the diagnosis of gynecologic malignancies in recent years, enabling a more accurate pretherapeutic evaluation of tumor extent than is possible with traditional roentgenographic methods. However, it is important to know the advantages and disadvantages of computed tomography (CT) and magnetic resonance imaging (MRI) relative to endosonography so that the roles of these modalities can be more accurately defined.

Whereas the lymphogenous spread of cervical carcinoma cannot be evaluated by endosonographic methods, CT and MRI show relatively high accuracy in the assessment of pelvic and para-aortic lymph node involvement (Brenner et al. 1982; Christ et al. 1983; Bandy et al. 1985; Ladner 1987). MRI is superior to CT in its ability to demonstrate involvement of the regional lymphatic drainage. However, while CT and MRI can disclose lymph node enlargement, they cannot demonstrate internal structural features that would distinguish a normal node from one involved by cancer. Also, it appears that lymph nodes located deep in the pelvis are more difficult to evaluate, and cannot be evaluated at as early a stage as high iliac or para-aortic nodes. Accordingly, comparative studies rate conventional lymphography as being equal or superior to CT and MRI for evaluating the lymphatic spread of carcinoma (Hohlweg-Majert et al. 1983; Zapf et al. 1987).

The evaluation of continuous cancer spread, so relevant to accurate staging, is not well accomplished by either CT or MRI; and neither modality can contribute to therapeutic decision making. MRI is slightly superior to CT in this regard owing to its capacity for arbitrary plane selection, which allows better differentiation of parametrial tissue from pelvic fat (Wischnik et al. 1986). CT and MRI are both inferior to endosonography for detecting infiltration of the rectum and bladder wall.

In summary, endosonography can evaluate the continuous spread of cervical carcinoma to better advantage than CT or MRI, and it can provide a correspondingly more accurate staging of disease. CT and MRI, on the other hand, are superior for the pretherapeutic staging of lymph node involvement.

1.3 Follow-up of Inoperable Cervical Carcinoma

Parametrial invasion by cervical carcinoma is crucial not only for staging and treatment planning (Bernaschek et al. 1984a, 1986) but also for monitoring therapeutic response when cancer is advanced and inoperable. Parametrial infiltration has traditionally been evaluated by rectal palpation, but rectosonography can eliminate the subjectivity of the rectal examination and provide a more objective assessment. The use of transaxial probes offers the additional advantage of documenting both parametria on the same scan so that lateral disparities and cervical displacement can be appreciated. A particular advantage of rectosonography is the ability to evaluate high-sited parametria that may be inaccessible to digital palpation.

With rectosonography it is possible not only to measure the width of the parametria and the proximity of the infiltration to the pelvic wall, but also to demonstrate the configuration of the parametria (Bernaschek and Janisch 1983; Yuhara 1987). Like all noninvasive diagnostic methods, rectosonography is unable to show whether a given parametrial lesion is benign or malignant (inflammatory or carcinomatous). That determination can be made only by a directed biopsy (which may be endosonographically guided) and subsequent histologic examination.

To substantiate the value of rectosonography in evaluating treatment response in advanced cervical carcinoma, we performed rectosonography in a series of ten patients before and after a standard course of radiotherapy. Our goal was to document cervical tumor regression and a reduction of parametrial invasion (Deutinger et al. 1989). The clinical diagnoses in this series, confirmed by biopsy in all cases, were FIGO stage II AB in two patients, stage III AB in seven patients, and stage IV in one patient.

In eight cases the parametria were widened on both sides, and in two cases parametrial involvement was unilateral. The radiotherapy consisted of concurrent telecobalt therapy and brachytherpay, the latter being fractionated into applications of a vaginal plaque, a uterine tube, and another plaque, each 2 weeks apart. The patients received an average of 3130 rad by intravaginal application and 2400 rad by the single intrauterine application. The patients additionally received telecobalt irradiation in two opposing fields with an axial dose averaging 4200 rad. The therapy was administered over a period of 6–9 weeks.

When rectal scans are taken for pre- and postirradiation comparisons, it is essential that the handle of the rectal probe be lowered until the long axis of the uterus is parallel to that of the scanner; this ensures a perpendicular beam incidence upon the cervix and parametria. The thickness of the involved parametria is then measured at their widest point 1 cm from the lateral cervical margin.

In patients with advanced cervical carcinoma, scans reveal an enlarged, nonhomogeneous cervix in addition to widened, shortened parametria that have an irregular configuration or show blunted lateral extensions. Usually the cervix is shifted toward the side of heavier parametrial invasion. In our investigations the width of the affected parametria averaged 24 mm before the start of radiotherapy and 17 mm after its conclusion. Bilateral reduction of infiltration was documented in six of eight cases (Fig. 19), and unilateral reduction was noted in the remaining two. In the two patients with unilateral involvement the parametrial width decreased by 1 cm and 1.2 cm by the end of therapy. Prior to therapy the cervix averaged 4.0 cm in its anteroposterior diameter and 4.2 cm transversely; these values decreased with therapy to 3.1 cm and 3.3 cm, respectively (Fig. 20).

Every gynecologic oncotherapist knows the striking effect that radiotherapy can have on the palpation findings in inoperable cervical carcinoma. In the present study we have furnished pictorial evidence of this familiar effect of brachy- and teletherapy. It was very instructive to contrast the pretherapeutic and post-therapeutic rectosonogra-

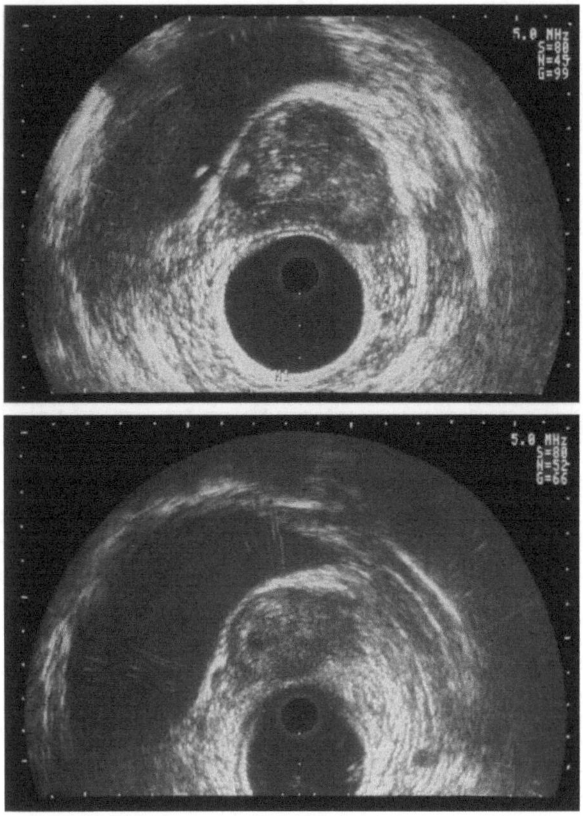

Fig. 19. Stage III AB cervical carcinoma. *Top:* transverse rectal scan before the start of radiotherapy. *Bottom:* transverse rectal scan following a course of radiotherapy, showing shrinkage of cervical tumor and marked regression of parametrial infiltration

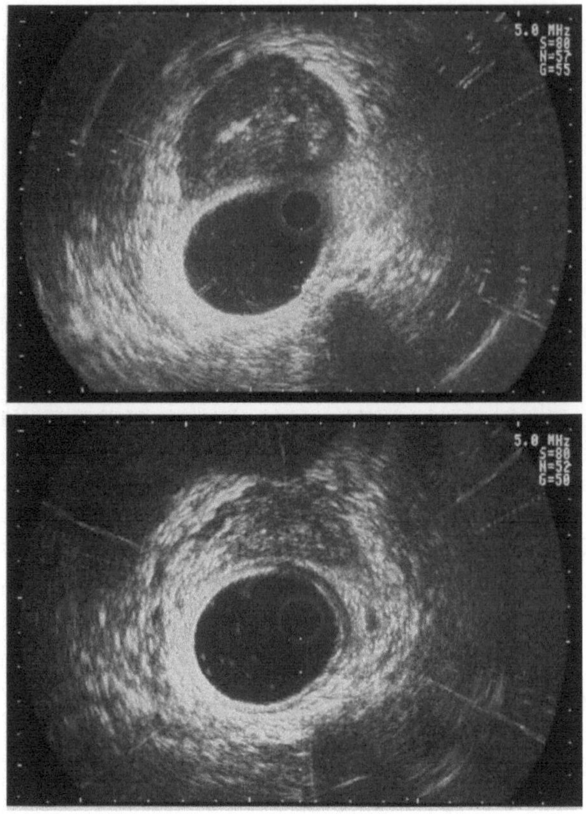

Fig. 20. Stage IV cervical carcinoma. *Top:* transverse rectal scan before the start of radiotherapy. *Bottom:* transverse rectal scan following a course of radiotherapy, showing a marked reduction in cervical size

phic findings and compare them with the results of later follow-up examinations (Deutinger et al. 1989). Rectosonography is a valuable adjunct to palpation in that it can be used to monitor treatment response and objectively document that response in patients with inoperable cervical cancer. The prognostic significance of sonographically demonstrable regression remains unclear at present due to the small number of cases studied.

2 Corpus Carcinoma

2.1 General

In recent years there has been a rise in the incidence of carcinoma of the uterine corpus, due in part to the increased life expectancy of women. Currently the estimated ratio of cervical to corpus carcinoma is approximately 1:1, compared with about an 8:1 ratio five decades ago (Michalica 1981). This change can be attributed largely to outstanding success in the early detection of cervical neoplasms with the aid of cervicovaginal smears. By contrast, the success rate for the cytologic detection of corpus carcinoma is no better than 50%. Results have been improved by the development of special intrauterine sampling techniques, but they can still be hampered by technical and methodologic errors, especially in patients with cervical stenosis. Vaginosonography could help to support or improve the early detection of endometrial thickening in menopausal women, at least in problem cases.

The staging of endometrial carcinoma follows FIGO guidelines. Stage I is subdivided into I A and I B according to the sounded length of the uterine cavity. However, numerous studies have shown that this has little if any prognostic signficance, since there is only a minimal correlation between uterine length and 5-year survival. This classification, then, results largely from a deficiency of diagnostic techniques able to provide satisfactory pretherapeutic differentiation. Meanwhile, the fact that currently about three-fourths of women with endometrial cancer was assigned to stage I demonstrates the urgent need for a more reliable means of pretherapeutic staging.

A study conducted at the Second University of Vienna Women's Hospital (Breitenecker et al. 1984) showed that, besides histologic grading, myometrial invasion is an extremely important prognostic indicator in corpus carcinoma. Thus, it was found that the survival curves of women undergoing primary operative treatment declined markedly with increasing depth of myometrial invasion. While hysterography and hysteroscopy are useful for localizing intracavitary disease and defining its extent, abdominal sonography first made possible the more accurate staging of endometrial cancer (Requard et al. 1981). The relatively complicated procedure of hysterosonography was subsequently used to estimate the depth of infiltration, but the far simpler procedure of vaginosonography will attain even greater importance in the future for measuring the depth of myometrial invasion.

2.2 Early Detection

The main advantage of vaginosonography over hysterosonography for endometrial screening is its relative ease of use, a basic requirement for an application of this kind.

As the peak incidence of endometrial carcinoma is between 50 and 70 years of age, screening would be appropriate for a high-risk population in that age group. It must be considered, however, that approximately 12% of women in this age group are not good candidates for vaginosonography (Rudelstorfer et al. 1988) because of stenotic changes. Vaginal stenosis, which most commonly affects the proximal third, can prevent insertion of the finger-thick vaginal probe. Obliteration of the fornices can also complicate the evaluation, as this causes the uterus to "ride" directly on the probe so that only the cervix can be adequately assessed.

The major value of vaginosonography as a screening examination for corpus carcinoma is its ability to detect endometrial thickening before clinical symptoms have appeared. An abnormal pattern is defined as uniform endometrial thickening that exceeds one-third the uterine wall thickness on the longitudinal scan. Circumscribed changes - usually incipient polypous lesion - are also regarded as pathologic. Special attention should be paid to myomas, which, especially when submucous or intramural, produce mass effects that can hinder the evaluation of small endometrial changes.

To investigate the value of vaginosonography in the detection and evaluation of endometrial changes, we performed a study in 77 women whose presenting complaint was postmenopausal bleeding (after more than 6 months' amenorrhea), and we correlated the results of vaginosonography with histologic findings in curettage material. Fifty-seven of the women had a normal endometrial scan according to the definition stated above. In 53 cases this correlated with a negative histologic outcome. So-

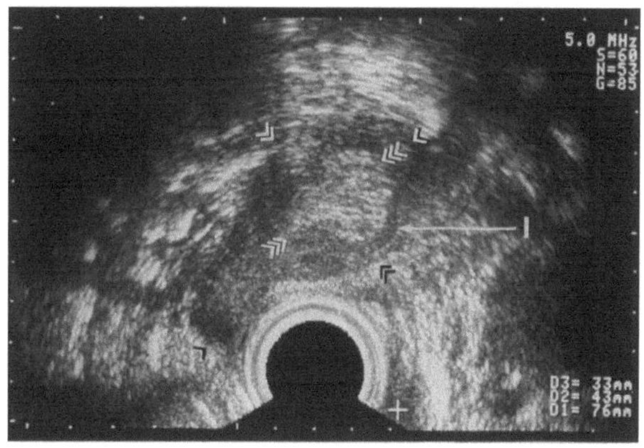

Fig. 21. Longitudinal vaginal scan through the uterus in a patient with postmenopausal bleeding. Widened, nonhomogeneous endometrial echo with hypoechoic myometrial infiltration *(I)* in the region of the uterine anterior wall (histology: endometrial carcinoma extending to the middle third of the myometrium). *Single arrowheads,* longitudinal axis of uterus; *double arrowheads,* anteroposterior axis of uterus; *triple arrowheads,* endometrium

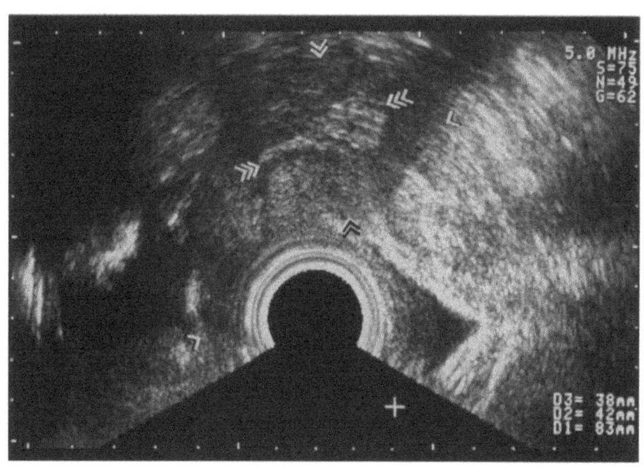

Fig. 22. Longitudinal vaginal scan through the uterus in a patient with postmenopausal bleeding. Widening of the endometrium close to the fundus (histology: benign polyp of the corpus mucosa). *Arrowheads,* see explanation in Fig. 21

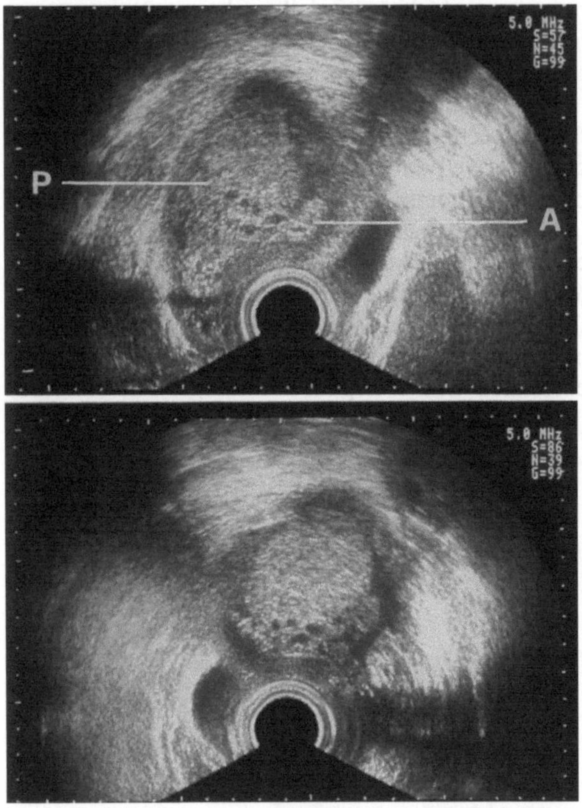

Fig. 23. *Top:* longitudinal vaginal scan through the uterus in a patient with postmenopausal bleeding. Significant endometrial widening with small cystic areas. Movement of the vaginal probe elicits typical motion of endometrial tissue on the anterior uterine wall *(A)* relative to the endometrium on the posterior uterine wall *(P)*. *Bottom:* horizontal vaginal scan (histology: cystic glandular hyperplasia of the endometrium)

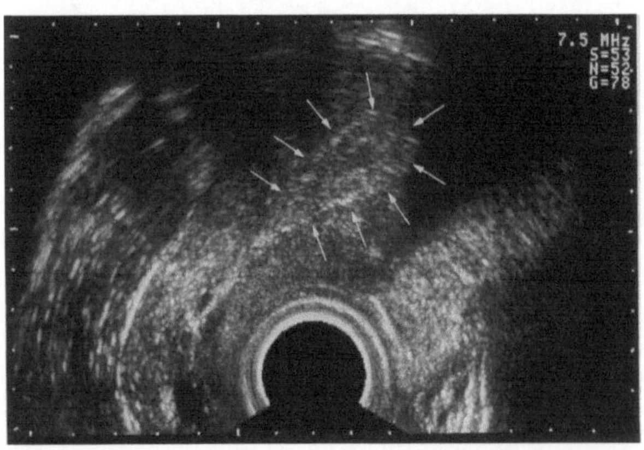

Fig. 24. Longitudinal vaginal scan through the uterus in a patient with postmenopausal bleeding. Uniform widening of the endometrium *(arrows)* with an intact central lumen and typical mobility (histology: endometrial hyperplasia)

nography missed one small endometrial carcinoma and three polyps ranging from 3 to 4 mm in size. In all the false-negative cases there were coexisting uterine myomas, which were readily identified by vaginosonography. In the remaining 20 women in our series, vaginosonography revealed endometrial thickening. All cases in this group correlated with abnormal histologic findings, regardless of whether or not there were coexisting myomas. In nine cases the endometrial widening was caused by endometrial carcinoma (Fig. 21), in seven cases by polyps (Fig. 22), and in four cases by hyperplasia (Fig. 23). Although vaginosonography did not allow a primary differentiation between benign and malignant changes, retrospective scrutiny did identify distinguishing features on some of the scans: Endometrial carcinomas, which occasionally are accompanied by mucometra, are characterized by markedly nonhomogeneous endometrial structures which distinguish them from polyps. A relatively hypoechoic area representing myometrial infiltration can provide an additional aid to differentiation. With endometrial hyperplasia, on the other hand, sonography demonstrates a central uterine cavity echo. Also, because of the glassy surface texture of hyperplastic endometrium (apparent on gross examination), movement of the vaginal probe elicits a characteristic motion of the anterior uterine wall relative to the posterior wall (Figs. 23, 24).

Our experience with the use of vaginosonography for the early detection of endometrial carcinoma can be summarized as follows:

1. Vaginosonography is able to demonstrate pathologic widening of the endometrial echo.
2. The changes must attain at least macroscopic size before they are detectable by this method.
3. Vaginosonography cannot positively differentiate between benign and malignant changes as a cause of endometrial widening (Fig. 25).
4. The presence of myomas interferes with evaluation of the endometrium.
5. As endometrial widening is a gradual process, it could be detected by vaginosonography before the onset of clinical symptoms. This gives vaginosonography potential value as a screening method for endometrial carcinoma in high-risk patients.

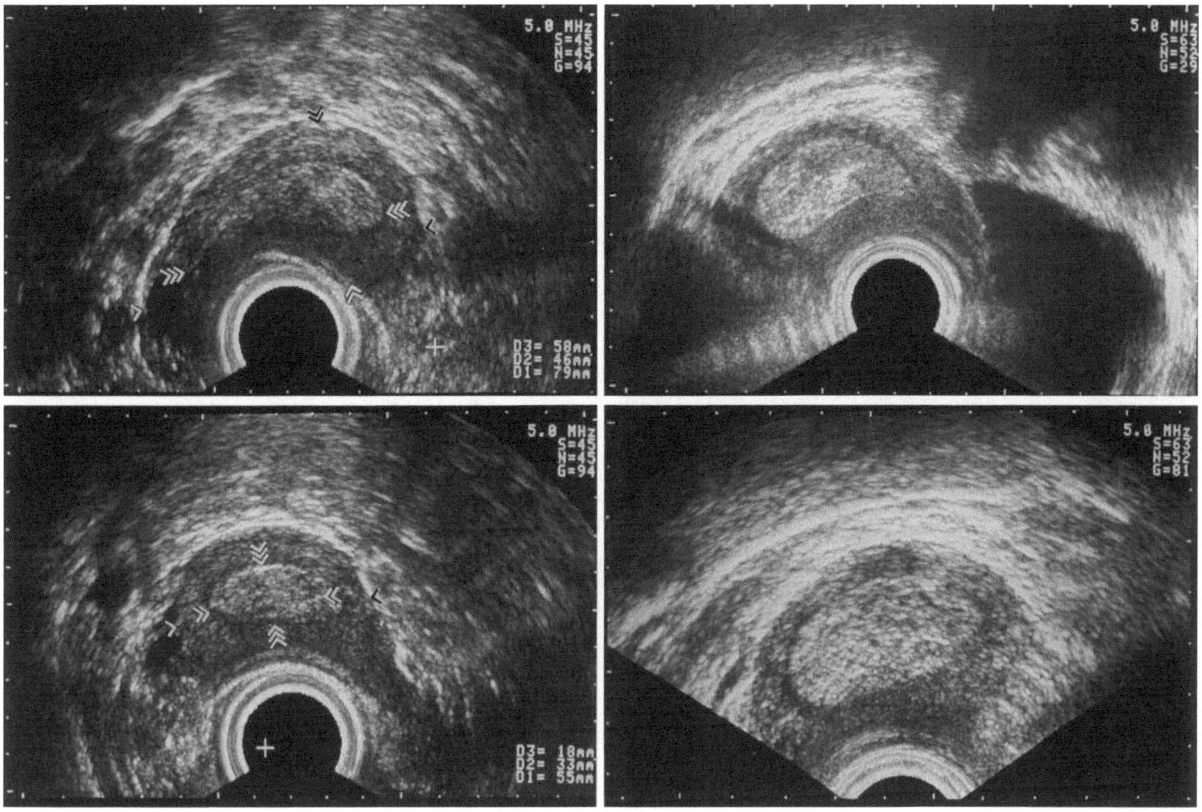

Fig. 25. *Left:* endometrial carcinoma not involving the myometrium: *top,* longitudinal scan; *bottom,* transverse scan through the uterus. *Right:* benign polyp of the corpus mucosa: *top,* longitudinal scan of the retroverted uterus; *bottom,* transverse scan through the uterus

2.3 Staging

2.3.1 Hysterosonography

Hysterosonography is generally performed with a transaxial probe of the type used for cystosonography. The major disadvantage of hysterosonography is the necessity of grasping the vaginal part of the cervix and dilating the cervical canal, requiring the use of general anesthesia. Thus the examination is usually performed as part of a diagnostic curettage.

One advantage over vaginosonography is that the central placement of the probe in the uterine cavity makes it possible to localize carcinomas that are confined to the endometrium. A transducer frequency of 7 MHz provides optimum resolution of details but limits the complete delineation of large uteri, especially when there is coexisting myomatous change. Given the limited depth of penetration of the high-frequency beam, hysterosonography cannot establish the extent of tumor spread outside the uterus.

2.3.1.1 Carcinoma Confined to the Endometrium

Hysterosonography is a method for demonstrating macroscopic endometrial disease. Intraepithelial changes are not accessible to hysterosonography and are indistinguishable from normal endometrial findings. The normal endometrium appears as an echogenic rim of variable thickness which bounds the echo-free uterine cavity. However, it can be difficult to demonstrate the normal uterine mucosa in cases where the probe completely fills the uterine cavity and is in direct contact with the endometrium.

Exophytic carcinomas appear hyperechoic because of the many interfaces in the tumor mass. The site or origin of smaller carcinomas is easily identified on intrauterine scans, and the extent of disease is easily determined (Fig. 26). With larger exophytic growths, echogenic structures appear to fill the uterine cavity on all sides of the transducer, which itself may be embedded in the soft tumor mass (Becker and Hötzinger 1986). The origin of an exophytic

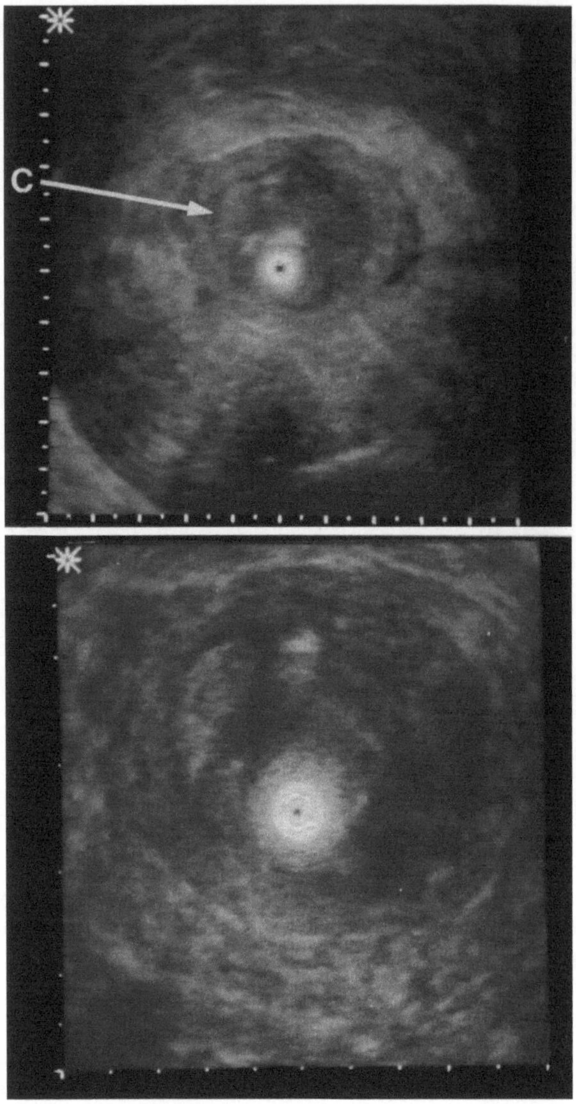

Fig. 26. Hysterosonography with a transaxial rotary scanner in a patient with histologically confirmed endometrial carcinoma. *Top:* transverse scan through the corpus showing a relatively echogenic area of carcinoma *(C)* confined to the endometrium. *Bottom:* enlarged view

growth usually appears as a relatively echogenic zone, and a careful search should be made in that area for signs of myometrial invasion.

2.3.1.2 Myometrial Invasion

In contrast to intracavitary tumor, cancer invading the myometrium appears markedly hypoechoic, serving to distinguish the infiltrated area from healthy myometrium. The paucity of echoes is due mainly to the less compact consistency of the tumor tissue relative to the myometrium. Areas of myometrial invasion are distinguished from myomas by their ill-defined borders, which lack the capsule that is characteristic of myomas. Hysterosonography can reveal only the macroscopic depth of myometrial invasion, which is determined on transaxial scans at the site of maximum extension and usually is expressed in relation to the total myometrial thickness. The usual method is to classify infiltration as extending to the inner, middle, or outer third of the myometrium.

2.3.1.3 Extension to the Cervix

The longitudinal extent of exophytic or infiltrating disease can be determined from the position of the probe relative to the external os and read on the scanner shaft.

Studies comparing hysterosonography and macroscopic sections (Hötzinger et al. 1984; Becker and Hötzinger 1986) in terms of maximum depth of invasion and longitudinal extent of endometrial carcinoma in a total of 60 cases indicate high correlation coefficients of 0.996 and 0.997, respectively. This confirms the high accuracy of endosonography in evaluating the two parameters stated above. As might be expected, the results of Obata et al. (1985), who compared the hysterosonographic and histologic depths of invasion, were somewhat less favorable. The hysterosonographic assessment of depth of myometrial invasion in terms of inner-, middle-, and outer-third involvement agreed with histologic findings in 26 of 32 cases (81.3%). In the remaining six patients hysterosonography underestimated the depth of myomatrial invasion, i. e., a deeper level of invasion was demonstrated histologically. However, in three of these cases myometrial involvement was in the fundal region, which cannot be scanned using a 90° transaxial probe. This leads Hötzinger et al. (1985) to recommend using a probe with a variable scan angle of 90° transaxial to 135° forward.

Hysterosonography also can help improve the results of radiotherapy in endometrial carcinoma by directing the individual planning of intracavitary therapy (Hötzinger 1984). Specifically, information on cavity length, cavity size, uterine size and position, and extent of tumor area is provided by a procedure that, from an anatomic standpoint, recapitulates the placement of an intrauterine radiation source. This provides an accurate definition of the target volume for intracavitary therapy. In addition, a computer program has been developed that plots the outer contour of the uterus in relation to the ultrasound probe and tumor area from the hysterosonogram and superimposes optimum isodose curves onto the individual scans. A quasi-three-dimensional isodose plot can also be generated (Engelmeier et al. 1985).

2.3.2 *Vaginosonography*

Vaginosonography can likewise be applied to the staging of endometrial carcinoma (Bernaschek 1987; Schurz et al. 1987). As in hysterosonography, the proximity of the transducer to the endometrium permits the use of high scanning frequencies, but visualization of the fundal region presents fewer problems. On the other hand, vaginosonography can localize endometrial carcinoma and define its origin only if there is demonstrable myometrial invasion. Exophytic cancer confined to the endometrium usually cannot be evaluated because the probe does not expand the uterine cavity as it does in hysterosonography.

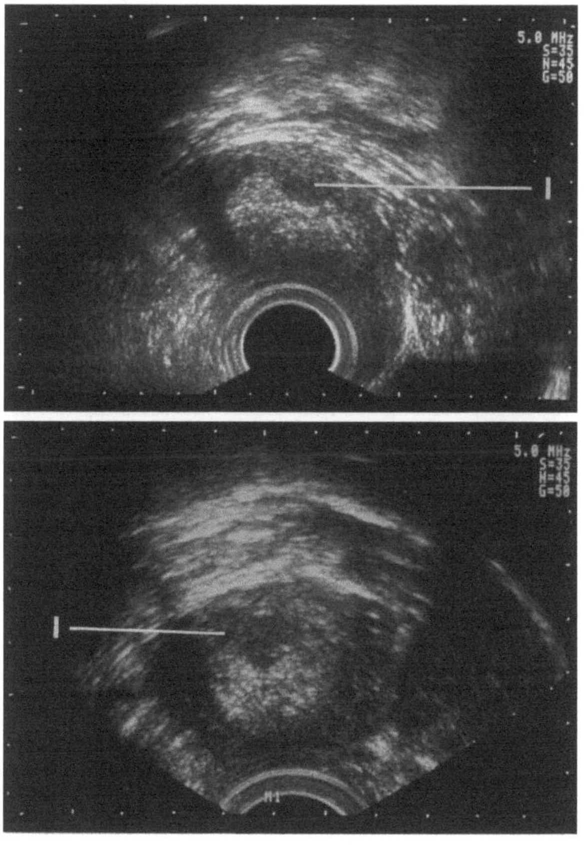

Fig. 27. Vaginosonography in a patient with postmenopausal bleeding due to histologically confirmed endometrial carcinoma. *Top:* longitudinal scan through the retroverted uterus showing a widened, irregular endometrium and a hypoechoic area representing myometrial infiltration *(I)*. *Bottom:* transverse scan through the corpus in the same case

The length of the uterus and its cavity are easily determined on the longitudinal vaginal scan. Carcinomas confined to the endometrium appear as irregular, hyperechoic areas that fill the uterine cavity and may or may not be accompanied by a fluid collection signifying mucometra. With deeper infiltration of the myometrium (superficial infiltration is accessible only to histology), the affected area appears hypoechoic to its surroundings (Bernaschek 1987). Usually the boundary between infiltrated and healthy uterine muscle is poorly defined (Fig. 27). Besides defining the depth of invasion, vaginosonography can also establish its longitudinal extent and thus can show possible involvement of the cervical region, enabling a pretherapeutic classification of stage II disease. In this regard vaginosonography shows the longitudinal extent of carcinoma to better advantage than hysterosonography, which employs a transaxial probe.

2.4 Summary

1. Experience to date indicates that vaginosonography is useful for the early detection of endometrial carcinoma, at least in high-risk populations. Widening of the endometrium can be recognized at an early stage.
2. Vaginosonography and especially hysterosonography can direct diagnosis and treatment planning by providing a pretherapeutic, macroscopic assessment of tumor extent. Sites of myometrial invasion can be identified, and spread to the cervical region can be detected.
3. The preoperative assessment of myometrial invasion and cervical spread enables the surgically active gynecologist to determine the necessary extent of resective surgery prior to the operation. Where radiotherapy is indicated, hysterosonography is able to define the treatment volume for intracavitary irradiation.

3 Ovarian Carcinoma

3.1 General

Given the lack of early symptoms and effective methods for early detection, more than 50%–70% of ovarian carcinomas are not diagnosed until they have reached an advanced stage. Some years ago this fact prompted the use of ultrasonography as a method for the early detection of ovarian carcinoma. Goswamy et al. (1982), for example, used abdominal ultrasound to determine ovarian volumes in postmenopausal women and compared both ovaries in an effort to detect early unilateral change. Another sign suggestive of cancer in this age group is ovarian structural change, especially the presence of cystic components, since approximately 95% of all ovarian malignancies contain cystic elements. A series of 1077 screening examinations performed in postmenopausal women by abdominal sonography (Goswamy et al. 1983) disclosed 15 abnormal ovarian findings, nine of which were identified as potentially malignant by subsequent histologic evaluation (serous cysts with adenofibromatous nodules, mucinous and papillary serous cystic adenomas, and dermoid cysts). In cases where abdominal sonography failed to demonstrate postmenopausal ovaries, the authors cited extreme obesity, adhesions, extreme ovarian atrophy, and inadequate bladder filling as the causes.

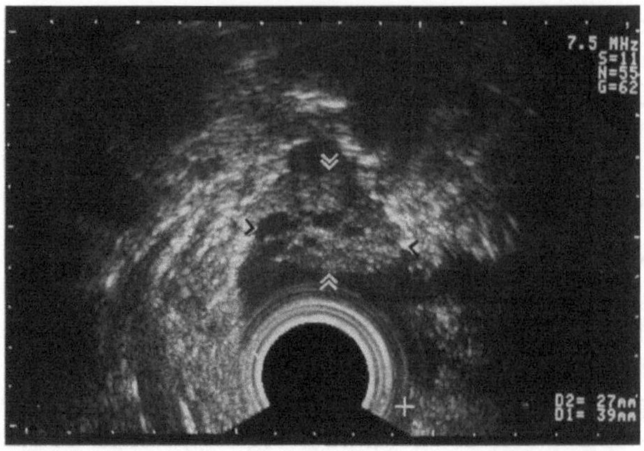

Fig. 28. Horizontal vaginal scan through a slightly enlarged ovary in a postmenopausal patient. Scan demonstrates free fluid in the abdomen (histologically confirmed ovarian carcinoma)

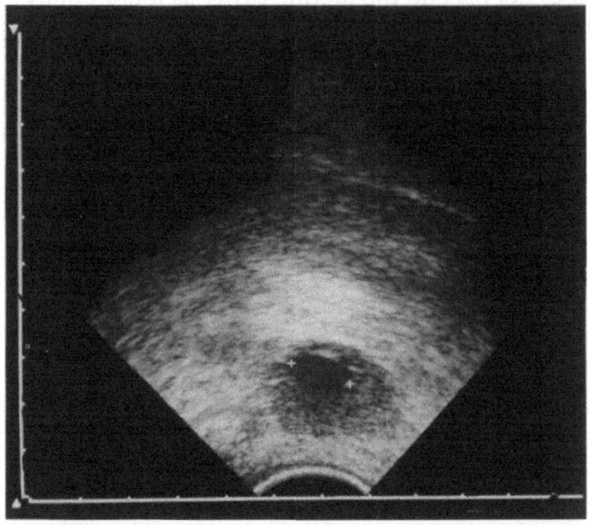

Fig. 29. Horizontal vaginal scan through a postmenopausal ovary that felt normal on palpation. Scan reveals a small cystic mass 13 mm in diameter (histology: benign cystoma)

3.2 Early Detection

The high resolving capacity of vaginosonography and the absence of most of the uncertainty factors that compromise transabdominal visualization of the ovaries enable the early detection of small changes whose size is at least in the macroscopic range (Fig. 28). Although ultrasound cannot establish whether a lesion is benign or malignant, it is still possible to identify suspicious areas at an early stage and refer them for histology evaluation - this at a time when palpation, inspection, and prior history are all negative (Fig. 29). Although it would be unrealistic at present to advise general screening by vaginosonography, due to costs and limited availability of equipment, such screening offers significant advantages in populations at risk and is superior to abdominal sonography. While several authors have reported on the possibility of endosonographic screening in postmenopausal women (Granberg and Wikland 1987; Popp et al. 1985), only Rodriquez et al. (1987) have published numerical results on vaginosonographic screening of the ovary, although their case numbers are still small. The relatively small percentage of demonstrable postmenopausal ovaries (75%) could probably be increased with vaginal ultrasound and concurrent palpation. Although their screening study failed to detect ovarian cancer, there were five cases (12%) in which microcystic ovarian changes noted sonographically were confirmed by histologic examination of the resection specimen.

It is reasonable to expect that vaginosonographic screening for pathologic ovarian changes will yield a significant percentage of suspicious findings. While histologic evaluation of the suspicious area is indicated in all cases, the method of obtaining the tissue sample remains a subject of controversy. Options include laparotomy, laparoscopy, and vaginosonographically guided needle biopsy.

4 Vaginal Carcinoma

The endosonographic diagnosis of a small vaginal carcinoma follows basically the same principles as the detection of vaginal infiltration by cervical carcinoma. Scanning may be performed with a transaxial rotary scanner which, when placed within a fluid-filled balloon, can provide 360° cross-sectional views of the vaginal wall (Fig. 30). One advantage of this technique is that the balloon expands the vaginal wall and creates conditions similar to those for the endosonographic assessment of rectal carcinoma. Another advantage is the bilateral comparison enabled by the transaxial scans, making it easier to recognize extension to the paracolpium. Scans may also be performed with a linear-array transducer, which can accurately depict the longitudinal extent of vaginal carcinoma. The excellent resolution of this scanner, even in the near-field range, makes it unnecessary to use a fluid-filled balloon. Carcinomas that have caused circumferential stricturing of the vagina are demonstrated better by rectosonography. The vaginal lumen is defined with greatest clarity by inserting a balloon (condom) into the vagina and filling it with liquid before performing the rectal scans.

Experience with the endosonographic diagnosis and staging of vaginal carcinoma is still very limited. Our own experience is also insufficient to permit final conclusions to be drawn.

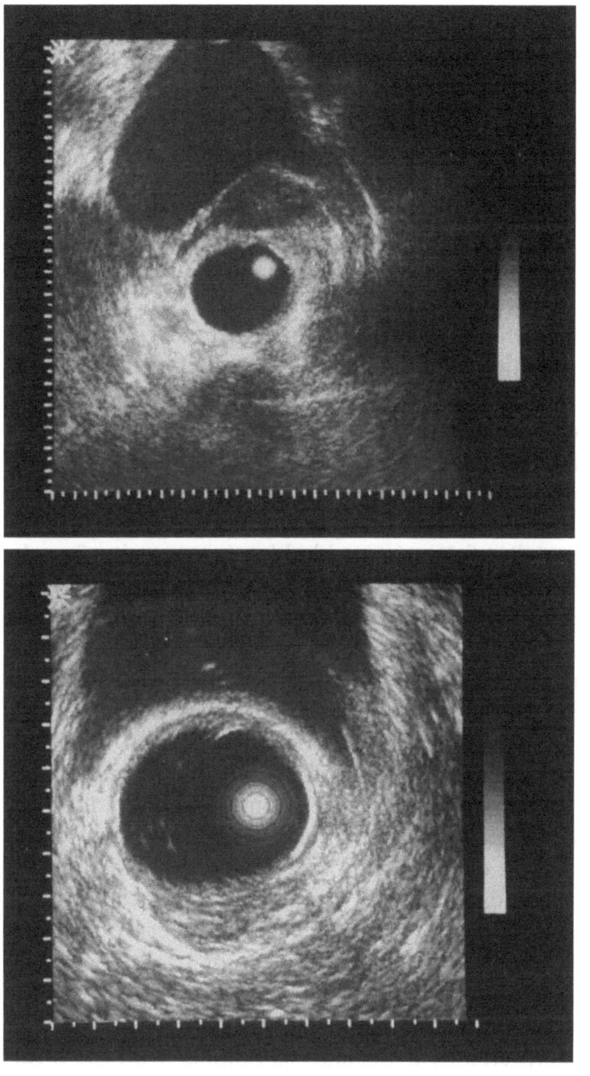

Fig. 30. Tumor in the rectovaginal septum metastatic to corpus carcinoma. *Top:* transverse rectal scan. The rectal wall appears intact, and the tumor cannot be accurately localized. *Bottom:* vaginal scan with a transaxial rotary probe and a fluidfilled endovaginal balloon around the probe. Scan demonstrates the tumor in the posterior vaginal wall, which appears disrupted in that area

5 Diagnosis of Recurrent Carcinoma

5.1 General

The primary goal of oncologic follow-up is the early detection of recurrent cancer so that appropriate treatment can be instituted. The diagnostic mainstays of this follow-up are inspection and palpation. The gynecologist has an advantage over other specialists in this regard, as the organs of the lesser pelvis are accessible to the palpating finger. Evaluation becomes difficult, however, in the presence of firm scar-tissue bands, pronounced muscular guarding, and especially in patients with high-sited lesions that are inaccessible to the examiner's finger. These difficulties have prompted efforts to find other alternatives for the diagnosis of recurrent gynecologic cancer.

Options for evaluating suspected lymph node metastases include CT, MRI, and abdominal ultrasound, which can demonstrate enlarged lymph nodes with a diameter greater than 2 cm both in the lesser pelvis and in the para-aortic region (Kratochwil et al. 1974). The advantage of these methods over X-ray lymphography and isotope scanning is that they do not rely upon the intact function of the lymphatic vascular system. This has led some authors to question the value of the latter two methods, even as preoperative studies (Frischbier 1969; Kindermann et al. 1970), and the disruption of lymph flow caused by prior radical surgery and primary or secondary radiotherapy casts doubt on their usefulness in the detection of recurrent cancer (Kofler and Kratochwil 1983).

Femoral angiography was formerly used for the direct visualization of pelvic wall recurrences, which are manifested by anomalous or disrupted vascular pattern (Lang and Greer 1969; Jakobs 1973). Today this invasive procedure is used only in exceptional cases owing to the availability of the less invasive techniques of CT and MRI. However, the technical complexity and high cost of these examinations are a major reason why primary consideration should be given to ultrasound in the search for recurrent cancer. Abdominal sonography can demonstrate larger pelvic-wall tumors as homogeneous areas with low-level internal echoes. It is less effective for detecting small recurrences located behind the full urinary bladder (a necessity for abdominal scanning), because artifacts caused by the impedance jump between the liquid and adjacent solid tissue can mask the presence of recurrent tumor (Kratochwil and Bernaschek 1981). Although this problem can be lessened somewhat by adjusting the gain setting (sensitivity-graded method), misinterpretations can still occur due to inadequate evacuation of the bowel. Diminished bladder capacity in older women, especially following radiotherapy, can also limit the value of the method.

Rectosonography promises to advance the diagnosis of recurrent female genital cancer by eliminating most of the disadvantages of traditional diagnostic procedures. The views provided by the transaxial rectal probe are easily appreciated and interpreted by gynecologists familiar with the traditional rectal presentation of recurrences. Besides visualizing palpable changes, the probe also represents an "extension" of the palpating finger. The examination is cost-effective and well tolerated. Rectosonography improves upon abdominal scanning by its ability to delineate the structures of the lesser pelvis even in the presence of adhesions or insufficient bladder capacity. These two factors are an especially frequent cause of poor-quality transabdominal images in patients evaluated for recurrence of malignant disease.

5.2 Detection of Recurrence

The sonographic detection of tumor recurrence in the lesser pelvis requires an adequate field of view, so paranoramic scanners should be used. Since in most cases the vagina is significantly shortened or stenosed following the extirpation or radiotherapy

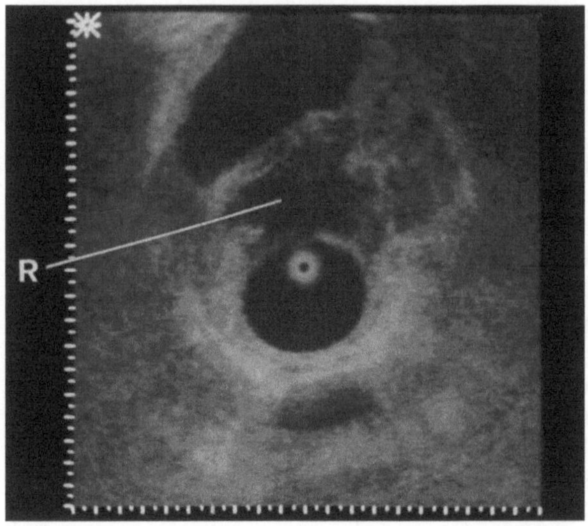

Fig. 31. Transverse rectal scan through the end of the vaginal blind pouch. There is an extensive local recurrence *(R)* 2 years after Wertheim's radical hysterectomy with a Boari reconstruction on the left side. A contour defect signifies invasion of the bladder wall

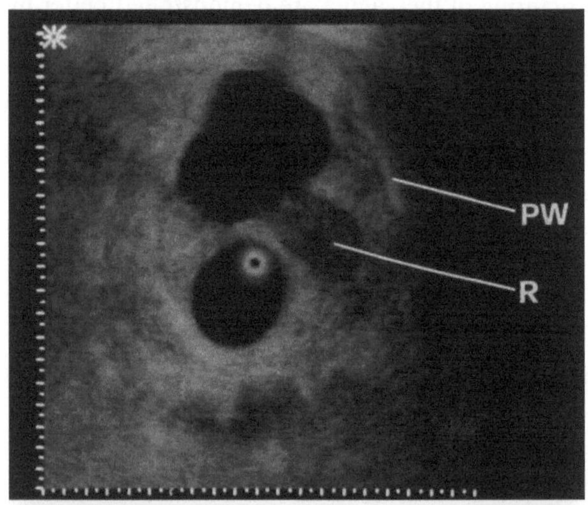

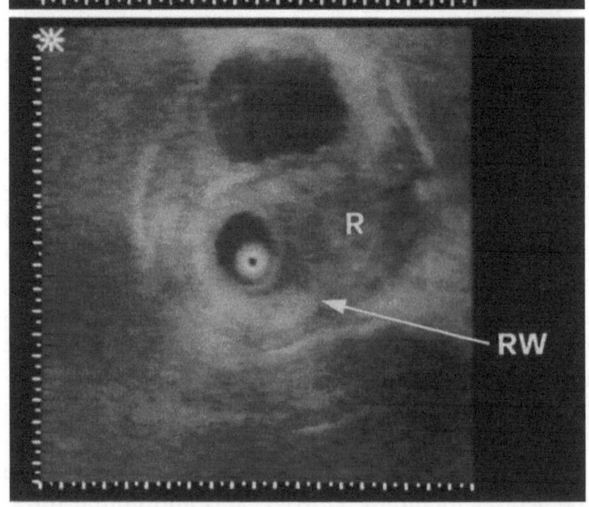

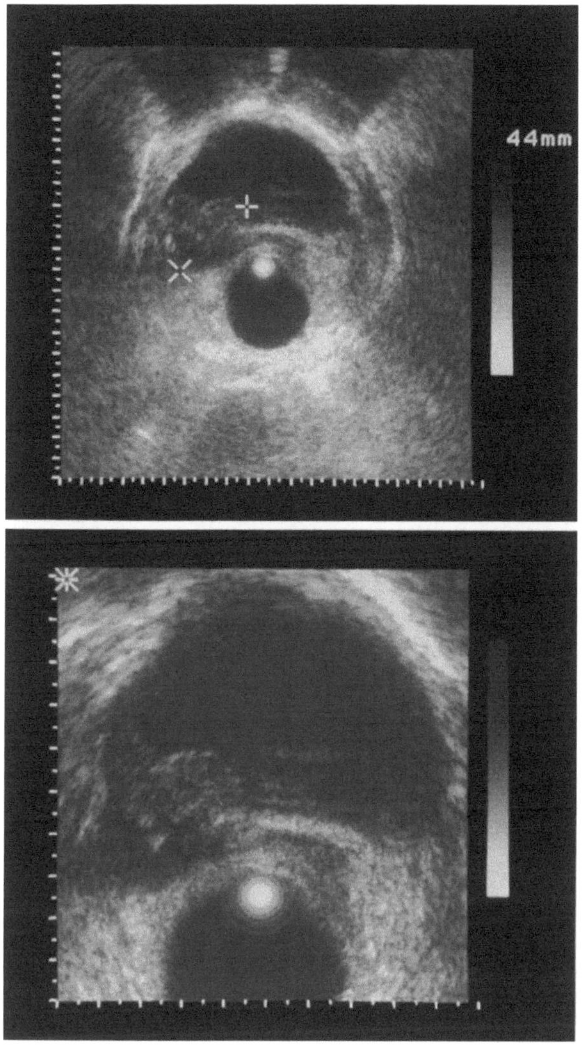

Fig. 33. Rectosonography with a transaxial rotary scanner. *Top:* Recurrence in the right paracolpium with bladder wall penetration 2 years after Wertheim's radical hysterectomy. *Bottom:* Enlarged view

◀ **Fig. 32.** Rectosonography with a transaxial rotary scanner. *Top:* pelvic wall recurrence *(R)* 1 year after Wertheim's radical hysterectomy and postoperative irradiation; pelvic wall *(PW)*. *Bottom:* same case 1½ years after radical hysterectomy. Rectal infiltration is manifested by a defect in the rectal wall contour *(RW)*. The irregular bladder wall contour is consistent with bullous edema

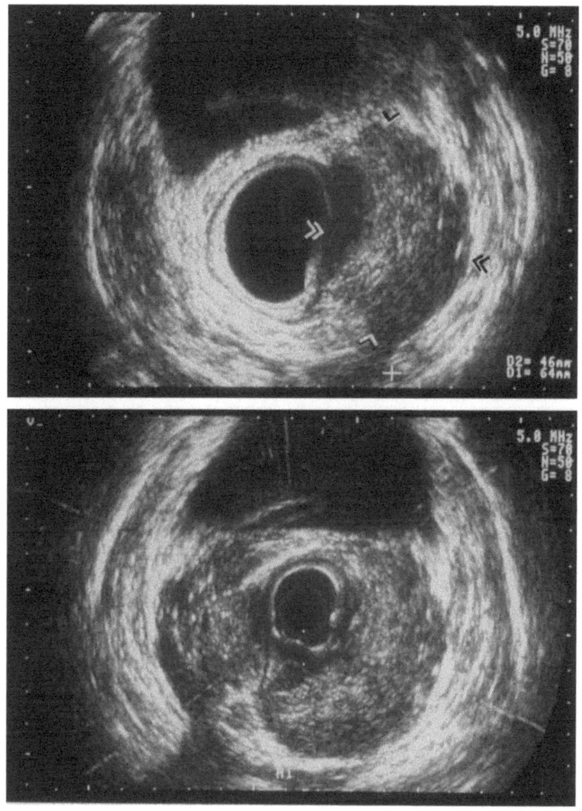

Fig. 34. Rectosonography 1½ years after Wertheim's radical hysterectomy and postoperative irradiation. *Top:* transverse scan with the probe inserted 6 cm into the rectum shows a pelvic-wall recurrence on the right side *(arrowheads)*. *Bottom:* transverse scan at 9 cm shows a large, bilateral pelvic-wall recurrence with circumferential rectal stenosis

of uterine cancer, rectosonography with a 360° transaxial probe is most effective for the detection of recurrence. In contrast to linear rectal probes, the transaxial probe also allows comparison of both sides, and the findings correlate optimally with the rectal examination (Bernaschek et al. 1984b).

Local recurrences typically appear on rectosonograms (and on vaginosonograms in suitable candidates) as hypoechoic areas with irregular margins (Fig. 31). Scans can also demonstrate invasion of the urinary bladder and rectum (Fig. 32), which can then be further evaluated by cystosonography. In addition, the location of a recurrent tumor can be accurately defined, making it possible to distinguish a central recurrence from a growth on the pelvic wall or in the paracolpium (Fig. 33). A central recurrence typically appears as a hypoechoic, usually irregular area on the vaginal stump whose echo pattern usually resembles that of the primary tumor. Suspicious areas on the pelvic wall usually have more sharply defined borders, especially when the lesion is a carcinomatous lymph node. Scars are not demonstrated by sonography; even firm, palpable bands or plaques are not definable as hypoechoic areas.

A major application of rectosonography is in the diagnosis of pelvic wall recurrence (Fig. 34), especially when the lesion is too high to be accessible to

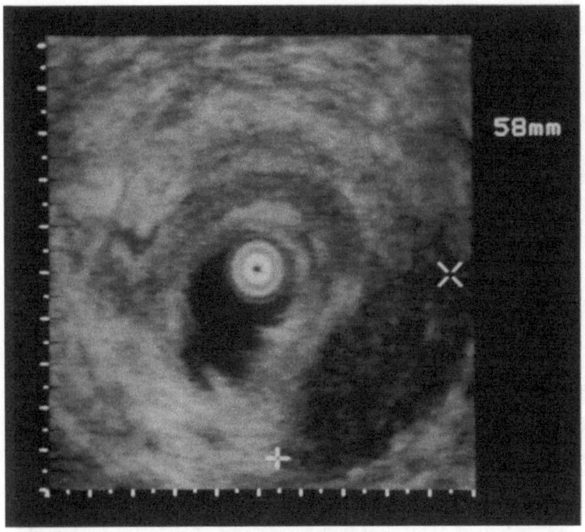

Fig. 35. Rectal scan shows a recurrence high on the left pelvic wall 2 years after Wertheim's radical hysterectomy. The full urinary bladder is not visualized

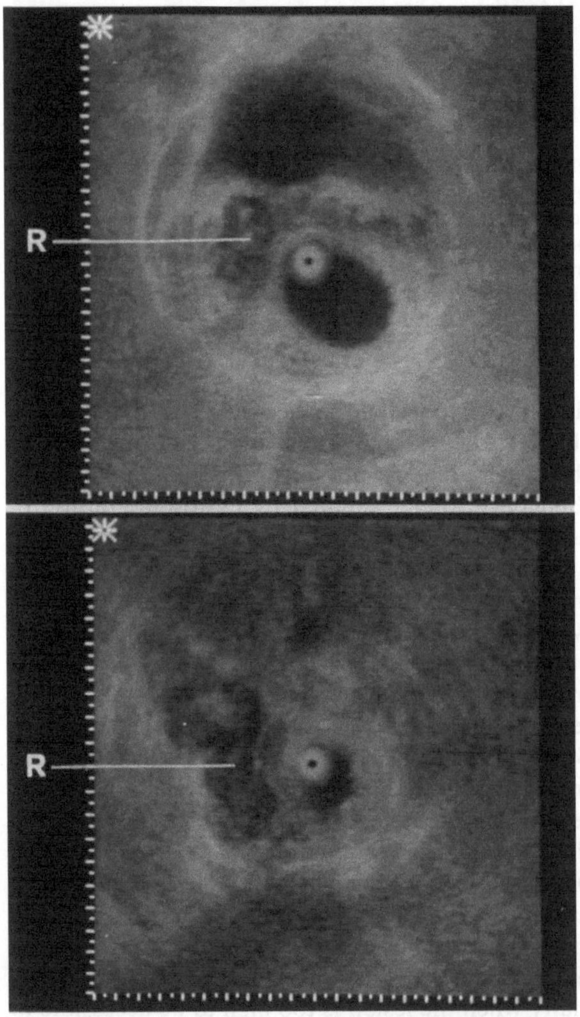

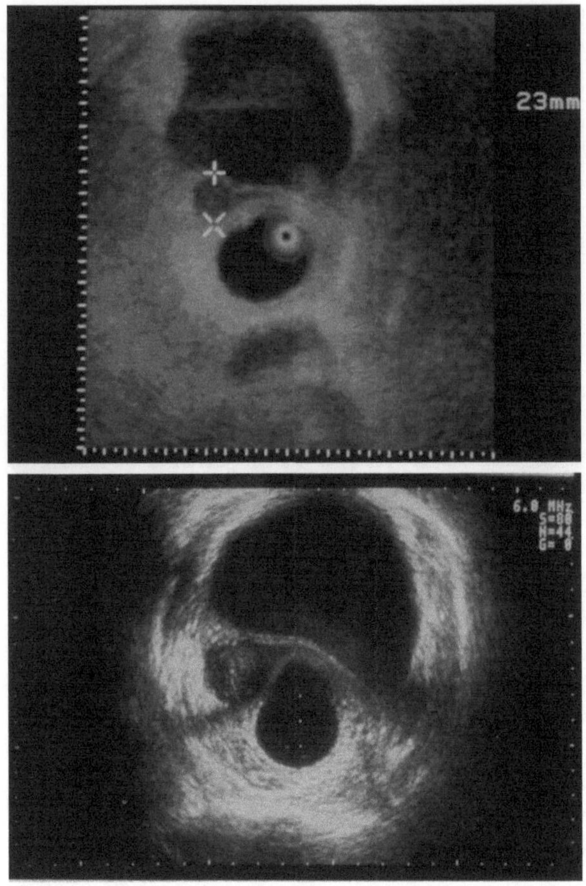

Fig. 37. Rectal scans of a histologically confirmed recurrence on the right side following primary surgery for ovarian carcinoma and subsequent irradiation and chemotherapy. *Top:* 3 years after the primary operation. *Bottom:* virtually no change 4½ years after the primary operation

◀ **Fig. 36.** Rectal scans of an early recurrence *(R)* 8 weeks after radical hysterectomy for uterine leiomyosarcoma. *Top:* transverse scan with the probe inserted 8 cm into the rectum. *Bottom:* transverse scan at 12 cm showing greater tumor size

rectal palpation (Fig. 35). Ordinarily the term "recurrence" signifies renewed tumor growth at the site of the primary neoplasm or very close to it, while "metastasis" denotes tumor occurrence separate from the primary lesion. These terms are often used loosely in gynecology, which is why a lymph node metastasis on the pelvic wall is commonly referred to as a "pelvic wall recurrence." If the "recurrence," or lymph node metastasis, is located at an extremely high level (i. e., in the lumbar region), it is seldom possible to visualize the growth with a rectal scanner, and other modalities such as CT or MRI would be preferred.

Rectosonography can also be used to determine whether a recurrence with a palpable lower pole

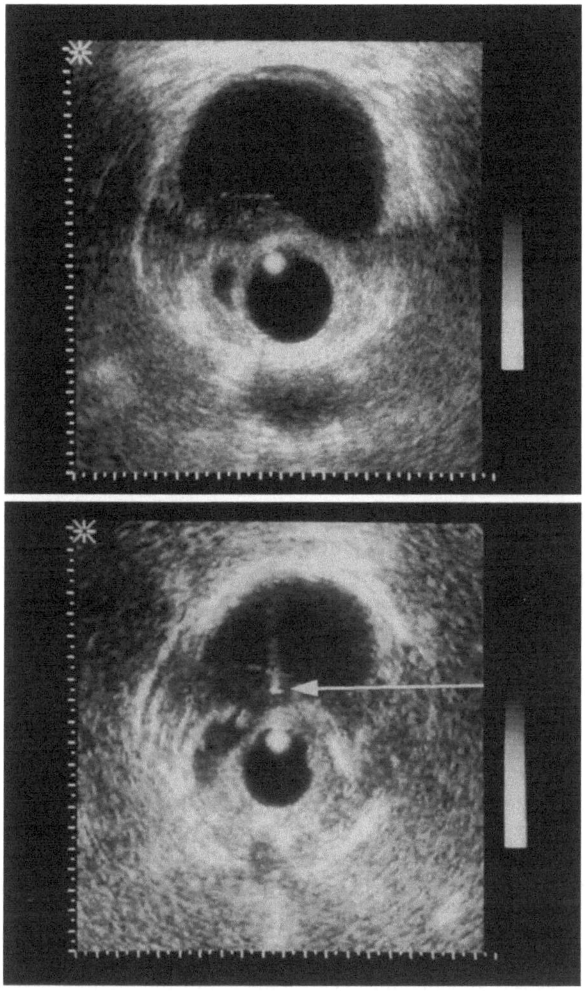

Fig. 38. Rectosonography: *Top:* local recurrence 1 year after Wertheim's radical hysterectomy for a stage II B cervical carcinoma. *Bottom:* ultrasound-guided biopsy of the local recurrence; needle *(arrow)*

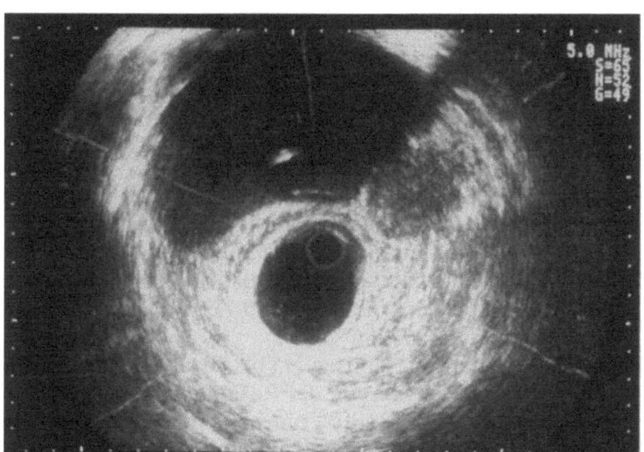

Fig. 39. Rectal scan 1 year after primary operation for ovarian carcinoma; left-sided recurrence (histologically confirmed)

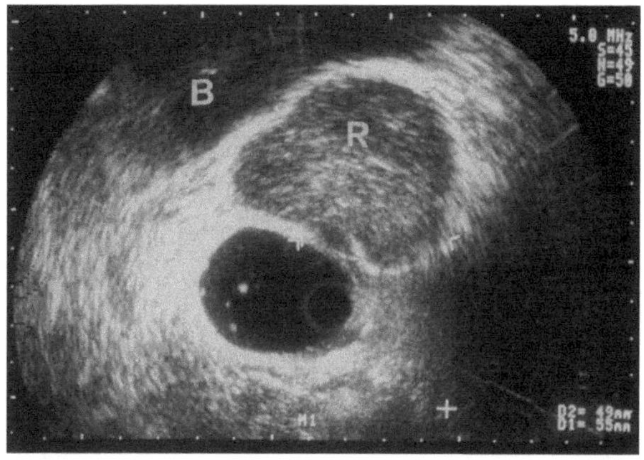

Fig. 40. Transverse rectal scan above the end of the vaginal blind pouch. Recurrence 2 years after primary operation for ovarian carcinoma

has a corresponding cranial extent (Fig. 36). In equivocal cases, and especially when clinical symptoms are absent, the situation can be clarified by watching for enlargement of the suspicious area on follow-up scans (Fig. 37) or by performing an ultrasound-guided needle biopsy (Fig. 38). Like all noninvasive diagnostic methods, rectosonography can only furnish a high index of suspicion of recurrent cancer. Recurrence can be proved only by histologic examination. But while only palpable tumors could be biopsied in the past, rectosonography and vaginosonography offer a means of sampling even high-sited recurrences with a transvaginal needle inserted under direct sonographic guidance.

The advantages of rectosonography for detecting recurrences in the lesser pelvis are apparent not only after surgery for uterine and ovarian malignancies (Figs. 39, 40) but also in patients who have undergone primary radiotherapy (Fig. 41). Uterine enlargement noted on serial examinations, especially when accompanied by the presence of hypoechoic to nonhomogeneous areas within the uterus, is consistent with a recurrence. Uterine enlargement may also be caused by mucometra secondary to occlusion of the cervical canal following intracervical or intrauterine radium insertion. This condition is easily identified by rectosonography, appearing as a predominantly cystic mass with a thin wall (Fig. 42).

Rectosonography also provides a means for the selective, ultrasound-guided interstitial irradiation of local recurrences of gynecologic cancer (Weigert et al. 1986), analogous to the procedure used for prostatic carcinoma. The initial therapy to be considered for recurrent cancer is reirradiation. Because external beam therapy alone usually cannot delivere a tumoricidal dose, the permanent implan-

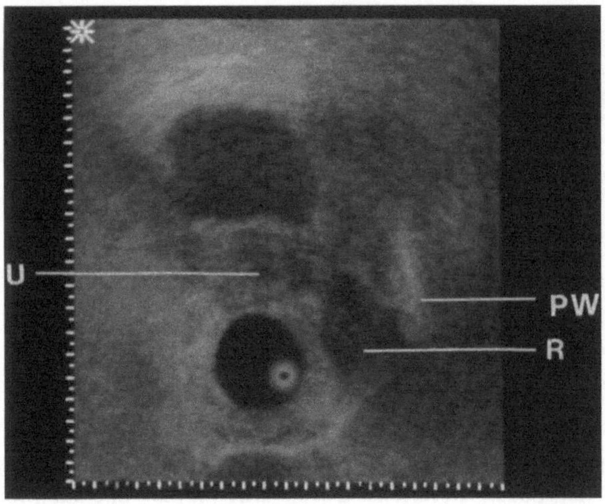

◀ **Fig. 41.** Transverse rectal scan with the probe inserted 12 cm into the rectum shows a recurrence *(R)* high on the left pelvic wall 1½ years after primary radiotherapy for stage III B cervical carcinoma. Ultrasound and clinical findings agreed, but rectal palpation was negative. Uterus *(U)*, pelvic wall *(PW)*

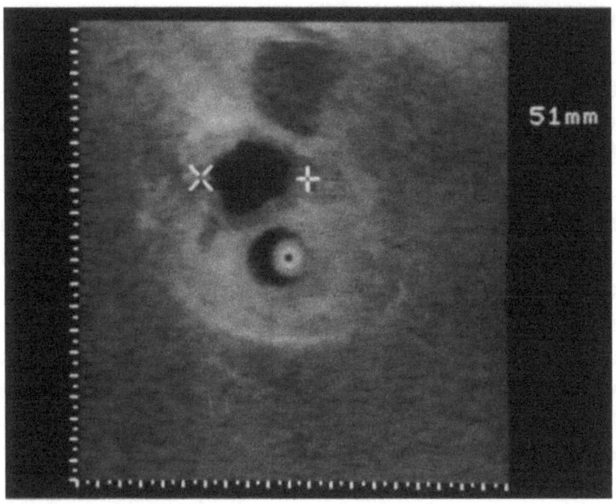

Fig. 42. Transverse rectal scan through a uterus affected by mucometra following primary radiotherapy for stage III cervical carcinoma

tation of iodine-125 seeds or the temporary implantation of a hollow needle for afterloading should be considered to achieve dose saturation. Rectosonography enables visual monitoring of the transvaginal implantation and allows the position of the implants to be adjusted according to the tumor location.

In summary, there are sound reasons to integrate rectosonography into the routine surveillance of patients who have been treated for uterine or ovarian cancer. Rectosonography offers obvious advantages in the detection of local recurrences, regardless of the nature of the primary therapy.

In patients considered at high risk for recurrence, endosonography has a place in post-therapeutic screening. Immediately following inspection and palpation, which are limited by anatomic constraints, the rectal probe can be introduced as an "extension" of the palpating finger, allowing the whole lesser pelvis to be explored. Suspicions can be tested histologically by performing a transvaginal needle biopsy under endosonographic guidance. Owing to its ease of use and powerful capabilities as a diagnostic tool, endosonography could easily take precedence over more costly and complicated techniques such as CT, MRI, and angiography, which may be necessary only if the results of endosonography are negative or equivocal.

References

Averette HE, Dudan RC, Ford JH (1972) Exploratory celiotomy for surgical of cervical cancer. Am J Obstet Gynecol 113: 1090-1096

Averette HE, Sevin BU, Girtanner RE, Ford HJ (1981) Prätherapeutische Staging-Laparotomie beim Zervixkarzinom. Gynäkologe 12: 164-169

Baltzer J, Goepcke W (1979) Tumor size and lymph node metastases in squamous cell carcinoma of the uterine cervix. Arch Gynecol 227: 271-278

Bandy LC, Clarke-Pearson DL, Silverman PM (19859 Computed tomography in evaluation of extrapelvic lymphadenopathy in carcinoma of the cervix. Obstet Gynecol 65: 73-76

Becker H, Hötzinger H (1986) Die Hysterosonographie und ihre Bedeutung in der Diagnostik des Endometriumkarzinoms. Geburtshilfe Frauenheilkd 48: 693-696

Bernaschek G (1987) Die endosonographische Diagnostik bei Uterusmalignomen. Ultraschall 8: 160-165

Bernaschek G, Janisch H (1983) Eine Methode zur Objektivierung des Parametrienbefundes beim Zervixkarzinom. Geburtshilfe Frauenheilkd 43: 498-500

Bernaschek G, Schaller A (1983) Operieren oder Bestrahlen des Zervixkarzinoms im Stadium II B. Geburtshilfe Frauenheilkd 43: 755-758

Bernaschek G, Bartl W, Janisch H (1984 a) Erste Ergebnisse eines echographischen Staging beim Zervixkarzinom. Wien Klin Wochenschr 96: 286-289

Bernaschek G, Tatra G, Janisch H (1984 b) Die rektale Sonographie - eine Erweiterung der Rezidivdiagnostik zervikaler Neoplasien. Geburthilfe Frauenheilkd 44: 495-497

Bernaschek G, Deutinger J, Bartl W, Janisch H (1986) Endosonographic staging of carcinoma of the uterine cervix. Arch Gynecol 239: 21-26

Bernoth EM, Link M, Weise W (1984) Gynäkologie: Differentialdiagnose und Klinik. Karger, Basel, p 326

Breitenecker G, Bartl W, Endler M, Gring H (1984) Bedeutung morphologischer Parameter bei Endometriumkarzinomen. Onkologie 7: 222-235

Brenner DE, Whitley NO, Prempree T, Villasanta U (1982) An evaluation of the computed tomographic scanner for the staging of the cervix. Cancer 50: 2323

Buchsbaum HJ (1972) Para-aortic lymph node involvement in cervical carcinoma. Am J Obstet Gynecol 113: 942-947

Christ F, Claussen C, Brandt H (1983) Die Wertigkeit der Computertomographie bei der präoperativen Diagnostik des Zervixcarcinoms. Arch Gynecol 235: 146-147

Deutinger J, Bernaschek G, Tatra G (1989) Rektosonographie vor und nach Strahlentherapie des inoperablen Zervixkarzinoms. Tumordiagnostik Ther 10: 119-122

Engelmeier UH, Hecker R, Hötzinger H, Thiel H (1985) Automatische Kontursuche und Segmentierung von sonographischen Transversalschnittbildern des Uterus hinsichtlich einer optimierten individuellen Isodosengestaltung bei Corpus-Ca. Strahlentherapie 161: 275-280

Frischbier HJ (1969) Grenzen der Metastasendiagnostik bei Lymphographie und Cavographie. Fortschr Rö [Beiheft] 199

Goswamy RK, Campbell S, Whitehead MI (1982) Establishment of normal ranges for ovarian volumes and identification of enlarged ovaries by real-time mechanical sector sonar in postmenopausal women. In: Ultrasound 82. Proceedings of third meeting of World Federation for Ultrasound in Medicine and Biology. Pergamon, Oxford, pp 615

Goswamy RK, Campbell S, Whitehead MI (1983) Screening for ovarian cancer. In: Campbell S (ed) Clinics in obstetrics and gynecology. Saunders, Philadelphia, pp 621

Granberg S, Wikland M (1987) Comparison between endovaginal and transabdominal transducers for measuring ovarian volume. J Ultrasound Med 6: 649-653

Heyman J (1938) Illustrating the division of cancer of the uterine cervix into four stages. League of Nations Health Organisation, Stockholm

Hohlweg-Majert P, Pohl R, Georgi M, Strauss L (1983) Vergleichende Untersuchungen zur Aussagekraft von Lymphographie und Computertomographie bei Kollumkarzinomen. Arch Gynecol 235: 145

Hötzinger H (1984) Durch Intrauterinsonographie optimierte intrakavitäre Strahlentherapie bei Korpuskarzinomen. Strahlentherapie 160: 600-604

Hötzinger H, Becker H, Becker V (1984) Intrauterine Ultraschalltomographie (IUT): Vergleich mit makroskopischen Präparatschnitten. Geburtshilfe Frauenheilkd 44: 219-224

Hötzinger H, Engelmeier KH, Ries G, Wischnik A, Sbornik E (1985) Zum Problem der Darstellung des Fundus uteri bei der Intrauterinsonographie. Ultraschall 6: 189-193

Jakobs H (1973) Möglichkeiten und Grenzen angiographischer Verfahren in der Rezidivdiagnostik bei gynäkologischen Tumoren. Röntgenblätter 26: 22-25

Kademian MT, Bosch A (1977) Staging laparotomy and survival in carcinoma of the uterine cervix. Acta Radiol 16: 314-324

Kindermann G, Gerteis W, Weishaar J (1970) Was leistet die Lymphographie in der Erkennung von Metastasen beim Zervixkarzinom? Geburtshilfe Frauenheilkd 30: 444-452

Kölbl H, Bernaschek G (1987) Der Stellenwert der Zystosonographie beim Staging von gynäkologischen Malignomen. Tumor Diagnostik und Therapie 8: 177-180

Kofler E, Kratochwil A (1983) In: Michalica W (ed) Gynäkologische Onkologie und Onkotherapie, vol 2. Maudrich, Vienna, p 159

Kottmeier HL (ed) (1982) Annual report on the results of treatment in gynecological cancer, vol 18. Radium hemmet, Stockholm (published under patronage of FIGO)

Kratochwil A, Bernaschek G (1981) Die Bedeutung der Sonographie in der Rezidivdiagnostik gynäkologischer Malignome. Wien Klin Wochenschr 93: 178-183

Kratochwil A, Stöger H, Zeibekis N (1974) Darstellung von Beckenwandrezidiven im Ultraschalltomogramm. Geburtshilfe Frauenheilkd 34: 742-752

Ladner HA (1987) Wertigkeit moderner bildgebender Verfahren in der Primärdiagnostik und Nachsorge von Patientinnen mit Genitalkarzinomen. Gynäkologe 20: 212-221

Lagasse LD, Ballon SC, Berman MI, Watring WG (1979) Pretreatment lymphangiography and operative evaluation in carcinoma of the cervix. Am J Obstet Gynecol 134: 219-224

Lang EK, Greer JL (1969) The value of pelvic arteriography for the staging of carcinoma of the cervix. Radiology 92: 1027-1034

Lange P (1960) Clinical and histological studies on cervical carcinoma. Praecancerosis, early metastasis, and tubular structures in the lymph nodes. Acta Pathol Microbiol Scand Suppl 143: 9-162

Michalica W (1981) Gynäkologische Onkologie und Onkotherapie, vol 1. Maudrich, Vienna

Nelson JH, Macasaet MA, Lu KT, Bohorqez JF, Smart GE, Nicastri AD, Walton LA (1974) The incidence and significance of paraaortic lymph node metastases in late invasive carcinoma of the cervix. Am J Obstet Gynecol 118: 749-756

Obata A, Akamatsu N, Sekiba K (1985) Ultrasound estimation of myometrial invasion of endometrial cancer by intrauterine radial scanning. J Clin Ultrasound 13: 397-404

Piver MS, Barlow JJ (1977) High dose irradiation to biopsy-confirmed aortic node metastases from carcinoma of the uterine cervix. Cancer 39: 333

Piver MS, Chung WS (1975) Prognostic significance of cervical lesion size and pelvic node metastasis in cervical carcinoma. Obstet Gynecol 46: 507-510

Popp LW (1986) Vaginosonographische Befunde bei gynäkologischen Malignomen. Möglichkeiten zur Früherkennung des Ovarialkarzinoms. In: Popp LW (ed) Gynäkologische Endosonographie. Klemke, Quickborn, pp 193-201

Popp LW, Lemster S, Müller-Holve W, Martin K (1985) Vaginosonography: an approach to a screening method for early ovarian cancer. XIth FIGO World Congress, Berlin, p 186

Requard CK, Wicks JD, Mettler FA Jr (1981) Ultrasonography in the staging of endometrial adenocarcinoma. Radiology 140: 781-785

Rodriquez M, Greensporn J, Platt L (1987) The use of transvaginal sonography in the evaluation of the postmenopausal ovary. 32nd Annual Convention of the American Institute of Ultrasound in Medicine. New Orleans, p 36

Rudelstorfer R, Schieder K, Bernaschek G (1988) Vaginosonographic measurements of endometrial thickness for early detection of carcinoma of the endometrium. J. E. M. U. 9: 244-246

Schurz B, Eppel W, Kucera H, Reinold E (1987) Die Darstellung des Endometriumkarzinoms mit der Vaginosonographie. Ultraschall 8: 166-170

Weigert F, Hötzinger H, Atzinger A (1986) Die rektosonographische Differentialdiagnose lokaler Rezidive gynäkologischer Malignome von Narben und die gesteuerte Afterloadingtherapie von Rezidiven im kleinen Becken. In: Popp LW (ed) Gynäkologische Endosonographie. Klemke, Quickborn, pp 63-66

Wharton JT, Jones HW, Day TG, Rutledge FN, Fletcher GH (1977) Preirradiation celiotomy and extended field irradiation for invasive carcinoma of the cervix. Obstet Gynecol 49: 333-337

Wischnik A, Hötzinger H, Bohrer M, Höss I, Hogo R v, Jacobs V (1986) Die gynäkologische Endosonographie im Vergleich mit anderen bildgebenden diagnostischen Verfahren. In: Popp LW (ed) Gynäkologische Endosonographie. Klemke, Quickborn, pp 203-209

Yuhara A, Akamatsu N, Sekiba K (1987) Use of transrectal radial scan ultrasonography in evaluating the extent of uterine cervical cancer. J Clin Ultrasound 15: 507-517

Zander J, Baltzer J, Lohe KJ, Ober KG, Kaufmann C (1981) Carcinoma of the cervix: an attempt to individualize treatment. Am J Obstet Gynecol 139: 752-759

Zapf S, Halbsguth A, Schweden F, Klose K, Lochner B, Beck T, Friese K, Kreienberg R (1987) Zum Problem des prätherapeutischen Staging des Carcinoma colli uteri. Geburtshilfe Frauenheilkd 47: 838-842

Diagnostic Evaluation of Urinary Incontinence

1 General

The variety of causes of female urinary incontinence demands an adequate and comprehensive preoperative diagnostic evaluation. Techniques used for the assessment of female urinary incontinence range from the simple gynecologic examination and cystoscopy to cystometry, sphincterometry, simultaneous urethrocystometry, cystourethrography, electromyography of the sphincter system, uroflowmetry, and even videographic cystourethrography with simultaneous pressure-flow measurement, the latter being the most modern and comprehensive study but also the most costly. At many centers the main selection criterion for a continence-restoring operation is the lateral cystourethrogram, which makes it possible, among other things, to accurately measure the posterior urethrovesical angle. In the past, the selection of candidates for surgical treatment has chiefly relied on a variety of roentgenographic techniques; and while these techniques have been significantly improved by the application of various catheters and contrast media, radiographic methods expose the patient to radiation and increase the risk of urinary tract infection.

Abdominal sonography was tried as an alternative to roentgenography for the assessment of urinary incontinence (Bernaschek et al. 1981), but it has not come into routine clinical use because of its disadvantages. Although the urethrovesical angle can be measured on abdominal scans with about the same accuracy as on X-ray films, serious difficulties can arise in obese patients and in women with pronounced genital prolapse and associated descent of the bladder neck behind the symphysis. Moreover, abdominal scans can at best demonstrate the proximal portion of the urethra, or the urethral axis must be inferred from the intravesical position of the catheter. Generally, however, the sonographic method is advantageous owing to the lack of need for contrast medium and the ability to prolong the examination as required.

2 Vaginosonography and Rectosonography

Since the development of intracorporeal probes for endosonography, experts have advocated their use for measuring the urethrovesical angle and observing dynamic processes in that region (Brown et al. 1985; Debus-Thiede et al. 1985). Most frequent use is made of linear-array scanners, which are best able to demonstrate the entire urethra and bladder neck area. The advantage of vaginosonography and rectosonography over abdominal sonography is their ability to demonstrate the region of interest with consistently good image quality and without interference from the symphysis or subcutaneous fat.

The main disadvantage of endosonography relates to the size and shape of the probe, which can cause displacements of anatomic structures. The close proximity of the urethra and bladder neck to the probe enables the use of higher frequencies, providing better resolution of details, but the space available for the examination is very confined. Another problem is the potential for probe displacement on straining. This frequently leads to artifacts, making it difficult to accurately assess positional relationships. Straining not only displaces the probe axially but can also cause lateral displacement of the urethra. Adjusting the probe to realign it with the urethra in this situation results in further displacement and loss of accuracy. Rectosonography is somethat less susceptible to these problems than vaginosonography, but it is also less acceptable to the patient.

3 Perineal and Introital Sonography

Given the difficulties of demonstrating the urethrovesical angle by conventional "semi-invasive" vaginal and rectal scanning, new methods have been devised for the sonographic visualization of the bladder neck region (Grischke et al. 1986; Kohorn et al. 1986; Kölbl et al. 1988).

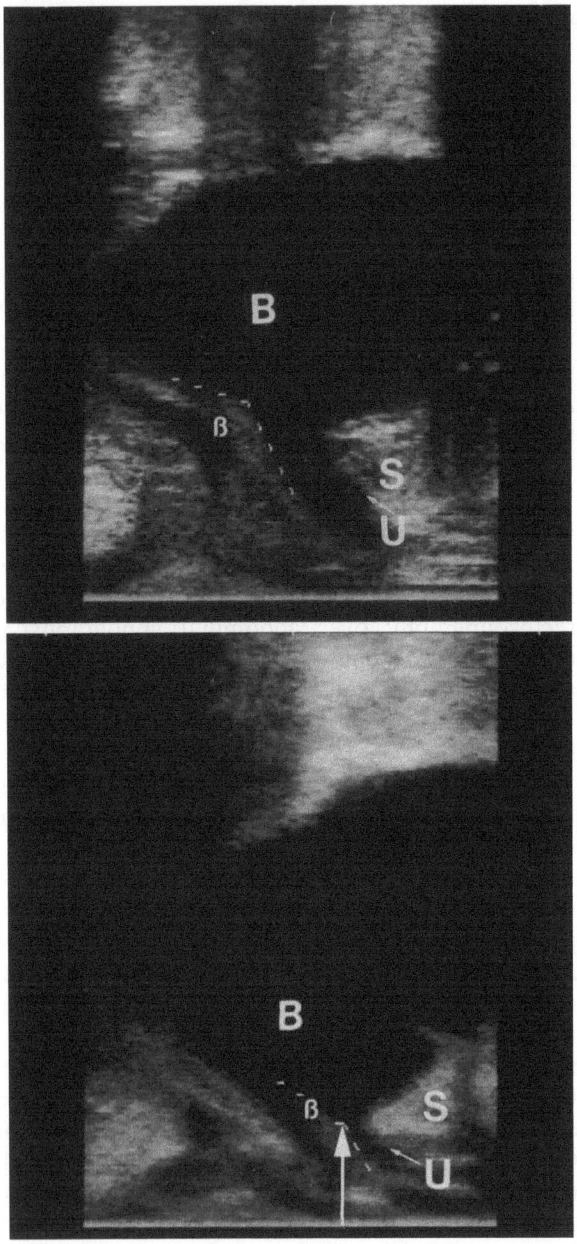

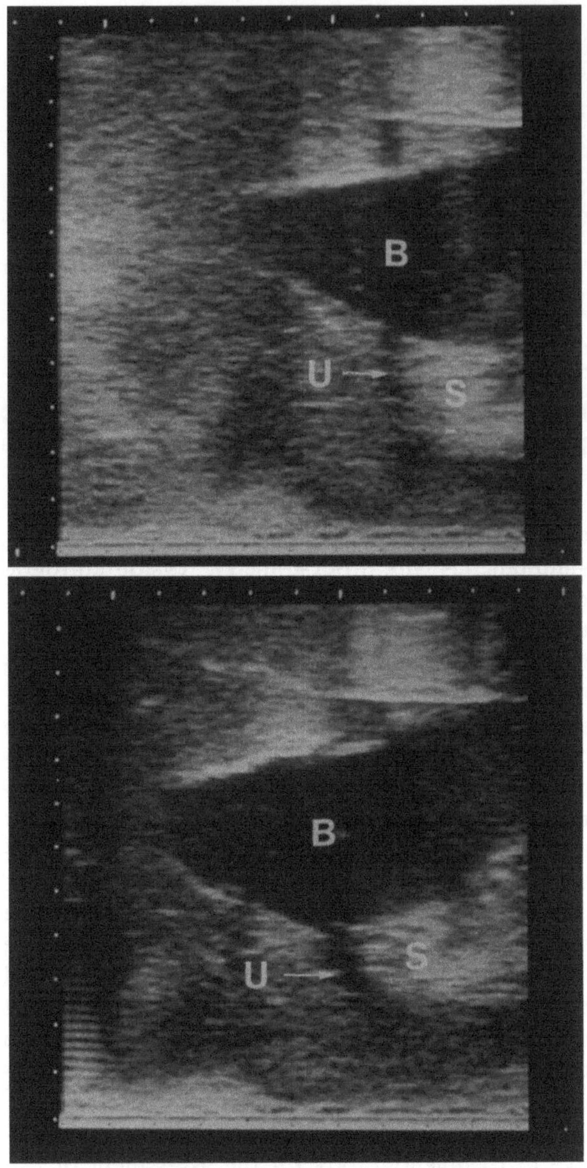

Fig. 2. Perineal scan of a 39-year-old woman with true type I stress incontinence in an upright position at rest *(top)* and during a Valsalva maneuver *(pottom)*, showing the bladder *(B)*, symphysis *(S)*, and urethra *(U)*

Fig. 1. Direct measurement of the posterior urethrovesical angle *(β)* by perineal sonography, showing the bladder *(B)*, the symphysis *(S)*, and the urethra *(U)* in a 34-year-old woman with true type II stress incontinence in an upright position. *Top:* at rest; *bottom:* during a Valsalva maneuver. Note the shortened distance between the internal urethral orifice and the transducer surface and also the increase in the urethrovesical angle *(arrow)*

Perineal sonography is usually performed with a linear-array scanner applied to the perineal region. As first described by Kohorn et al. (1986), some training and experience are required in order to identify the structures in the bladder neck region. For the beginner, orientation can be aided by the passage of a disposable catheter, although high-resolution equipment can easily delineate the urethra without catheterization (Fig. 1). Unlike vaginosonography and rectosonography, perineal scanning does not alter topographic relationships, and the

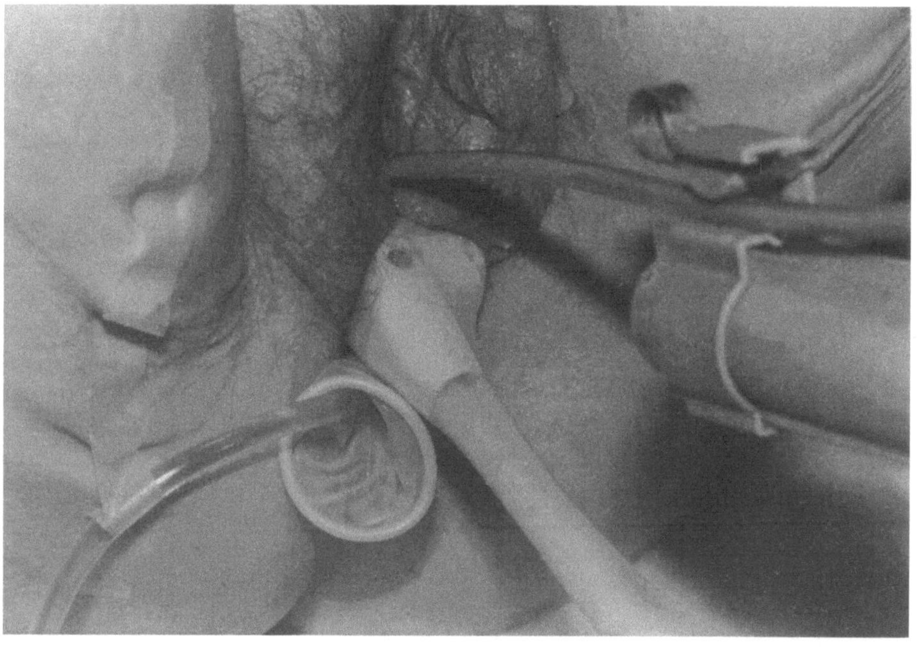

Fig. 3. Demonstration of introital sonography and simultaneous pressure-flow measurement with the patient in a sitting position

transducer position is not affected by straining. In addition, the need for a vaginal or abdominal operation to restore continence is quickly and easily assessed by determining the level of the urethrovesical junction in relation to the symphysis.

Bladder descent on straining also can be measured by applying the perineal transducer (Fig. 2). This study provides an additional, objective parameter for evaluating the success of a continence-restoring operation and of nonoperative treatments for stress incontinence. An initial comparison with the results of X-ray examinations indicates good agreement (Kölbl et al. 1988).

Difficulties arise when cystourethrometry is performed concurrently with perineal scanning due to the long "footprint" of the linear-array scanner, which tends to cause catheter displacement. The scanner may also interfere with voiding when simultaneous uroflowmetry is attempted. In an effort to eliminate these problems, Kölbl and Bernaschek (1989) devised the technique of introital sonography, which is intermediate between perineal scanning and vaginosonography. Though the technique is performed with a standard front-viewing vaginal scanner, the probe is not applied to the perineum or inserted into the vagina (Fig. 3). Rather, it is applied

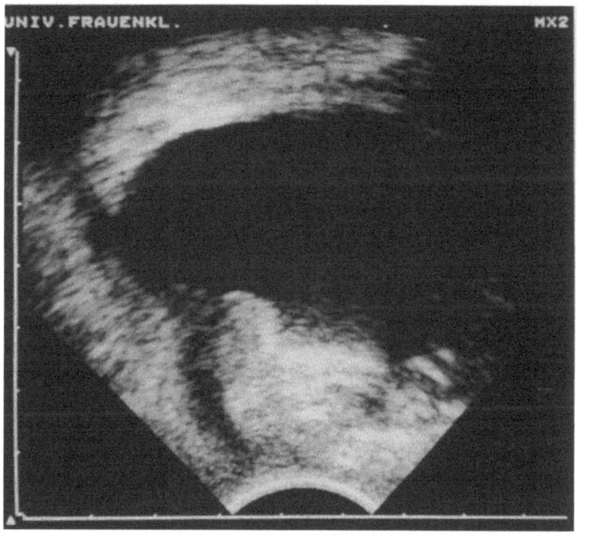

Fig. 4. Introital sonography at rest in a continent patient

over a small area in the introital region just behind the urethral orifice (Fig. 4). The long scanner shaft allows optimum freedom of motion for the examiner and patient, and the small transducer contact area does not interfere with voiding or urethral catheterization. Meanwhile the technique preserves all the advantages of perineal scanning. Thus, introital sonography provides an ideal method for ultrasound video cystourethrography. Simpler and less costly than the X-ray method, introital sonography has a comparable range of indications, including

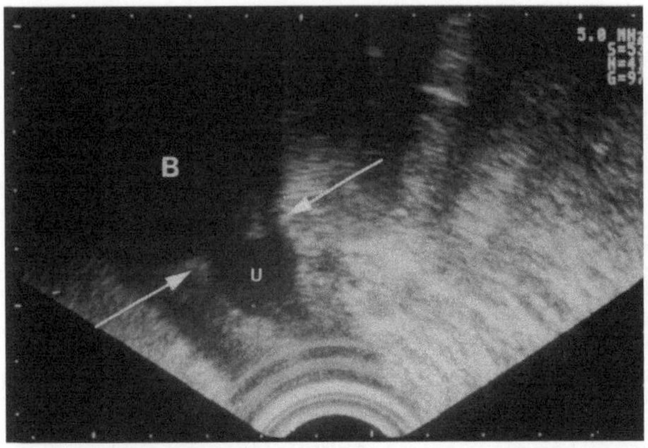

Fig. 5. Introital scan in a nulliparous woman with detrusor instability combined with urodynamic assessment. Scan shows marked opening of the bladder neck *(arrows)*; urethra *(U)*, bladder *(B)*

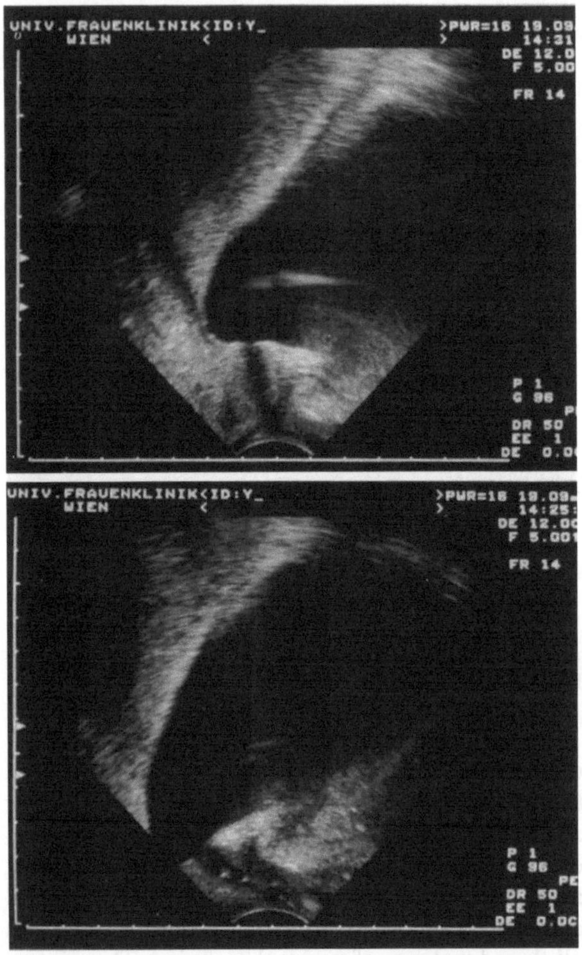

Fig. 6. Introital scan during rest *(top)* and during coughing *(bottom)* in a patient with genuine stress incontinence combined with rotatory bladder descent. Descent of the urethrovesical junction during coughing and marked increase of the retrovesical angle and of the angle of inclination

the diagnostic evaluation of recurrent incontinence, detrusor instability (Fig. 5), etc. The absence of radiation exposure allows extended examining times, and the examination may be repeated as often as necessary until a diagnosis is established.

Introital scans are equivalent to videographic records in their ability to demonstrate dynamic processes in bladder function studies. The examination is noninvasive, easy to learn and perform, and is well tolerated by patients.

In summary, introital sonography, based on vaginosonography, is preferable to other sonographic methods for the evaluation of female urinary incontinence. It can demonstrate static changes in the bladder neck region and is of particular value for the study of dynamic processes (Fig. 6). Owing to its efficacy and ease of use, introital sonography could provide a serious alternative to X-ray video cystourethrography.

References

Bernaschek G, Spernol R, Wolf G, Kratochwil A (1981) Vergleichende Bestimmung des Urethra-Blasenwinkels bei Inkontinenzfällen mittels Ultraschall und lateralem Urethrozystogramm. Geburtshilfe Frauenheilkd 41: 339-342

Brown MC, Sutherst JR, Murray A, Richmond DH (1985) Potential use of ultrasound in place of X-ray fluoroscopy in urodynamics. Br J Urol 57: 88-90

Debus-Thiede et al. (1985) Erste Erfahrungen mit der vaginalen Sonographie von Urethra und Blase im Rahmen der Inkontinenzdiagnostik. Geburtshilfe Frauenheilkd 45: 891-894

Grischke EM, Dietz HP, Jeanty P, Schmidt W (1986) Eine neue Untersuchungsmethode: Perineal Scan in der Geburtshilfe und Gynäkologie. Ultraschall 7: 154-161

Kohorn EI, Scioscia AL, Jeanty P, Hobbins JC (1986) Ultrasound cystourethrography by perineal scanning for the assessment of female stress urinary incontinence. Obstet Gynecol 68: 269-272

Kölbl H, Bernaschek G (1989) A new method for sonographic cystourethrography and simultaneous pressure-flow measurements. Obstet Gynecol 74: 417-422

Kölbl H, Bernaschek G, Wolf G (1988) A comparative study of perineal scan and urethrocystography in patients with genuine stress incontinence. Arch Gynecol Obstet 244: 39-45

Infertility

1 General

Ultrasonography has earned an established place in the evaluation and management of female infertility. The history of this application began when Kratochwil et al. (1972, 1973) first evaluated the ovaries with ultrasound. Later, improvements in image resolution enabled the sonographic detection of cyclic changes in the female genital organs. Building on the work of Kratochwil, Hackelöer and Hansmann were able in 1976 to demonstrate follicular growth and involution with ultrasound and to show a strong correlation between follicular size and endocrine parameters (Hackelöer et al. 1979).

Before the application of ultrasound to follicular monitoring, it was necessary to rely on clinical and hormonal parameters. Today folliculometry is commonly used in conjunction with hormonal parameters to determine the most favorable time for coitus or artificial insemination in both spontaneous and induced cycles. Precise monitoring is essential for successful oocyte retrieval, especially in anovulatory patients undergoing costly stimulation therapy or when extracorporeal fertilization is planned. When combined with hormone assays, sonographic follicular monitoring provides a valuable adjunct to stimulation therapy as a prelude to oocyte collection and in vitro fertilization (De Crespigny et al. 1981 a, b; Buttery et al. 1983; Hull et al. 1986).

Sonography is also valuable for diagnosing causes of infertility that do not relate to follicular maturation or ovulation, such as uterine anomalies (Confino et al. 1985) and tubal changes (Timor-Tritsch and Rottem 1987; Timor-Tritsch et al. 1988).

A major application of sonography for in vitro fertilization, besides folliculometry, is to guide follicular aspiration. The use of an vaginal probe provides an important simplification for in vitro fertilization programs. Owing to its ease of use and high patient acceptance, vaginosonography is finding increasing use for follicular measurement and aspiration, and it has significantly reduced the role of laparoscopy for oocyte retrieval. Once follicular aspiration and oocyte fertilization have been accomplished, embryo transfer likewise can be carried out under sonographic guidance.

Vaginosonography can be used along with biochemical parameters (Deutinger et al. 1984, 1986 a) for monitoring the corpus luteum phase and for very early pregnancy detection owing to its ability to demonstrate an intrauterine conceptus as small as 2 mm in diameter (Bernaschek 1987 a, b; Bernaschek and Deutinger 1988). Meanwhile, the vaginal probe has become a routine instrument for many other investigations in gynecologic patients.

2 Evaluation of the Menstrual Cycle

2.1 General

Since introduction of the in vitro fertilization program at the Second University of Vienna Women's Hospital, a transabdominal real-time sector scanner has been chiefly used for the daily ultrasound monitoring of follicular growth (Feichtinger et al. 1981). A linear scanner is less suitable for transabdominal folliculometry, as it is unable to demonstrate both ovaries in 20% of cases (Hackelöer 1986). At one time compound scanners were used routinely for folliculometry, but they are rarely used today because real-time instruments yield better results in less time (Leerentveld et al. 1984; Hackelöer 1986). But despite the use of high-resolution equipment, the transabdominal visualization of the ovaries and their cyclic changes has continued to be a problem area in gynecologic sonography (Martin 1986). One problem is the need for a full urinary bladder, which straightens the uterus from its normal anteflexed, anteverted position and also elevates the uterus and adnexa out of the lesser pelvis. This is necessary to make the organs accessible to ultrasound evaluation and obtain a meaningful result. The distended bladder also displaces gas-filled loops of bowel, which may come to overlie structures of interest in the adnexal region. An overdistended bladder may also press the adnexa against the pelvic side wall, where they become difficult to evaluate with ultrasound.

The results of transabdominal sonography are not entirely reproducible due to the inability to standardize bladder distention. Eissa et al. (1985) found 95% confidence limits of ±2.4 mm for transabdominal ultrasound measurements of follicular diameters. Usually, folliculometry is performed in the morning, meaning that the patient has to undergo the particular discomfort of traveling to the examination facility early without emptying her bladder. Often it is found that the bladder is inadequately distended, causing loss of time for the patient and examiner. Indeed, some patients are unable to achieve adequate bladder distention. Another potential problem in transabdominal scanning is a full rectum, which may overlie and obscure the adnexal structures of interest. Finally, it may be difficult or impossible to evaluate the adnexa with transabdominal ultrasound in markedly obese patients.

Since an infertility patient referred for folliculometry is required to undergo daily ultrasound examinations, it was natural that the problems of transabdominal sonography would stimulate interest in vaginosonography as a routine alternative. This technique does not require a full bladder, so it eliminates a major source of discomfort while avoiding loss of time due to an inadequately filled bladder. Even extreme obesity does not pose an obstacle to successful scanning.

Initial vaginosonographic examinations were performed as early as 1969 (Kratochwil 1969). The vaginal approach provides excellent visualization of all organs of the lesser pelvis (Schwimer and Lebovic 1984) and gives substantially improved resolution of the ovaries (Meldrum et al. 1984). This is because the transducer can be positioned closer to the organ of interest, which in turn allows the use of higher transducer frequencies that provide better resolution than is attainable with traditional abdominal ultrasound (Kratochwil 1969). This is particularly helpful in defining small ovarian follicles in the early follicular phase (Deutinger et al. 1987 a).

Female infertility patients often have experience with transabdominal as well as vaginal sonography. After a brief explanation virtually all patients opt for the vaginal examination, considering the distended bladder to be far more objectionable than the insertion of a vaginal probe (Deutinger et al. 1986 b). Other advantages such as improved resolution, greater accuracy, visualization of the ovaries even under difficult anatomic conditions, and a more rapid result have made vaginosonography the method of choice for daily follicular monitoring. It has been found that vaginosonography improves results in 87% of patients in whom follicles cannot be demonstrated transabdominally (Yee et al. 1987).

The development of a vaginal scanner with a 240° forward view angle ("panoramic scanner") has led to the exclusive use of vaginosonography for follicular monitoring at the Second University of Vienna Women's Hospital (Deutinger et al. 1986 b). The large view angle of the scanner makes orientation easy, since the bladder, the whole uterus, and the symphysis all appear as landmarks on the same image. This represents a substantial improvement over the small field of sector scanners. In addition, the transducer frequency is adjustable from 5 to 7.5 MHz, making it possible to obtain high-resolution images of structures close to the scanner (such as the endometrium or very small follicles on ovaries located in the cul-de-sac).

Before vaginal folliculometry is carried out, the patient is told to empty her bladder. The examination itself is performed in the lithotomy position, which allows the greatest freedom of scanner manipulation. Coupling gel is applied to the transducer to improve contact with the vaginal vault. Then the transducer is covered with a condom-like sleeve. This is done for hygienic reasons and also because various gels have been found to adversely affect sperm motility in vitro and to decrease the oocyte cleavage rate - a potentially serious disadvantage in infertility patients (Schwimer et al. 1984; Sheean et al. 1986).

The scanner for vaginal folliculometry is first inserted into the vagina so that a sagittal image is obtained. This provides a long-axis view of the uterus that is helpful for orientation. The endometrium can also be evaluated at this time (Hackelöer 1986). Next the scanner is rotated 90° to obtain a horizontal image. In many cases both ovaries can be seen concurrently on this plane.

Unaltered fallopian tubes usually are not delineated with ultrasound and are seen only in the immediate postovulatory period when there is free fluid in the cul-de-sac (Timor-Tritsch and Rottem 1987).

2.2 Visualization and Measurement of the Follicles

Since the introduction of real-time scanning, daily monitoring of the number and size of the ovarian follicles has become a routine clinical procedure (Hoffmann 1987). Follicular monitoring is especially important when stimulation methods are applied that induce superovulation; such monitoring is necessary in order to relate the daily determinations of hormone levels (estradiol, progesterone, LH) to the stimulation-induced changes in the ovary. Because the increase in follicular size is linear, a mathematical model can be used to predict the time of ovulation at a very early stage (Doody et al. 1987).

The ovaries are ellipsoid in shape and normally are located between the uterus and the lateral pelvic wall. However, they may be abnormally positioned as a result of postoperative adhesions, inflammatory disease, or endometriosis and therefore difficult to identify. If the ovaries are at different levels, it may still be possible to demonstrate them on one image by rotating the scanner slightly; otherwise they must be evaluated separately. Very often the ovaries are located close to the internal iliac artery and vein, although they do not maintain a consistent relationship with those vessels. The ovary usually appears as a hyperechoic structure. Its volume is calculated from its mean longitudinal, transverse, and sagittal dimensions, where the greatest diameter visible on each plane is measured. The volume of the normal ovary is less than 6 cm^3, but that increases considerably in the follicular phase due to the size of the mature follicle (average diameter 25 mm).

The plane on which the ovary can be demonstrated concurrently with the infundibulopelvic ligament, which transmits the ovarian vessels, has been suggested as the standard reference plane for reproducible folliculometry (Hackelöer and Nitschke-Dabelstein 1980). This ligament provides an anatomic landmark which helps to ensure that other cystic structures such as a hydrosalpinx or bowel loop (recognizable on real-time scans by peristaltic motion) are not mistaken for follicles (Hill et al. 1986). Both ovaries should always be imaged for the evaluation of follicular growth, even though it is common for ovulation to occur in the same ovary in several consecutive cycles (Werlin et al. 1986) (Fig. 1). It should be noted that hormonally inactive ovaries can be extremely difficult to visualize, even by vaginal scanning (Granberg and Wikland 1987).

The primordial follicles are too small to be delineated with ultrasound. Approximately 99.9% of all primordial follicles formed during fetal development are lost to atresia; only a very small number achieve ovulation. In the spontaneous, nonstimulated menstrual cycle, a variable number of small cystic areas 3-5 mm in diameter, the developing follicles, can be demonstrated on the surface of the ovaries between days 5 and 7 of the cycle. One of these small follicles is selected as the dominant follicle, and under the influence of follicle-stimulating hormone (FSH) the granulosa cells begin to synthesize estrogens. By day 8-12 of the cycle the dominant follicle can be distinguished sonographically by its size from the remaining follicles, which do

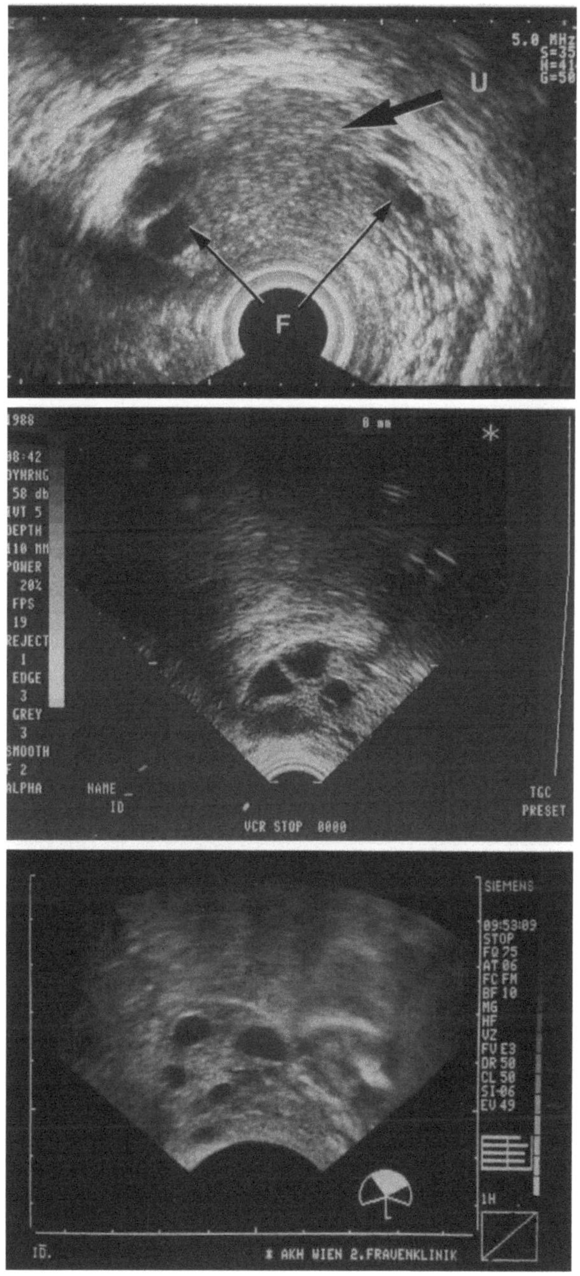

Fig. 1. *Top:* vaginal scan showing the uterus and bilateral ovarian follicles in the early follicular phase. *U*, uterus; *F*, follicles. *Center:* early follicular phase. Three follicles are delineated on this scan. *Bottom:* vaginal scan of follicles in the right ovary on day 7 of a stimulated cycle

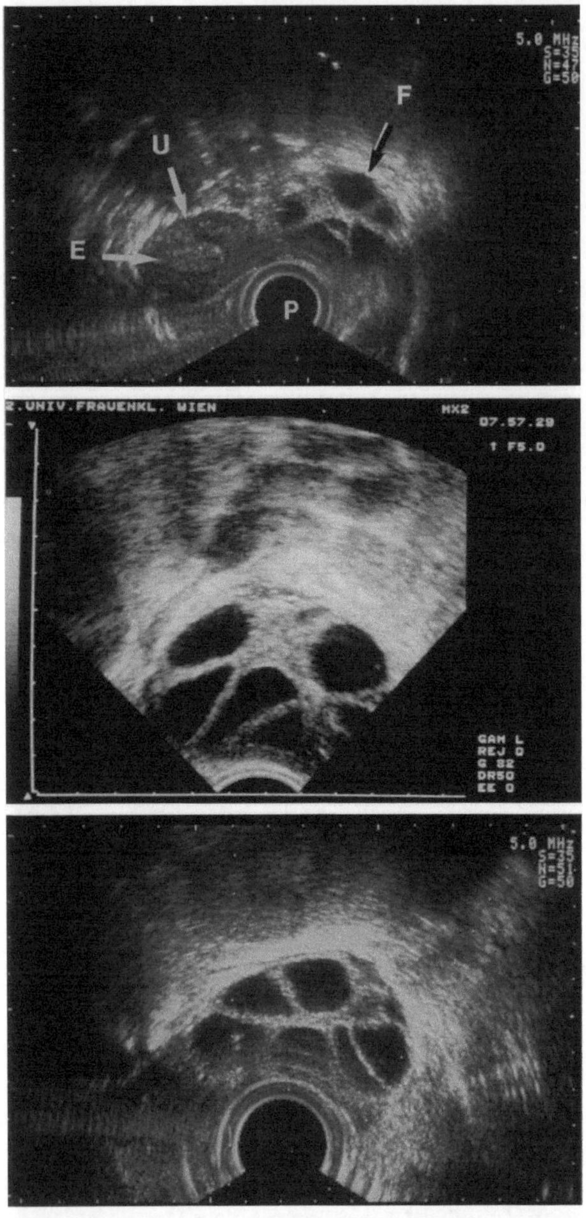

Fig. 2. Vaginal scan of stimulated ovaries in three women just prior to aspiration. *Top:* uterus *(U)* with a widened endometrium *(E)*, left ovary with follicles *(F)*, and vaginal probe *(P)*

not develop further and rarely reach a size of 14 mm before becoming atretic.

Consequently, a follicle larger than 14 mm in diameter is destined to ovulate (Ritchie 1985). The dominant and subordinate follicles are distributed in both ovaries. In more than 80% of ovaries with a dominant follicle, a second follicle can be demonstrated in the same or opposite ovary. Three or four growing follicles are found in 20%-50% of cycles, and two dominant follicles may develop in 5%-11% (O'Herlihy et al. 1980b). In the last 4-5 days preceding ovulation the dominant follicle grows at a rate of 2-3 mm per day, reaching a diameter of 18-28 mm by the time of ovulation (Hackelöer et al. 1977; Picker et al. 1983; Buttery et al. 1983).

The follicle appears as an echo-free or hypoechoic area surrounded by an echogenic border (Fig. 2). Once the ovary has been visualized on the scan, the follicles are individually defined and measured on the plane of their greatest diameter. The follicular diameter is always measured from inside to inside (Hansmann et al. 1985). For most size determinations it is sufficient to measure the greatest follicular diameter, which correlates both with the ovarian volume and with endocrine parameters (Vargyas et al. 1982). The volume of the follicle can be determined from its three greatest dimensions by using the formula for calculating the volume of an ellipsoid. Additionally, laparoscopic studies have shown good agreement between sonographic measurements and the volume of fluid aspirated from the cul-de-sac (O'Herlihy et al. 1980a).

The distended bladder in transabdominal sonography can displace the follicles and distort them into a more oblong shape, introducing some error into the measurement of the greatest follicular diameter. Nevertheless, there does not appear to be a significant difference between transabdominal and vaginal determinations of the total volume of all the follicles in the early follicular phase (Schwimer and Lebovic 1985). In any case, measurement of the greatest follicular diameter is problematic with a full urinary bladder.

Only the prior sonographic detection of a cumulus oophorus or its dissociation from the surrounding granulosa tissue confirms the presence of an oocyte or a corresponding stage of follicular maturity (Fig. 3) (Fleischer et al. 1981; Hackelöer 1984; Roberts et al. 1985). However, this feature can be identified in only about 20% of cases by traditional abdominal scanning (Ritchie 1985). This is an area where vaginosonography offers significant advantages, for it can demonstrate a cumulus oophorus in two-thirds of cases when a 6.5-MHz transducer is used (Drugan et al. 1988). Further improvements may be expected from the use of higher ultrasound frequencies. Ovulation occurs within about 36 h of the spontaneous LH surge or the administration of hCG.

Following ovulation there is a rapid reduction in follicular size and an escape of fluid to the perifollicular area (De Crespigny et al. 1981b; Nitschke-Dabelstein et al. 1981; Werlin et al. 1986). Within

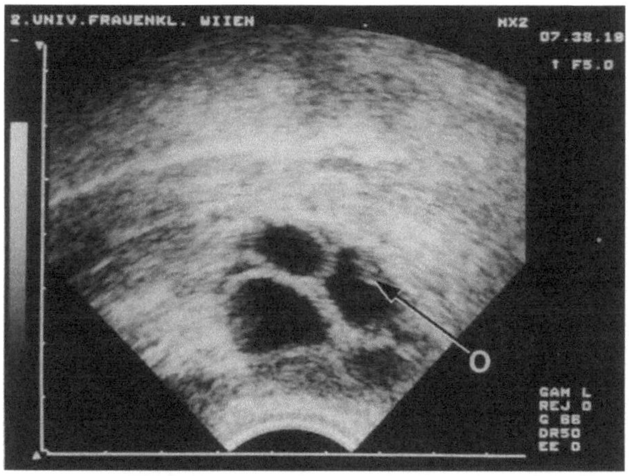

Fig. 3. Appearance of a cumulus oophorus *(O)* on a vaginal scan

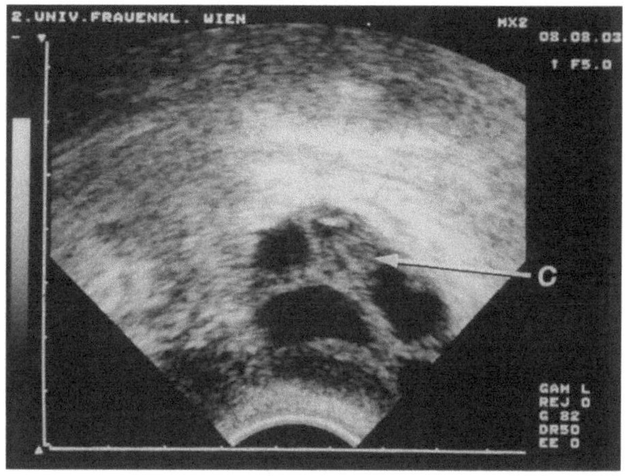

Fig. 4. Appearance of a corpus luteum *(C)* in a stimulated ovary. The essentially echo-free preovulatory follicle is transformed into a more echogenic structure

45 min the follicular fluid empties completely, and a corpus hemorrhagicum develops which appears sonographically as a cystic structure of variable echogenicity. In about 25% of ovulatory cycles, free fluid can be detected in the cul-de-sac following ovulation (Davis and Gosink 1986). A spoked-wheel pattern can frequently be demonstrated within the corpus hemorrhagicum. In some cases a corpus luteum cyst develops and resolves spontaneously with the onset of the next cycle.

Vaginosonography can provide a much better resolution of the internal ovarian structure (Granberg 1987). An ovary fixed in the cul-de-sac is also easier to demonstrate by the vaginal approach. In addition, ovulation can be established by the detection of free fluid in the cul-de-sac, and this is accomplished much more easily by the vaginal technique.

2.3 Corpus Luteum

Ovulation is succeeded by a fall in the serum estradiol level. Meanwhile, once luteinization has taken place, the theca cells of the developing corpus luteum begin to synthesize large amounts of progesterones. The mean length of the luteal phase in humans is 14 days (range 11.3–17.0 days) (Cooke 1988). The length of the luteal phase can be determined precisely with the aid of ultrasound monitoring. Once ovulation has occurred, a rapid change takes place in the sonographic appearance of the follicle. Sometimes the follicle can no longer be identified; alternatively, it may collapse and simulate the presence of two follicles (Hansmann et al. 1985). Frequently, internal echoes forming a spoked-wheel pattern can be demonstrated within the follicle itself (Hackelöer and Robinson 1978; Nitschke-Dabelstein et al. 1981). By day 16 of the physiologic cycle, increased solid components can be seen within the follicle, and by day 19 the area no longer appears cystic at all, its internal echoes resulting from the presence of blood, collagen, and new blood vessels (Fig. 4).

The corpus luteum is best appreciated sonographically in the midluteal phase and in hormone-induced cycles, appearing as an elliptical, hypoechoic feature with an echogenic rim. If pregnancy does not occur, the corpus luteum will resolve at least by the onset of menstruation. Persistence of the corpus luteum may be an early sonographic sign that nidation has occurred.

2.4 Cyclic Endometrial Changes

An important aspect of cycle monitoring besides folliculometry is the cyclic endometrial changes that are detectable by ultrasound (Fig. 5). Vaginosonography is superior to transabdominal sonography for assessment of the endometrium. Abdominal ultrasound fails to demonstrate the endometrium in 5% of examinations, usually because of marked obesity or a retroverted uterus (Fleischer et al. 1984). The vaginal technique, however, is not affected by uterine version (Fig. 6). A panoramic scanner has advantages over other scanner types for evaluating the endometrium. Only the panoramic scanner has a field of vision wide enough to portray a complete long-axis view of the uterus (Fig. 7).

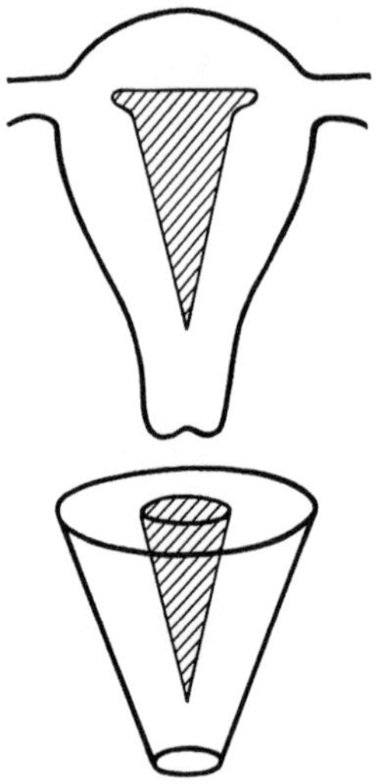

Fig. 5. *Top:* schematic representation of the sonographic appearance of the endometrium. *Bottom:* in reality the endometrium has the approximate shape of an inverted cone

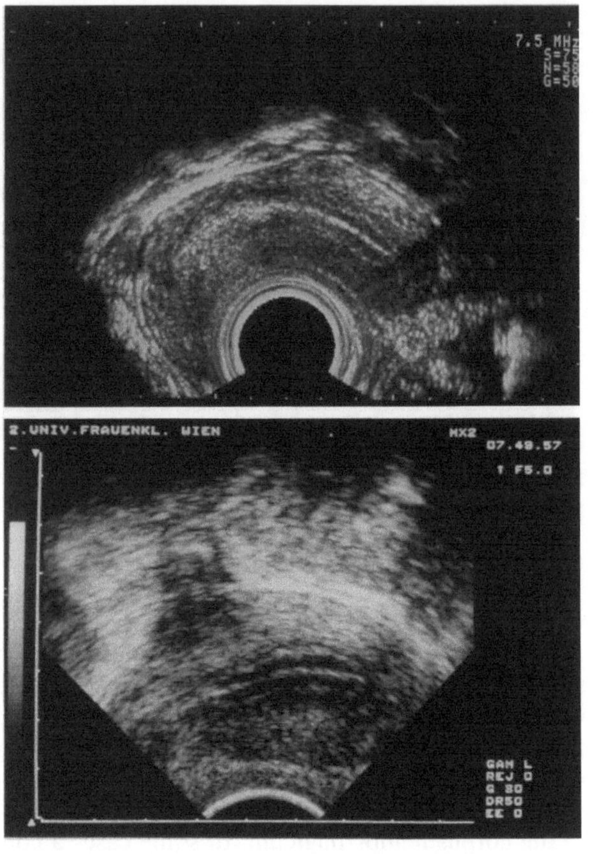

Fig. 7. Longitudinal scans through the uterus. The upper scan was made with a panoramic scanner, the lower scan with a curved array scanner. The complete long-axis view of the uterus clearly demonstrates the advantage of the wide-angle probe. *Top:* the endometrium appears streaklike in the early follicular phase. *Bottom:* near the time of ovulation the endometrial echo is widened, and the anechoic areas in the uterine cavity signal the onset of the secretory phase

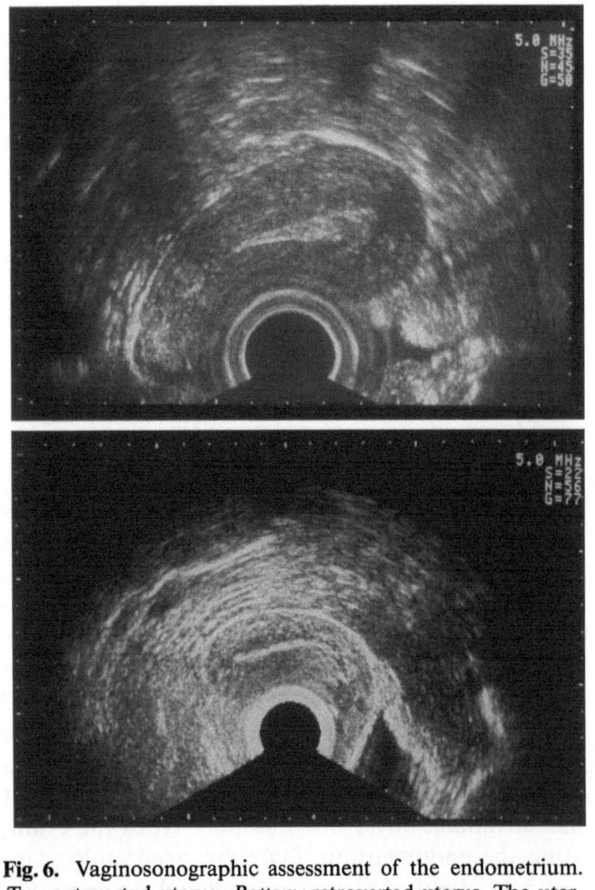

Fig. 6. Vaginosonographic assessment of the endometrium. *Top:* anteverted uterus. *Bottom:* retroverted uterus. The uterine version is not an obstacle to vaginal scanning

The endometrium is seen to undergo a proliferative and a secretory phase during the menstrual cycle. During the follicular phase the endometrium undergoes a thickening that is easily observed on daily sonographic folliculometry. The endometrium is composed chiefly of glands, secretions, and connective-tissue elements. Sonography early in the cycle shows a thin streak that disappears near the end of the first half of the cycle. Endometrial thickness and echogenicity increase until the postovulatory phase, when the endometrium becomes markedly less echogenic. A mucinous material is secreted that appears as a central, hypoechoic feature. Patients who undergo trials of various stimulation therapies in multiple cycles show a similar endometrial response pattern in each cycle (Fleischer et al. 1984). Because the endometrial response to different stimulation therapies is individualized, the protocol for

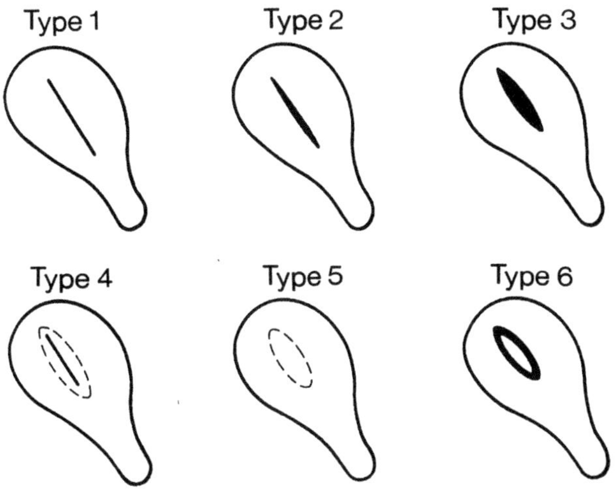

Fig. 8. The six endometrial types that can be seen in the nonstimulated monofollicular cycle (after Bald 1983). These types have been found to correlate with serum estradiol and serum progesterone levels

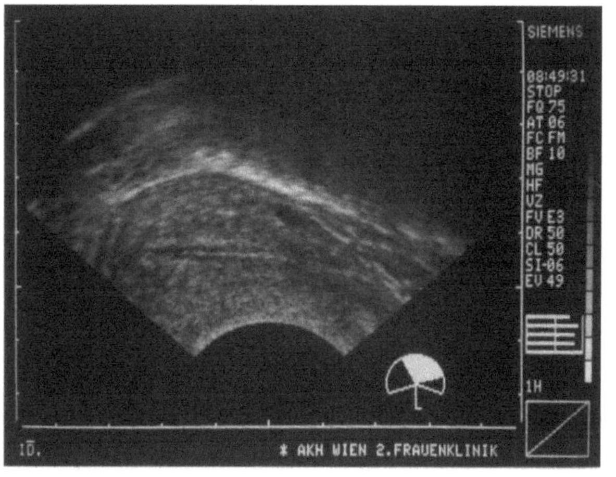

Fig. 9. Endometrium in the early cycle phase in a retroverted uterus (corresponding to type 1 in Fig. 8)

Fig. 10. Endometrium in the periovulatory phase, with a widened endometrial echo and the appearance of a loop figure. *Top:* longitudinal scans through the uterus; *bottom:* corresponding transverse scans. The images at left were made with a panoramic scanner, which demonstrates the entire uterine length ▼

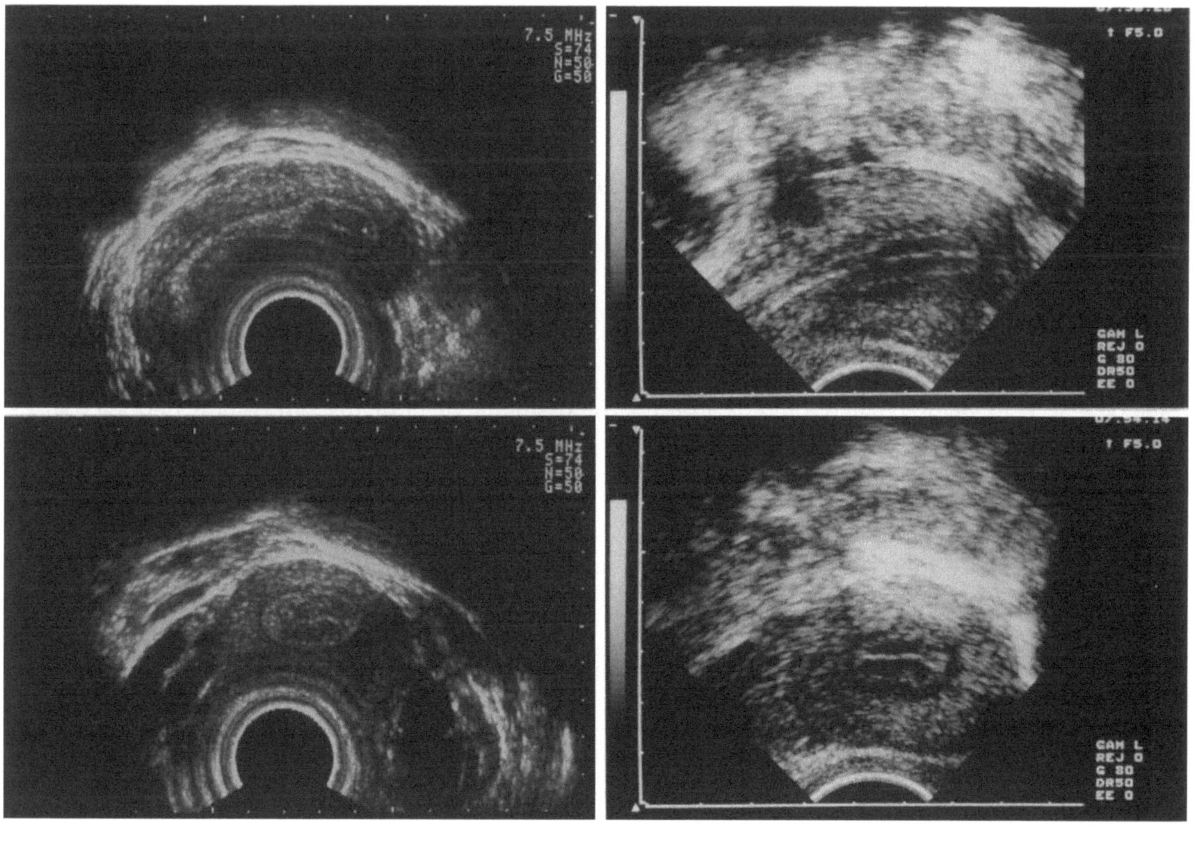

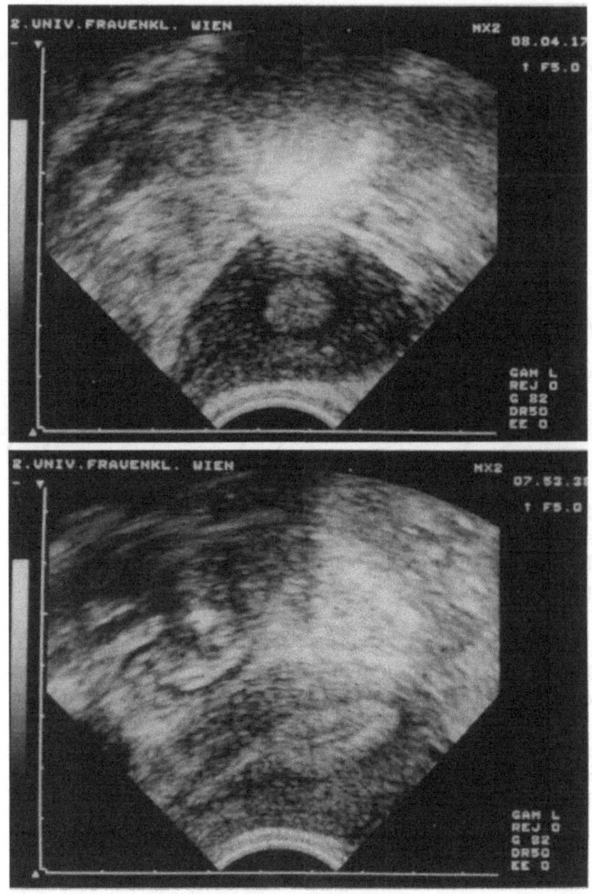

Fig. 11. Case in which pregnancy ensued following embryo transfer. *Top:* transverse scan trough the uterus prior to follicular aspiration. *Bottom:* longitudinal scan through the uterus

luteal phase support should be tailored to the individual patient. While a definitive assessment of the endometrium can be accomplished only by endometrial biopsy (Dallenbach-Hellweg 1988), most examiners prefer a less invasive approach based on serum progesterone determinations (Crosignani 1988) combined with ultrasound visualization of the endometrium, with particular attention given to the endometrial thickness.

Scanning of the endometrium should precede examination of the ovaries so that a knowledge of follicular sizes will not bias assessment of the endometrial phases (Hansmann et al. 1985). Data in the literature are contradictory regarding the sonographic appearance of the endometrium in relation to hormone levels. Thus, while some authors found no correlation between the thickness of the endometrial echo and hormone values (Fleischer et al. 1984, 1986; Glissant et al. 1985), other studies have shown a correlation with serum estradiol as well as serum progesterone (Bald 1983; Deichert et al. 1986). Different forms of stimulation might be responsible for these discrepancies.

As many as six different patterns in the sonographic appearance of the preovulatory endometrium have been described in the nonstimulated, monofollicular cycle (Fig. 8) (Bald 1983). These changes correlate both with follicular size and with serum estradiol and progesterone levels. Type 1 is characterized by a narrow endometrial line (Fig. 9) that widens during the proliferative phase (types 2 and 3) and finally expands into a loop figure (type 4). The ring-like echo appearing in the peri- and postovulatory phases (types 5 and 6) could be caused by mucinous secretions (Fig. 10). This ring (type 6) resembles the pseudogestational sac of ectopic pregnancy, is highly unstable, and may represent the earliest sign of ovulation (Hansmann et al. 1985). This hypothesis is strengthened by the fact that no such ring has been seen in patients with LUF syndrome (see Sect. 3.2 below), where there is a failure of ovulation.

Four different sonographic patterns of endometrial response have been described in stimulated cycles (Smith et al. 1984): Type A endometrium shows a greater echogenicity than the myometrium and consequently appears brighter on the gray-scale image. Type B endometrium is comparable to the myometrium in its echogenicity, with an initial hypoechoic halo separating the endometrium from the myometrium. Type C endometrium appears darker (less echogenic) than the myometrium, and type D is characterized by an echo-free region with a prominent, central midline echo. The "quality" of the endometrial preparation for nidation decreases from A to B, although so far there is no evidence of a correlation with implantation rates. The type B pattern is most common (approximately 75%) and is associated with higher serum progesterone levels.

While there has been considerable interest in the possibility of applying ultrasound endometrial assessment to predicting the success of embryo transfers, different endometrial configurations have not been found to correlate with implantation rates (Fleischer et al. 1986; Thickman et al. 1986).

The most important aspect of endometrial assessment is the measurement of endometrial thickness. This parameter increases continuously during the follicular phase until about the midluteal phase (Fig. 11). If the endometrial thickness does not reach at least 0.8 mm during stimulation, it appears that the likelihood of subsequent implantation is reduced (Mettler 1988). Endometrial thickness decreases from the midluteal phase onward. The increase in endometrial thickness is least pronounced following stimulation with clomiphene citrate (Fleischer 1988).

When implantation occurs, the thickness of the endometrial echo continues to increase even beyond the midluteal phase (Imoedemhe et al. 1987).

All studies performed to date indicate that assessment of the endometrium by ultrasound is of limited value as an adjunct to cycle monitoring (Giorlandino et al. 1987). It is possible that ultrasound scanning of the endometrium will assume greater importance through the improved resolution of vaginosonography. Preliminary data obtained by means of vaginosonography suggest that the amount of endometrial growth during ovarian hyperstimulation and the endometrial thickness on the day before oocyte retrieval could serve as possible parameters for implantation (Gonen et al. 1989).

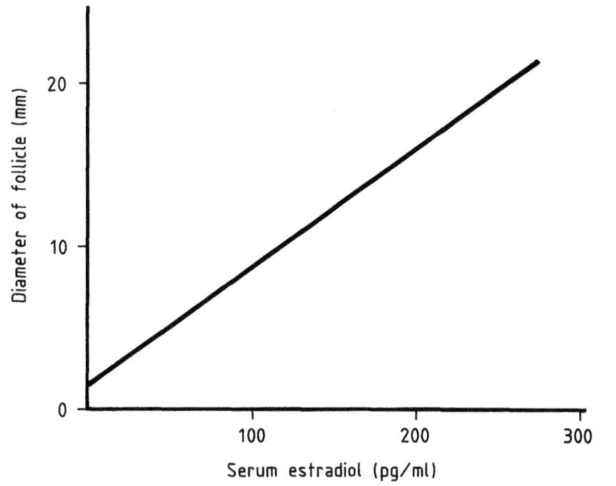

Fig. 12. Correlation of growth of the lead follicle with serum estradiol levels. (After Hackelöer et al. 1979)

2.5 Correlation of Folliculometry with Hormonal Parameters

In nonstimulated cycles there is a linear relationship between the size of the dominant follicle and the day of the cycle. In a 28-day cycle the diameter of the leading follicle is approximately 24 mm at the time of ovulation. The size of the leading follicle also correlates with the serum estradiol level (Hackelöer et al. 1979) (Figs. 12, 13), and even in stimulated cycles a link has been demonstrated between hormone levels and follicular size (Fig. 14) (Vargyas et al. 1982).

A comparison of the results of vaginal ultrasound folliculometry with serum assays of LH and estradiol, LH urinary assays, daily basal body temperature (BBT) measurements, and postovulatory serum progesterone assays showed that in cases where a serum LH value of 40 mU/ml was measured, ovulation occurred the next day in 84% of cases (Vermesh et al. 1987). In 97% the serum estradiol level peaked before the LH surge. A preovulatory rise of urinary LH was found in 87.5% of the women examined. However, daily vaginosonography was able to confirm the occurrence of ovulation in all cases. The dominant follicle could no longer be discerned. The great advantage of vaginosonography is that it can identify the exact day on which ovulation occurs, whereas the rise of serum progesterone or BBT can detect ovulation only retrospectively.

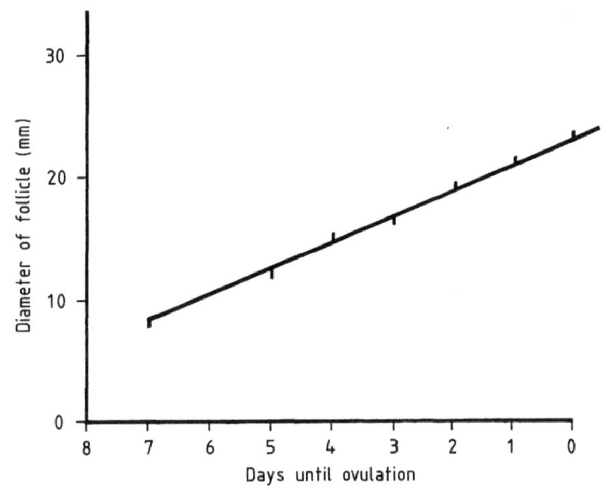

Fig. 13. Growth of the lead follicle at the time of ovulation. (After Bryce et al. 1982)

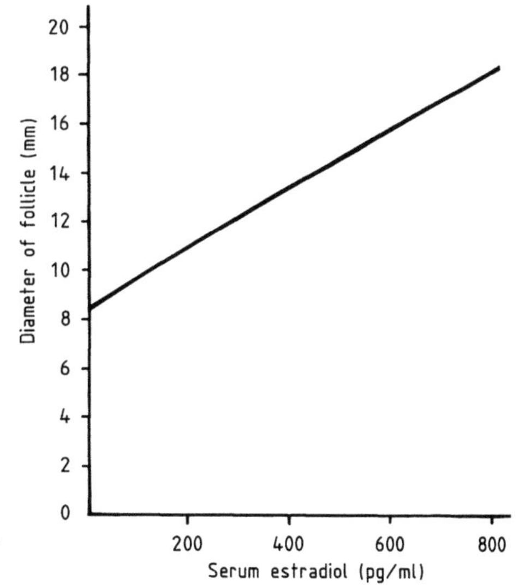

Fig. 14. Correlation of follicular size and serum estradiol levels in the stimulated cycle. (After Vargyas et al. 1982) ▶

2.6 Doppler Measurements of Blood Vessels in the Lesser Pelvis

The calibers of the ovarian blood vessels may change significantly during the course of the menstrual cycle (Hansmann et al. 1985). There are isolated reports on transabdominal flow measurements of the uterine artery (DuBose et al. 1985) and vaginal flow studies of the ovarian and uterine arteries in nonpregnant women (Taylor et al. 1985). So far only continuous-wave transducers have been used for vaginal Doppler investigations (Schulman et al. 1986).

With the recent development of a vaginal probe for pulsed Doppler scanning, it has become possible to detect and record cycle-dependent changes of blood flow velocity in the vessels of the lesser pelvis, most notably the ovarian artery (Deutinger et al. 1987 c). Details on the technique of vaginal Doppler flowmetry are given in the chapter "Vaginal Doppler Techniques". In our own studies, we found that as the number of follicles over 14 mm increased, there was a decrease in the average A/B ratio in the ovarian artery from 3.85 with two follicles to 2.71 with five follicles. Rising serum estradiol levels are likewise associated with a decrease in the average A/B ratio (Deutinger et al. 1989). The fall of the A/B ratio is caused by an increase in the diastolic flow velocity, which appears to be caused in turn by hormonal influences. Proliferative processes during the follicular phase are associated with hyperemia in the affected organs. Hormonal stimulation induces the maturation of multiple follicular vesicles, causing the ovarian volume to expand. This not only alters the blood distribution within the ovary but also changes the rate of blood flow through the ovary as a whole. Both the expanded ovarian volume and the change in the physical parameters of the blood vessels lead to a change in the perfusion profile (Fig. 15). The decreased vascular resistance during the follicular phase leads to a decline in the A/B ratio due to an increase in the diastolic flow velocity. Given the

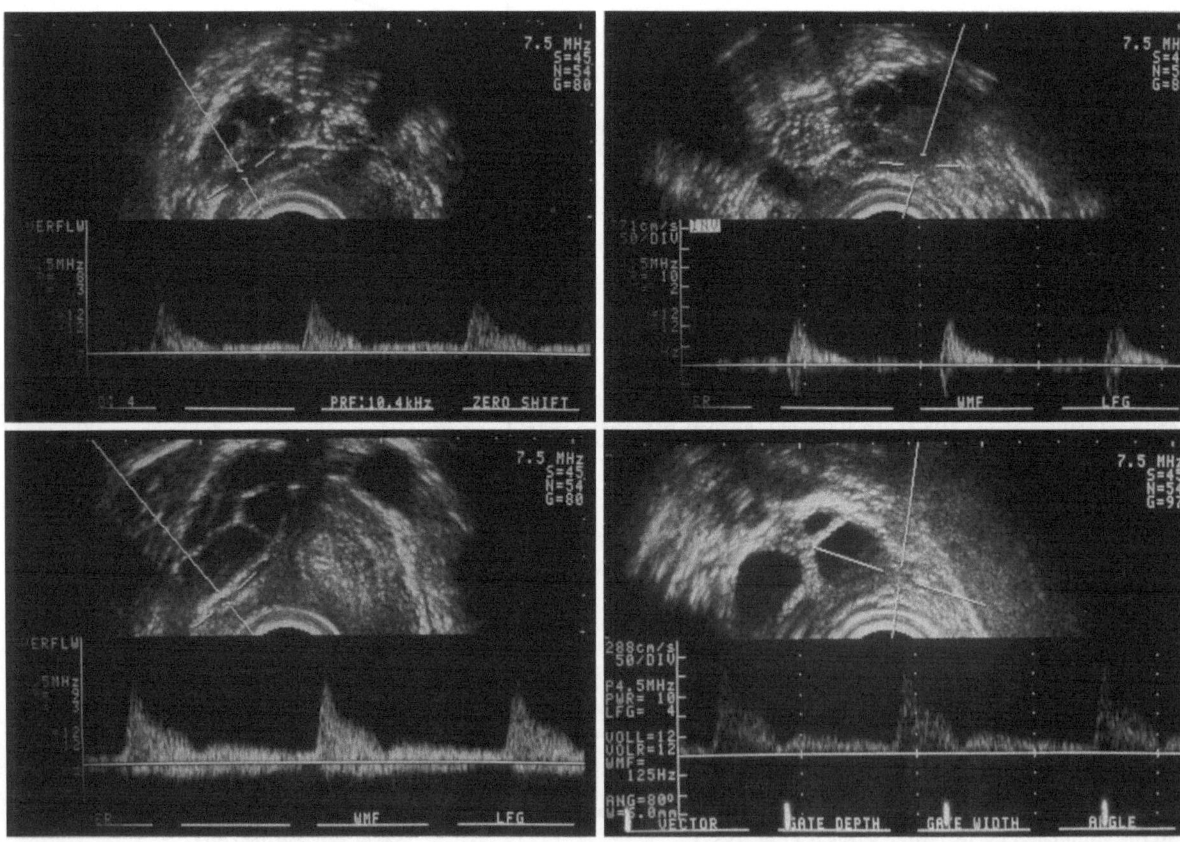

Fig. 15. *Top:* vaginal pulsed Doppler scanning of the ovarian artery on day 8 of a stimulated cycle. The right ovarian artery is shown on the left, and the left artery on the right. The diastolic blood flow velocity is very low or undetectable. *Bottom:* two days before follicular aspiration, the diastolic blood flow velocity increases

wide variation that we found in A/B ratios for the ovarian artery in stimulated cycles, the interpretation of individual findings can be difficult. Nevertheless, we were able to demonstrate a rise of the diastolic flow velocity close to ovulation in our study population, although we found no cycle-dependent flow changes in the iliac artery. The method of vaginal pulsed Doppler flowmetry of the ovarian artery clearly represents an additional option for cycle monitoring whose capabilities must still be verified by studies in larger populations posing a range of diagnostic problems.

3 Endocrine Disorders

3.1 Corpus Luteum Insufficiency

Luteal phase defects are manifested clinically by a short second cycle half (<11 days), decreased progesterone production, and infertility. The reduced progesterone production is expressed in low serum progesterone levels and/or a shortened period of progesterone synthesis. Endometrial biopsy is the classic way to diagnose corpus luteum insufficiency, as the low progesterone levels do not allow for adequate secretory transformation of the endometrium. Due to its invasiveness, this method may be used only if conception fails to occur. If implantation does not occur after in vitro fertilization, a luteal phase defect is causative in 28%-76% of cases (Graf et al. 1988).

Determination of the plasma progesterone level has become the standard technique for monitoring of the luteal phase. Corpus luteum insufficiency may be excluded if the progesterone level exceeds 10 ng/ml in a single assay or ranges from 12-13 ng/ml in serial assays performed from the 5th to 8th day after ovulation (Crosignani 1988).

The insufficient priming of the endometrium interferes with implantation and leads to a decreased pregnancy rate. If implantation does occur, early abortion may result due to the insufficient production of progesterone.

Ultrasound scanning may reveal ruptured follicles of normal size, ruptured small follicles, or luteinized unruptured follicles (LUF syndrome). The predominant feature, however, is an abnormal folliculogenesis: the development of multiple follicles of insufficient size could be demonstrated in 39% of patients with unexplained infertility (Ying et al. 1987). These small follicles are unable to produce sufficient estradiol or induce pituitary LH secretion, resulting in inadequate luteinization. The frequency of luteal phase defects is increased following clomiphene stimulation. Low serum progesterone levels and a high estradiol/progesterone ratio are associated with a low implantation rate following in vitro fertilization (Lejeune et al. 1986). The efficacy of administering progesterone in the second half of the cycle remains controversial. The nature of the previous cycle stimulation is a major factor determining the form, duration, and dosage of the replacement therapy (van Steirteghem et al. 1988). Other treatment options prior to ovulation are the administration of low doses of clomiphene or hMG, with clomiphene citrate showing particular efficacy in prolonging the corpus luteum phase (Dlugi et al. 1985). hCG also may be administered following ovulation. hCG prolongs progesterone production, although an increase in pregnancy rates has not been observed following in vitro fertilization (Buvat et al. 1988). Often the administration of bromocriptine through the cycle will lead to an ovulatory cycle. Sonography provides a means of checking the therapeutic response and altering the treatment regimen as may be required.

3.2 LUF Syndrome

Luteinized unruptured follicle (LUF) syndrome is said to be present when a follicle is transformed into a corpus luteum without ovulating (Fig. 16). The incidence of this syndrome is uncertain, with reports ranging from 5% to 29% (O'Herlihy et al. 1980b; Liukkonen et al. 1984). It is believed to be

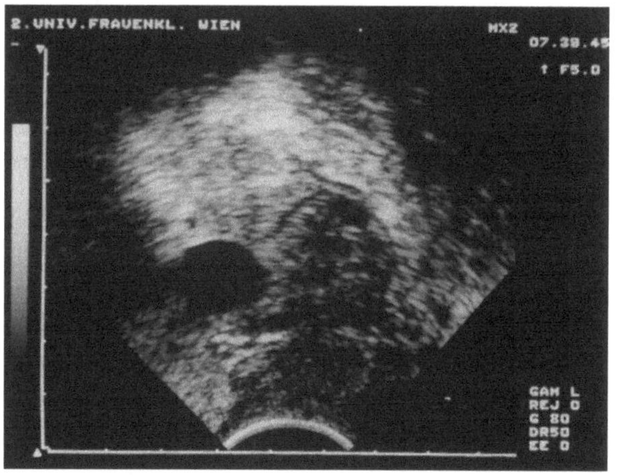

Fig. 16. Right ovary of a patient with infertility due to LUF syndrome on cycle day 18 (29-day cycle length). Vaginosonography at this time demonstrated a standing follicle

present in about 10% of women with unexplained infertility (Daly et al. 1985). LUF syndrome is diagnosed with greater frequency in patients with endometriosis and may account for the increased infertility rate in that condition (Donnez and Thomas 1982; Koninckx 1983). The BBT shows a biphasic course, while LH and progesterone production are normal, so that ovulation appears to have occurred based on standard endocrinologic criteria (Eissa et al. 1987). However, it can be shown laparoscopically that the follicle remains unruptured 4 days after the spontaneous LH surge, and that it has not undergone transformation to a corpus hemorrhagicum (Bernardus et al. 1983). The presence of LUF syndrome can be demonstrated sonographically (Liukonnen et al. 1984; Davis and Waskey 1987), but the frequency with which LUF is diagnosed by ultrasound alone appears to be low. Perhaps the detection of this syndrome can be improved by a twice-daily schedule of follicular monitoring (Janssen-Caspers et al. 1986 a).

Before the advent of sonography, suspicion of LUF syndrome could be confirmed only by the inability to demonstrate a ruptured follicle by laparoscopy 2-4 days after the time of ovulation. This procedure is not only invasive but also leads to a high rate of false-positive diagnosis (Daly et al. 1985).

In ultrasound scanning of the endometrium, a transient, central ring resembling the pseudogestational sac of ectopic pregnancy can be seen in the uterine cavity following ovulation. So far this central ring has not been demonstrated in anovulatory cycles and therefore may serve as an additional sonographic criterion for LUF syndrome (Hansmann et al. 1985).

When diagnosed, LUF syndrome is commonly treated by the administration of clomiphene citrate in a subsequent cycle. However, in vitro fertilization has become the therapy of choice (Liukonen et al. 1984). A newer approach is the GIFT procedure (gamete intrafallopian transfer), which appears preferable owing to the higher pregnancy rate.

3.3 PCO Syndrome

Polycystic ovarian (PCO) syndrome typically appears before 35 years of age. It is characterized clinically by oligo- or amenorrhea, hirsutism, obesity, and infertility, although 25% of patients present no clinical symptoms.

The disease has a primary form, known as Stein-Leventhal syndrome, or it may occur secondarily in the setting of Cushing's disease (congenital adrenocortical hyperplasia), hyper- and hypothyroidism, or as a sequel to hyperprolactinemia. Sonography does not contribute to differential diagnosis in the secondary forms (Parisi et al. 1984). Pathophysiologically the syndrome is based on a functional disturbance of the ovary that may or may not involve observable changes in gonadotropin and androgen metabolism. This underscores the value of direct sonographic diagnosis.

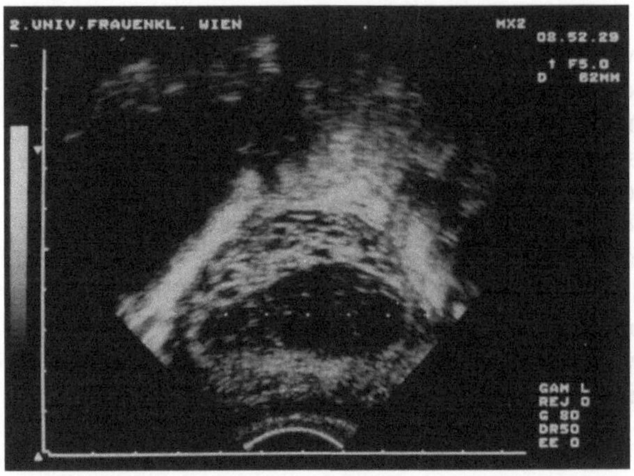

Fig. 17. Right ovary of a patient with PCO syndrome. The ovary is enlarged to 62 mm and presents a mixed hyperechoic and hypoechoic pattern

In 75% of cases sonography demonstrates markedly enlarged ovaries with numerous small cysts. The volume of the ovaries averages 14 cm^3, compared with the normal volume of 6 cm^3. The uterus is smaller than normal.

Four different sonographic ovarian patterns are observed in PCO syndrome: (1) enlarged ovaries with a predominantly cystic structure, (2) enlarged ovaries with a predominantly solid structure, (3) normal-sized ovaries with a predominantly cystic structure, and (4) normal-sized ovaries with a predominantly solid structure (Orsini et al. 1985). When the ovaries are enlarged, it is more common to find elevated testosterone and dehydroepiandrostenedione levels in conjunction with obesity, amenorrhea, hirsutism, and hyperprolactinemia (Fig. 17). When the ovaries are normal-sized, it is more common to find an elevated LH/FSH ratio as well as high androstenedione and estrogen levels in association with oligomenorrhea (El Tabbakh et al. 1986). The biologically active testosterone fraction is markedly elevated in obese patients (Laatikainen et al. 1983). The size of the cysts is extremely variable, and both microcystic and macrocystic forms of the disease are recognized. Most cysts are smaller than 6 mm, which is why transabdominal scanning is of little value. Other limiting factors in abdominal scanning are obesity and the ovarian location, which can vary considerably as a function of bladder distention and rectal fullness (Orsini et al. 1985).

Because the features of polycystic ovaries are so variable, sonography is poorly suited as a screening examination. Nevertheless, sonography has a definite role in the diagnosis of PCO syndrome (Kun and Bösze 1973; Swanson et al. 1981; Parisi et al. 1982; Nicolini et al. 1985; El Tabbakh et al. 1986), and vaginosonography could enhance that role further owing to the superior resolution of the 7.5-MHz transducer.

The treatment of PCO syndrome consists in hormonal treatment (administration of antiandrogens, cortisone, and clomiphene citrate) or in surgical wedge resection or partial laparoscopic resection. Response tends to be best in cases with ovarian enlargement. Treatment is frequently successful in producing ovulation, and it is common for ovulatory cycles to persist after therapy is discontinued (Venturoli et al. 1985).

4 In Vitro Fertilization

4.1 General

The birth of the first "test-tube" baby in 1978 attracted worldwide attention (Steptoe and Edwards 1978). Shortly thereafter the method of in vitro fertilization was introduced at many infertility clinics, and so far approximately 8000 babies have been successfully conceived by that method and carried to term. Initially the procedure was difficult and costly, but the individual steps involved in extracorporeal fertilization and embryo transfer were gradually simplified until today it is performed as an outpatient procedure at many centers.

The successful conduct of in vitro fertilization without the use of ultrasound is hard to imagine. Sonography is used to monitor follicular maturation in stimulated cycles. It is also used to confirm the presence and location of the follicles so that subsequent follicular aspiration can be successfully accomplished (De Crespigny et al. 1981 a). Whereas laparotomy and laparoscopy were used for oocyte retrieval in the early days of in vitro fertilization, sonographically guided follicular aspiration has now become the procedure of choice.

In vitro fertilization was introduced at the Second University of Vienna Women's Hospital in 1981 and led in 1982 to the birth of the first test-tube baby in Austria (Feichtinger et al. 1981). Since then, this form of infertility therapy has enabled many infertile couples to have their own children (Deutinger et al. 1985).

4.2 Stimulation Therapy

Modern infertility therapies include various forms and regimens of hormonal replacement (Hoult et al. 1981; Buttery et al. 1983; Reinthaller et al. 1985; Tal et al. 1985). The goal of all these treatments is to induce the maturation of multiple follicles in one cycle. Pregnancy rates can be improved by the aspiration of multiple oocytes and by the transfer of multiple embryos (Speirs et al. 1983). Ultrasound follicular monitoring is of special importance in patients receiving stimulation therapy, because the daily measurements of hormone levels (estradiol, progesterone, LH) are meaningful only when interpreted in relation to the number and size of the ovarian follicles. Reports vary on the presence of a dominant follicle and the size of the follicles in stimulated cycles. Regarding the presence of a dominant follicle, it is reasonable to assume that when increased hormone doses are administered for induction, it is incorrect to speak of a dominant follicle since the maturation of multiple follicles is induced. There is considerable disagreement in reports on follicular sizes, with some authors stating that induced follicles are smaller than spontaneous follicles and others claiming that there is no difference or that induced follicles are larger (Ylöstalo et al. 1979; O'Herlihy et al. 1980b; Marinho et al. 1982; Eissa et al. 1986). These discrepancies may result from different forms of induction therapy or measurements performed at different times.

When clomiphene citrate alone is used for induction, it is administered between cycle days 5 and 9 in the 28-day cycle. Sonographic monitoring should be initiated no later than cycle day 10 or 11. When the lead follicle has reached a diameter of 20-24 mm, ovulation can be induced by the injection of hCG, an analog of LH. An appropriate follicular size alone does not necessarily portend a successful oocyte retrieval, because stimulation with FSH can cause the appearance of follicle-like hypoechoic areas in the ovary that do not contain oocytes (Ashkenazi et al. 1987). Only the prior sonographic detection of a cumulus oophorus or its dissociation from the surrounding granulosa tissue can establish the presence of an oocyte of appropriate maturity (Fleischer et al. 1981; Hackelöer 1984;

Roberts et al. 1985). However, this can be accomplished with traditional abdominal ultrasound in only 20% of cases (Ritchie 1985). This is an area where vaginal scanning offers significant advantages, because a cumulus can be successfully identified with a 6.5-MHz vaginal probe in two-thirds of cases (Drugan et al. 1988). This rate can be improved even further by the use of higher transducer frequencies. Ovulation occurs approximately 36 h after the spontaneous LH surge or administration of hCG.

Potent alternatives to clomiphene citrate stimulation include the injection of hMG, pure FSH, or GnRH. These costly therapies call for very precise monitoring due to the greatly increased risk of hyperstimulation and multiple pregnancy. While the introduction of ultrasound follicular monitoring has not led to a significant increase in pregnancy rates, it has decreased the incidence of hyperstimulation and allowed reduction of the hMG dose, which in itself is a major advance in view of the high cost of this medication (Dornbluth et al. 1983). When hMG stimulation is used, daily determinations of serum estradiol, serum progesterone, urinary LH, and follicular measurements should be commenced 3 or 4 days after the start of therapy (i. e., on menstrual day 7 or 8), because follicular growth is initiated an average of 2 days earlier in stimulated cycles than in spontaneous cycles (Rossavik and Gibbons 1986). hCG is administered when the lead follicle reaches a diameter of 20 mm and the estradiol values per follicle have reached 400 pg/ml serum. This is followed 36 h later by coitus, insemination, or follicular aspiration as a prelude to in vitro fertilization, depending on the cause of the infertility.

In stimulated cycles the calculated total follicular volume in both ovaries shows a strong correlation with the preovulatory serum estradiol levels, whereas the size of the putative lead follicle shows no correlation with estradiol levels (Hull et al. 1986). The total follicular surface area also correlates with the estradiol level (Eissa et al. 1986). However, the calculation of follicular volume and surface area is too cumbersome for routine folliculometry.

Given the great variety of induction regimens that are used and the fact that stimulated follicles (especially those induced with hMG) can produce varying amounts of estradiol regardless of their size, the value of sonographic folliculometry in stimulated cycles may be characterized as follows: The number of follicles can be determined by ultrasound, but the "quality" of those follicles is reflected in the estradiol level (DeCherney et al. 1985). Daily determinations of both parameters - number of follicles and estradiol level - and the determination of serum progesterone and urinary LH values make it possible to assess the dynamics of the follicular phase and enable the experienced examiner to determine the optimum time for conception. Clinical parameters such as the cervical score should have only an adjunctive value when these costly induction procedures are used, and rarely should they be applied in isolation (Oelsner et al. 1986).

Ultrasound ovarian scanning is essential before induction therapy is reinitiated, as there is a 10%-15% incidence of residual cysts following cycle stimulation. These cysts always resolve spontaneously, but induction therapy should not be reinitiated while such cysts are present.

4.3 Determining the Time of Ovulation

Determining the time of ovulation is an important step in the management of female infertility. Before the advent of sonography, ovulation could be timed only indirectly by clinical means such as evaluating the cervical mucus or the width of the cervical os, or later by hormone assays.

Ultrasound scanning of the ovaries can be repeated as often as desired and thus provided a means for the direct, daily measurement of follicular growth in spontaneous and induced cycles (Hackelöer et al. 1977; Hackelöer and Robinson 1978; Queenan et al. 1980; Fleischer et al. 1981; Mantzavinos et al. 1983; Ritchie 1985). It is less costly and more convenient than endocrinologic monitoring. There is diversity of opinion, however, as to whether sonography is sufficient in itself to determine the time of ovulation, or whether it must be combined with hormone assays (Robertson et al. 1979; Hoult et al. 1981). In any case it is clear that sonographic folliculometry is superior to clinical examination and BBT charts in establishing the time of ovulation (Hackelöer et al. 1980; Paulson et al. 1984; Leader et al. 1985). Laparoscopy has confirmed the accuracy of sonography in the measurement of follicular diameters (O'Herlihy et al. 1980 a).

Because follicular growth is linear in the spontaneous, nonstimulated cycle, it is possible to predict the time of ovulation at a very early point in the cycle using a mathematical model (Doody et al. 1987). Frequently, however, the largest follicle is not the one that ovulates, a fact which can complicate prediction of the time of ovulation based on ultrasound findings alone (Vermesh et al. 1987). But while views differ regarding the value of cycle monitoring by ultrasound alone, it has been shown that pregnancy can be achieved following in vitro fertilization when this type of monitoring is used (Hoult et al. 1981). The best time for predicting ovulation in the spontaneous cycle is the point where the lead follicle has reached a diameter of 20 mm; ovulation will then occur in 1.4 ± 1.2 days. Bryce et al. (1982) found that concurrent daily assays of LH, FSH, and estradiol did not improve the accuracy of the ovulation prediction.

A comparison of the results of vaginosonographic folliculometry with determinations of serum LH and estradiol, urinary LH, BBT charting, and postovulatory serum progesterone has shown that in cases where a serum LH value of 40 mU/ml was measured, ovulation occurred the next day in 84% of cases (Vermesh et al. 1987). In 97% of cases the serum estradiol level peaked before the LH surge. A preovulatory rise of urinary LH was found in 87.5% of the women examined. However, daily vaginosonography was able to confirm in all cases that ovulation had occurred. The dominant follicle could no longer be visualized. The main advantage of this method is that it can identify the exact day on which ovulation occurs, whereas the rise in serum progesterone or BBT can detect ovulation only retrospectively.

Thus, when we consider the question of patient convenience and the costs of stimulation therapy, we can appreciate the need to provide the most effective cycle monitoring available. Ultrasound scanning and serum estradiol are equivalent and mutually complementary studies for determining the time of ovulation (Robertson et al. 1979).

4.4 Inadequate Stimulation Therapy

4.4.1 Hyperstimulation

A crucial aspect of ultrasound follicular monitoring is the avoidance of ovarian hyperstimulation syndrome. This refers to a complication or side effect of stimulation therapy that is most commonly observed in patients receiving gonadotropins. It is characterized by an uncontrolled growth of numerous follicles from 2 to 6 mm in diameter at the start of the cycle. Hyperstimulated ovaries quickly develop features similar to those observed in polycystic ovarian syndrome, i.e., the ovaries contain large numbers of variable-size follicles exhibiting cystic change (Orsini et al. 1985). In very severe cases there may be extreme ovarian enlargement with the development of ascites, hydrothorax, hemoconcentration, and oliguria. If a dominant follicle larger than 20 mm and several secondary follicles are present at the time of hCG administration in the induced cycle, there is a greatly increased likelihood that hyperstimulation syndrome will occur (Tal et al. 1985; Blankstein et al. 1987). On the other hand, the pregnancy rate declines significantly if the follicular diameter is less than 23 mm at the time of follicular aspiration.

Sonographically, the hyperstimulated ovary presents a typical spoked-wheel pattern with numerous follicles of varying size (Fig. 18). Again, vaginal scanning offers significant advantages for the monitoring of these patients since the empty-bladder technique eliminates the compression effects that can distort measurements. Before the advent of sonography, serum estradiol assays and palpation were the only means available for avoiding hyperstimulation, and before vaginosonography was introduced measurement of serum estradiol was still superior to ultrasound for this purpose (Haning et al. 1983). Despite the ability of sonography to establish the optimum timing for hCG administration, the incidence of hyperstimulation syndrome can be reduced but cannot be entirely eliminated (Blankstein et al. 1987). In milder forms the pregnancy rate is equivalent to that in patients responding normally (McArdle et al. 1983; Haning et al. 1985). Indeed, pregnancies have even been observed in pronounced forms of hyperstimulation syndrome.

Ordinarily hCG is administered in the gonadotropin-stimulated cycle when the serum estradiol level is between 200 and 500 pg/ml per mature follicle. However, this rule applies only to monofollicular cycles or cases in which there are dominant follicles of similar size. If there is also growth of multiple smaller follicles, it is not possible to estimate the level of estradiol production in the individual follicles.

Severe hyperstimulation syndrome is managed by tracking the cyst sizes with ultrasound and by monitoring renal function. Surgical intervention is seldom necessary and generally is not advised due to the risk of uncontrolled hemorrhage. Approximately 500 cycles have been stimulated with gonadotropins at the Second University of Vienna Women's Hospital during the past 5 years. Significant hyperstimulation occurred in four of these cases, and two of the patients had to be hospitalized. No patients required surgical intervention.

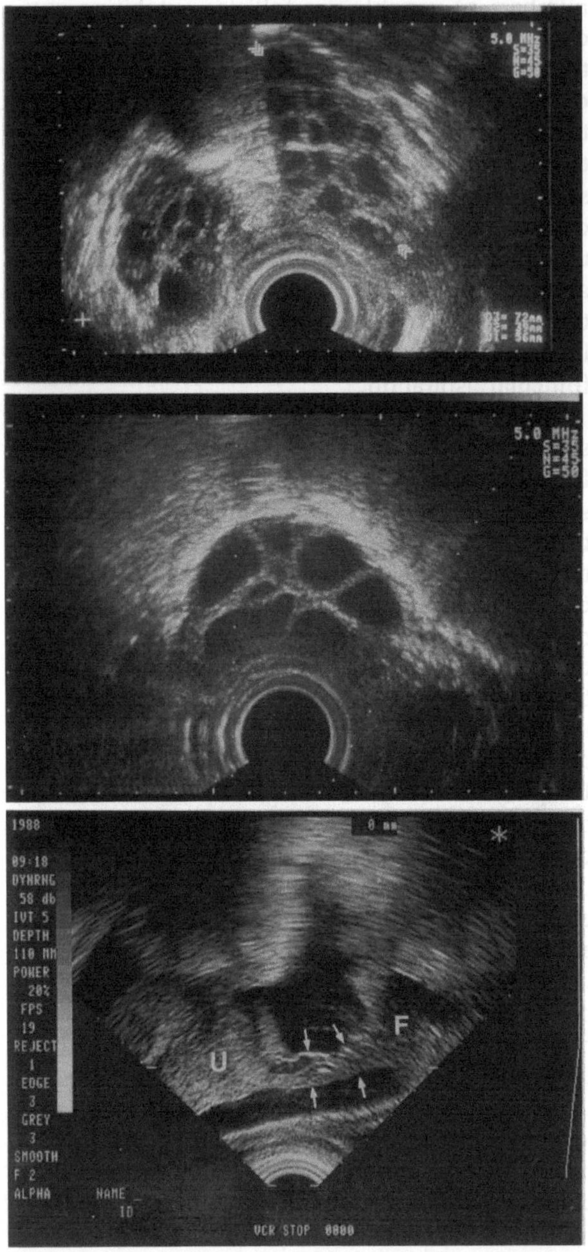

Fig. 18. *Top:* horizontal scan through both ovaries in hormonal hyperstimulation. The left ovary is enlarged to 72 × 56 × 39 mm. *Center:* cycle day 7 after hormonal stimulation (started on day 2 with two ampules of hMG daily). Multiple, apparently mature follicles are seen even at this early stage. *Bottom:* after hormonal hyperstimulation, free fluid is demonstrated in the cul-de-sac. The presence of the fluid allows visualization of the left fallopian tube (*U,* uterus; *F,* fibriae; *small arrows,* course of tube)

4.4.2 Insufficient Stimulation

The evaluation of intraovarian structures is a valuable aid in the diagnosis of insufficient or anovulatory cycles. Thus, it may be assumed that ovulation will take place only if a preovulatory cumulus oophorus can be demonstrated sonographically. In "low responders" who show an insufficient rise of serum estradiol levels, the daily follicular growth rate in reduced (Mantzavinos et al. 1983). The maximum follicular size in the insufficiently stimulated cycle is considerably smaller than in the spontaneous cycle (Geisthövel et al. 1983). Delayed follicular growth is associated with insufficient follicular maturation and with a significant decrease in the maximum diameter of the dominant follicle. As a result, the dominant follicle fails to ovulate and becomes atretic. This leads to an insufficient, shortened luteal phase combined with irregular BBT readings and reduced serum progesterone levels. Falling serum estradiol values and/or decreasing follicular diameters after the administration of hCG are poor prognostic indicators of successful oocyte recovery and fertilization (Ben-Rafael et al. 1986).

4.5 Follicular Aspiration

The earliest attempts to obtain an oocyte for in vitro fertilization were made in spontaneous cycles, but the number of oocytes recovered by that method was too small. There was considerable inconvenience for the patient due to the need for frequent LH determinations, and also for personnel, who had to be available on a round-the-clock basis. Today follicular aspiration is performed almost exclusively in stimulated cycles approximately 34–36 h after the administration of hCG.

4.5.1 Laparotomy and Laparoscopy

The oocytes for in vitro fertilization were originally collected by laparotomy and later by a modified mini-laparotomy (Lopata et al. 1974). Once laparoscopy had established itself as a diagnostic and therapeutic procedure in gynecology, it was natural to apply the technique for oocyte collection (Steptoe and Edwards 1970; Feichtinger et al. 1981; Wood et al. 1981). Manipulations are performed under vision, so bleeding is easily detected and controlled. However, because many infertility patients have had previous surgery, there is a risk of injury to adherent bowel loops during insertion of the laparoscope. Dense adhesive bands, whether caused by surgery or endometriosis, can so envelop an ovary that it cannot be visualized by laparoscopy. A further objection is the necessity of intubation anesthesia and the potentially adverse effect of intra-abdominal gas on the fertilization of the retrieved oocytes.

4.5.2 Ultrasound-Guided Aspiration Methods

The foregoing disadvantages are eliminated by sonographically guided transvesical aspiration, which requires a decreased duration and depth of anesthesia. Indeed, local anesthesia is sufficient in many cases (Lenz and Lauritsen 1982). The procedure is less invasive than laparoscopy since it does not involve entry into the abdominal cavity. Injuries are unlikely to occur even when there is extensive adhesion of bowel to the abdominal wall, because the aspiration needle is inserted through the full urinary bladder under sonographic guidance. Another advantage of ultrasound guidance is that it enables the number and location of the follicles to be monitored continuously throughout the procedure.

The currently used abdominal sector scanners are smaller than earlier transducers so that, even in thin patients, visualization of the lesser pelvis is not hampered by the pelvic bones. With high-lying ovaries it may not be possible to image the bladder and follicles on the same scan, and ovaries located in the cul-de-sac may be difficult to puncture with precision. Moreover, like all methods except for laparoscopy, sonography does not allow for the direct identification of bleeding sites. The follicles are often difficult to define with ultrasound in extreme obesity, which also increases the distance the needle must travel to reach the target. A full bladder is required in this method.

Our oocyte recovery rate with the transabdominal method (2.0 per aspiration) was lower than that achieved by laparoscopy (Deutinger et al. 1987 d). The main reason for this is that transabdominal aspiration was used consistently in cases where laparoscopy did not seem promising or was considered too hazardous because of extensive adhesions. Possible complications of transabdominal follicular aspiration include bleeding, cystitis, and micro- or macrohematuria. But aside from three patients who complained of mild lower abdominal discomfort for several days, we observed no complications.

The introduction of ultrasound-guided follicular aspiration lowered the cost and invasiveness of oocyte recovery and reduced the anesthetic risk. Soon the aspirations were performed under local anesthesia, leading to further cost reduction by allowing the procedure to be performed in ambulatory patients.

Recently, ultrasound-guided follicular aspiration has assumed greater importance since it was demonstrated that, with experience, the success rate can not only be maintained (Lewin et al. 1985) but even improved (Feichtinger and Kemeter 1984). This method of oocyte collection has relied mostly on traditional transabdominal sonography with the aspiration performed: (1) through the bladder (Lenz et al. 1981; Lenz and Lauritsen 1982; Wikland et al. 1983; Bernaschek et al. 1984; Feichtinger and Kemeter 1984; Lenz 1984; Lenz and Bang 1986; Taylor et al. 1986; Lewin 1986 a, b; Janssen-Caspers 1986 b; Mao et al. 1986), (2) through the urethra (Parsons et al. 1985; Fateh et al. 1986; Riddle et al. 1987, or (3) by culdocentesis (Gleicher et al. 1983; Dellenbach et al. 1984, 1985; Schulman 1985 a). A needle guide is generally used for transabdominally guided aspirations (Lenz and Lauritsen 1982; Wikland et al. 1983; Trotnow et al. 1985; Habermann et al. 1986). Vaginal aspiration using a freehand technique under transabdominal ultrasound guidance has not yielded satisfactory results (Feichtinger and Kemeter 1986 b, 1987). Transurethral aspiration was developed at a time when researchers were seeking to simplify the procedure, and vaginal scanners were not yet available. The decisive breakthrough for ultrasound-guided follicular aspiration came with the development of vaginal transducers with an attached needle guide (Popp 1986).

4.5.3 Vaginal Follicular Aspiration Guided by Vaginosonography

Several groups of workers have reported independently on the development and use of vaginal transducers for oocyte retrieval (Popp and Müller-Holve 1984; Schwimer and Lebovic 1984; Meldrum et al. 1984; Enk and Wikland 1985; Kemeter and Feichtinger 1986). The rationale for vaginal ultrasound guidance is that the transducer can be positioned closer to the ovary, and a higher frequency can be used to provide improved resolution (Kratochwil 1969; Popp and Müller-Holve 1984; Schwimer and Lebovic 1984; Deutinger et al. 1986 b; Bernaschek and Deutinger 1988). Since the ovaries are usually located in the cul-de-sac region very close to the vaginal wall, there is also a rationale for aspirating the follicles via the vaginal route, which eliminates the problems of a "long-distance" puncture. A number of groups have already published their results with this technique and have reported on the development of new technical aids (Feichtinger and Kemeter 1986 a, b, 1987; Cohen 1986; Deutinger et al. 1986 c, 1987 b, d, e; Wikland 1987; Lenz et al. 1987; Hammarberg et al. 1987; Seifer et al. 1988).

The best type of transducer for vaginal follicular aspiration is the panoramic scanner, which provides the operator with a clear, immediate anatomic overview (Popp et al. 1985). The transducer frequency is 5 MHz, and the 240° forward view angle affords coverage of the entire lesser pelvis (see section on Scanner Types for further technical details). The patient is placed in the lithotomy position for the follicular aspiration. Immediately before the procedure the patient empties her bladder, and the vagina is cleansed with physiologic saline. The aspiration may be performed under brief, light general anesthesia or under local anesthesia with mild sedation; some patients require no analgesia at all. Further details may be found in the chapter "Endosonographically Guided Punctures."

The scanner is inserted into the vagina and initially positioned so that a sagittal image appears on the monitor. The probe is then rotated 90° to obtain a transverse scan of the lesser pelvis, which in many cases will demonstrate both ovaries concurrently. The transducer is equipped with an attached, fixed needle guide. Calibration is necessary only for initial use of the guide, because the calibration lines can be stored and redisplayed at any time. The aspiration needle has an overall length of 35 cm, an outside diameter of 1.4 mm, and an inside diameter of 1.2 mm. It is connected to a vacuum pump that develops a negative pressure of 180 mm Hg. If the initial aspirate does not contain an oocyte, the follicle, which becomes more conspicuous on the monitor as it fills with blood, is reaspirated (Fig. 19). We do not flush the follicle with culture medium. Because vaginal follicular aspiration rarely takes longer than 10 min, we prepare heparinized test tubes so that the follicular fluid, which is sometimes blood-tinged, will not coagulate until the microscopic search for oocytes is completed.

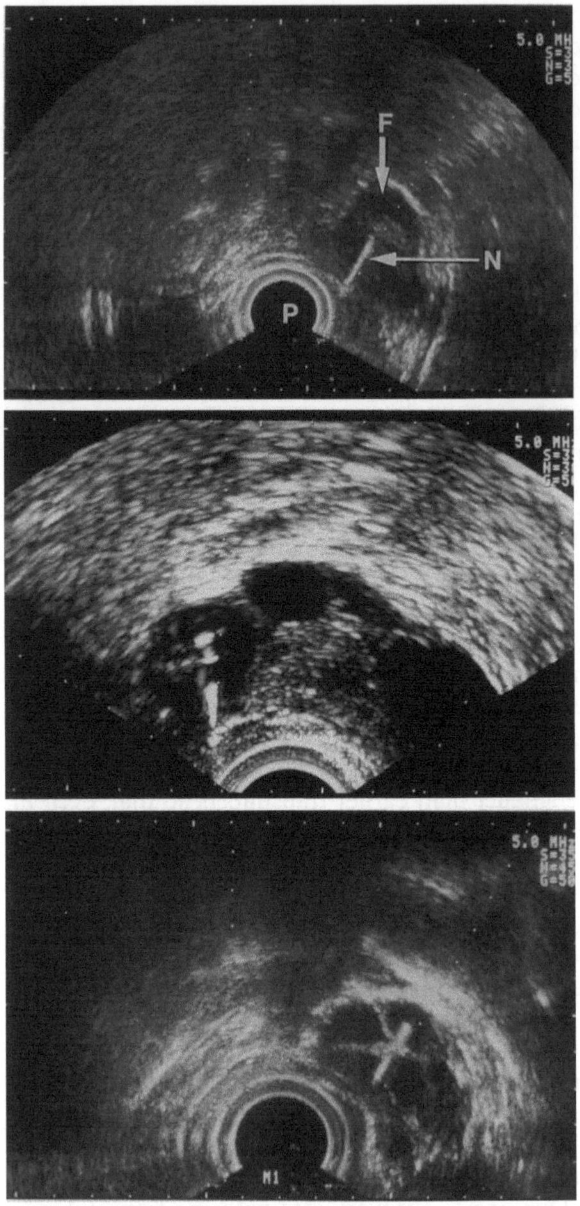

Fig. 19. Vaginal follicular aspiration guided by vaginosonography. *Top:* aspiration of the last follicle of the left ovary. *F,* follicle; *N,* needle; *P,* vaginal probe. *Center:* aspiration of the most laterally situated follicle of the right ovary. *Bottom:* aspiration of a follicle in the left ovary

This method of oocyte recovery saves time and is easily performed even in obese patients. In contrast to transabdominal aspiration, the urinary bladder is not punctured. Previous abdominal operations and adhesions do not interfere with the vaginal retrieval. The disadvantage of this method is the potential for bacterial contamination of the culture, since the vagina cannot be made completely aseptic. Nevertheless, we found that the cleavage rate of the oocytes in our patients exceeded that of oocytes collected by transvesical aspiration. Although the different induction regimens make direct comparisons difficult, it is clear that the vaginal route does not adversely affect the fertilization rate despite the lack of perfect asepsis (Deutinger et al. 1987 b).

So far we have observed no complications associated with the use of this method. Vaginal aspiration has replaced earlier methods of follicular aspiration and oocyte recovery and has made laparoscopy obsolete for the purpose of in vitro fertilization (Schulman 1985 b).

The pregnancy rate in patients undergoing GIFT is as high as 40%, depending on the cause of the infertility, and thus is greater than the rates achieved with in vitro fertilization (Mettler 1988). Attempts have been made to utilize the less invasive method of vaginal ultrasound guidance to harvest oocytes for the GIFT procedure and then undertake the gamete transfer under hysteroscopic vision (Lindemann 1988). In this method the pressure in the uterus is raised to a maximum of 25-80 mm Hg until the orifice of the selected tube is clearly visualized. The gametes are transferred into one tube only. So far the pregnancy rate with this method is approximately 30% (Mettler 1988).

4.6 Embryo Transfer

Once fertilization of the oocytes has been accomplished, embryo transfer is the final step of the in vitro fertilization procedure. The embryos are transferred in various stages of division. At most centers it has become standard practice to make the transfer in the four-cell stage, or about 48 h after follicular aspiration. The fertilization of at least one oocyte can be achieved in approximately 70%-80% of cases. Nevertheless, the pregnancy rate after a successful transfer is only about 25%. Implantation, then, is the weak point of in vitro fertilization and the phase about which we have the least information. It is known that the pregnancy rate is improved by the transfer of multiple embryos (Speirs et al. 1983). However, the pregnancy rate declines again when more than four embryos are transferred (Sharma et al. 1988). Thus, a careful transfer technique is currently the only way to influence the implantation rate.

The cervical canal is routinely catheterized on cycle day 2 (i. e., at the start of induction therapy), but in some cases it proves difficult to pass the flexible transfer catheter into the uterine cavity. Great care should be taken not to injure the endometrium during catheter insertion, because the uterus responds to any external stimulus by increasing its contractility, and this could prevent a successful implantation. This has prompted attempts to perform the embryo transfer under sonographic guidance so that the position of the transfer catheter can be accurately monitored (Strickler et al. 1985). However, this method has not gained wide acceptance because the transfer procedure normally presents no difficulties.

Laparoscopically guided embryo transfer has been attempted in cases where the cervical os is obliterated by adhesions and in patients with a history of difficult transfer. So far, however, this method has not produced a pregnancy (Leeton and Kerin 1984). Recently a new, successful method of embryo transfer was proposed for problem cases. The transfer is performed transmurally under vaginosonographic guidance using the guide apparatus designed for follicular aspirations (Popp 1986). Application of this method has already led to pregnancies (Parsons et al. 1987). An alternative method is transurethral transfer guided by abdominal ultrasound. A 14-gauge needle is inserted into the urethra and passed under vision into the uterine cavity, where the embryos are deposited. Pregnancies have been achieved using this method as well (Parsons et al. 1987).

Embryo transfer can be undertaken in approximately 70%-80% of cases following a successful follicular aspiration. The average pregnancy rate is 20%-25% of the transfers, demonstrating that the transfer phase or subsequent implantation phase is currently the weak point of in vitro fertilization. Besides performing the transfer under laparoscopic vision, recent attempts have been made to transfer embryos under hysteroscopic control. Pregnancies have already been achieved using this method (Mettler 1988).

5 Summary

Sonography has become indispensable to the effective evaluation and management of female infertility. Within a decade after the first reports on folliculometry, sonography is being used to diagnose uterine malformations and tubal lesions, and especially for the monitoring of follicular growth. Sonography has gained in importance through the simultaneous development of in vitro fertilization. Advances in infertility treatment and sonography are stimulated largely by the impetus that each imparts to the other. The development of vaginosonography represents another important milestone in the still short history of ultrasound. This method has been applied for folliculometry, follicular aspiration, embryo transfer, and luteal phase tracking in infertile women who desire children. Besides saving time and lowering costs, vaginal follicular aspiration has made in vitro fertilization widely accessible, extending its benefits to women with unexplained infertility as well as patients with absent or irreparably occluded tubes. Already this method has obviated the need for surgical treatment in many cases. Vaginal probes equipped for pulsed Doppler signal processing are one of the most recent innovations in the development of vaginosonography for infertility evaluation.

References

Ashkenazi J, Feldberg D, Shelef M, Dicker D, Goldman JA (1987) Empty follicle syndrome: an entity in the etiology of infertility of unknown origin, or a phenomenon associated with purified follicle-stimulating hormone therapy? Fertil Steril 48: 152-154

Bald R (1983) Studien über die sonographische Endometriumdarstellung. Dissertation, University of Marburg

Ben-Rafael Z, Kopf GS, Blasco L, Flickinger GL, Tureck RW, Strauss JF, Matroianni L jr (1986) Follicular maturation parameters associated with the failure of oocyte retrieval, fertilization, and cleavage in vitro. Fertil Steril 45: 51-57

Bernardus RE, van Dop PA, van Kessel H, Schoemaker J (1983) New evidence for the existence of the luteinized unruptured follicle syndrome as a cause of infertility: a case report. Fertil Steril 39: 376-378

Bernaschek G (1987a) Endosonography. In: Sabbagha RE (ed) Diagnostic Ultrasound, 2nd edn. Lippincott, Philadelphia, pp 572-582

Bernaschek G (1987b) Vorteile der endosonographischen Diagnostik in Gynäkologie und Geburtshilfe. Geburtshilfe Frauenheilkd 47: 471-475

Bernaschek G, Deutinger J (1988) Endosonographie in Geburtshilfe und Gynäkologie. In: Gockel HP (ed) Jahrbuch der Radiologie 1988. Regensberg & Biermann pp 105-119

Bernaschek G, Szalay S, Müller-Tyl E, Janisch H (1984) Ultraschallgeleitete Follikelpunktion zur Eizellgewinnung. In: Lutz H, Reichel L (eds) Ultraschalldiagnostik, vol 83. Thieme, Stuttgart, p 68

Blankstein J, Shalev J, Saadon T, Kukia EE, Rabinovici J, Pariente C, Lunenfeld B, Serr DM, Mashiach S (1987) Ovarian hyperstimulation syndrome: prediction by number and size of preovulatory ovarian follicles. Fertil Steril 47: 597-602

Bryce RL, Shuter B, Sinosich MJ, Stiel JN, Picker RH, Saunders DM (1982) The value of ultrasound, gonadotropin, and estradiol measurements for precise ovulation prediction. Fertil Steril 47: 42-45

Buttery B, Trounson A, McMaster R, Wood C (1983) Evaluation of diagnostic ultrasound as a parameter of follicular development in an in vitro fertilization program. Fertil Steril 39: 458-463

Buvat J, Marcolin G, Herbaut J-C, Dehaene J-L, Verbecq P, Fourlinnie J-C (1988) A randomized trial of human chorionic gonadotropin support following in vitro fertilization and embryo transfer. Fertil Steril 49: 458-461

Cohen J, Debache C, Pez JP, Junca AM, Cohen-Bacrie P (1986) Vaginal sonographically controlled ovarian puncture for oocyte retrieval for in vitro fertilization. J In Vitro Fert Embryo Transfer 3: 309-313

Confino E, Friberg J, Giglia RV, Gleicher N (1985) Sonographic imaging of intrauterine adhesions. Obstet Gynecol 66: 596-598

Cooke ID (1988) The corpus luteum. Hum Reprod 3: 153-156

Crosignani PG (1988) The detective luteal phase. Hum Reprod 3: 157-160

Dallenbach-Hellweg G (1988) The endometrium in natural and artificial luteal phases. Hum Reprod 3: 165-168

Daly DC, Soto-Albors C, Walters C, Ying Y, Riddick DH (1985) Ultrasonographic assessment of luteinized unruptured follicle syndrome in unexplained infertility. Fertil Steril 43: 62-65

Davis JA, Gosink BB (1986) Fluid in the female pelvis. Cyclic patterns. J Ultrasound Med 5: 75-79

Davis J, Waskey J (1987) Ultrasonic evaluation of the menstrual cycle: a new approach to diagnosing luteal phase defect (LPD) and luteinized unruptured follicle (LUF) syndrome. Proceedings of 32nd Annual Convention of American Institute of Ultrasound in Medicine, Oct 6-9, 1987, New Orleans, p 36

DeCherney A, Tarlatzis BC, Laufer N (1985) Follicular development: lessons learned from human in vitro fertilization. Am J Obstet Gynecol 153: 911-923

De Crespigny LC, O'Herlihy C, Hoult IJ, Robinson HP (1981a) Ultrasound in an in vitro fertilization program. Fertil Steril 35: 25-28

De Crespigny LC, O'Herlihy C, Robinson HP (1981b) Ultrasonic observation of the mechanism of human ovulation. Am J Obstet Gynecol 139: 636-639

Deichert U, Hackelöer BJ, Daume E (1986) The sonographic endocrinologic evaluation of the endometrium in the luteal phase. Hum Reprod 1: 219-222

Dellenbach P, Nisand I, Moreau L, Feger B, Plumere C, Gerlinger P, Brun B, Rumpler Y (1984) Vaginal sonographically controlled ovarian follicle puncture for egg retrieval. Lancet I: 1467

Dellenbach P, Nisand I, Moreau L, Feger B, Plumere C, Gerlinger P (1985) Vaginal sonographically controlled follicle puncture for oocyte retrieval. Fertil Steril 44: 656-662

Deutinger J, Müller-Tyl E, Bernaschek G (1984) Predictive value of early biochemical and ultrasound parameters for pregnancy detection after in vitro fertilization. Arch Androl 12: 168

Deutinger J, Reinthaller A, Müller-Tyl E, Riss P, Fischl F, Janisch H (1985) Vier Jahre erfolgreiche In vitro Fertilisierung an der 2. Universitätsfrauenklinik Wien. Österr Ärzteztg 40: 52-58

Deutinger J, Neumark J, Reinthaller A, Riss P, Müller-Tyl E, Fischl F, Bieglmayer Ch, Janisch H, (1986a) Early pregnancy-specific parameters in pregnancies after in vitro fertilization: prediction of the course of pregnancy. Fertil Steril 46: 77-80

Deutinger J, Reinthaller A, Riss P, Bernaschek G, Csaicsich P, Müller-Tyl E, Fischl F, Janisch H (1986b) Vaginosonography: an improvement for daily follicular measurement. Hum Reprod 1, Suppl 2: 7-8

Deutinger J, Reinthaller A, Bernaschek G, Riss P, Fischl F, Müller-Tyl E (1986c) Vaginal sonography and follicle puncture with a new vaginal scanner. 42nd Annual Meeting of the American Fertility Society, Sept 27-Oct 2, 1986, Toronto

Deutinger J, Reinthaller A, Riss P, Bernaschek G, Csaicsich P, Fischl F, Müller-Tyl E (1987a) Comparison of the results of vaginal and abdominal follicle scans. Arch Gynecol Obstet 241: 171-176

Deutinger J, Reinthaller A, Csaicsich P, Riss P, Fischl F, Bernaschek G, Müller-Tyl E (1987b) Follicular aspiration for in vitro fertilization: sonographically guided vaginal versus laparoscopic approach. Eur J Obstet Gynecol Reprod Biol 26: 127-133

Deutinger J, Rudelstorfer R, Bernaschek G (1987c) Vaginale gepulste Dopplerultraschalluntersuchungen an den Ovarialgefäßen. Ultraschall Klin Prax Suppl 1: 45-46

Deutinger J, Riss P, Reinthaller A, Bernaschek G, Csaicsich P, Fischl F, Müller-Tyl E, Janisch H (1987d) Vergleich von vaginosonografischer, transabdomineller sonografischer und laparoskopischer Follikelpunktion zur Gewinnung von Eizellen im Rahmen eines In-vitro-Fertilisierungsprogrammes. Wien Med Wochenschr 5/6: 108-112

Deutinger J, Bernaschek G, Csaicsich P, Reinthaller A (1987e) Vaginal sonography and follicle puncture with a new vaginal scanner. 32nd annual convention of the American Institute of ultrasound in medicine, Oct 6-9, 1987, New Orleans

Deutinger J, Rudelstorfer R, Bernaschek G (1989) Vaginale gepulste Dopplermessungen von Strömungsgeschwindigkeiten in Beckengefäßen nach Zyklusstimulation. Geburtshilfe Frauenheilkd 49: 33-36

Dlugi AM, Laufer N, Botero-Ruiz W, DeCherney AH, Polan ML, Haseltine FP, Mezer HC, Behrmann HR (1985) Altered follicular development in clomiphene citrate versus human menopausal gonadotropin-stimulated cycles for in vitro fertilization. Fertil Steril 43: 40-47

Doody MC, Gibbons WE, Zamah NM (1987) Linear regression analysis of ultrasound follicular growth series: statistical relationship of growth rate and calculated date of growth onset to total growth period. Fertil Steril 47: 436-440

Donnez J, Thomas K (1982) Incidence of the luteinized unruptured follicle syndrome in fertile women and in women with endometriosis. Europ J Obstet Gynecol Reprod Biol 14: 187-170

Dornbluth NC, Potter JL, Shepard MK, Balmacedo JP, Siler-Khodr TM (1983) Assessment of follicular development by ultrasonography and total serum estrogen in human menopausal gonadotropin-stimulated cycles. J Ultrasound Med 2: 407-412

Drugan A, Blumenfeld Z, Erlik Y, Timor-Tritsch IE, Brandes MJ (1988) The use of vaginal sonography in infertility. In: Timor-Tritsch IE, Rottem S (eds) Vaginal sonography. Heinemann Medical, London, pp 143-159

DuBose TJ, Hill LW, Hennigan HW Jr, Nichols DH, Mezaraups GG, Porter L, Marley L, Butschek CM, Karnaze GC, Walser E, Reyes K, Cunyus JA (1985) Sonography of arcuate uterine blood vessels. J Ultrasound Med 4: 229-233

Eissa MK, Hudson K, Docker MF, Sawers RS, Newton JR (1985) Ultrasound follicle diameter measurement: an assessment of interobserver and intraobserver variation. Fertil Steril 44: 751-754

Eissa MK, Obhrai MS, Docker MF, Lynch SS, Sawers RS, Newton JR (1986) Follicular growth and endocrine profiles in spontaneous and induced conception cycles. Fertil Steril 45: 191-195

Eissa MK, Sawers RS, Docker MF, Lynch SS, Newton JR (1987) Characteristics and incidence of dysfunctional ovulation patterns detected by ultrasound. Fertil Steril 47: 603-612

El Tabbakh GH, Lofty I, Azab I, Rahman HA, Southren AL, Aleem FA (1986) Correlation of the ultrasonic appearance of the ovaries in polycystic ovarian disease and the clinical, hormonal, and laparoscopic findings. Am J Obstet Gynecol 154: 892-895

Enk L, Wikland M (1985) Preliminary report on a vaginal ultrasound transducer for oocyte collection. 4th World Conference on In-Vitro-Fertilization, Melbourne, 1985. Brüel and Kjaer, Kopenhagen

Fateh M, Ben-Rafael Z, Blasco L, Tureck RW, Meloni F, Mastroianni L (1986) Comparison of ultrasonographic transurethral and laparoscopic guided oocytes retrieval. Fertil Steril 46: 653-656

Feichtinger W, Kemeter P (1984) Laparoscopic or ultrasonically guided follicle aspiration for in vitro fertilization? J In Vitro Fert Embryo Transfer 1: 244-249

Feichtinger W, Kemeter P (1986a) Transvaginal sector scan sonography for needle guided vaginal follicle aspiration and other applications in gynecological routine and research. Fertil Steril 45: 722-725

Feichtinger W, Kemeter P (1986b) Erste Ergebnisse der vaginalen Follikelpunktion mit Hilfe eines vaginalen Ultraschallsektorscanners mit Punktionsnadelführung. In: Popp LW (ed) Gynäkologische Endosonographie. Aktualisierter Kongreßband der 1. Arbeitstagung gynäkologische Endosonographie 1985 in Hamburg. Klemke, Quickborn, pp 147-151

Feichtinger W, Kemeter P (1987) Aspirationstechniken unter Ultraschallkontrolle. Verhandlungen der Deutschen Gesellschaft für Gynäkologie und Geburtshilfe. Arch Gynecol Obstet 242: 141-146

Feichtinger W, Szalay S, Beck A, Kemeter P, Janisch H (1981) Die Gewinnung reifer menschlicher Eizellen mittels Laparoskopie zum Zwecke der In-vitro-Fertilisierung. Geburtshilfe Frauenheilkd 41: 400-403

Fleischer AC (1988) Vaginal sonography of the endometrium during ovulation induction. 1st World Congress on vaginosonography in gynecology. June 9-12, 1988, Washington DC

Fleischer AC, Daniell JF, Rodier J, Lindsay AM, James AE (1981) Sonographic monitoring of ovarian follicular development. J Clin Ultrasound 9: 275-280

Fleischer AC, Pittaway DE, Beard LA, Thieme GA, Bundy AL, James AE, Wentz AC (1984) Sonographic depiction of endometrial changes occurring with ovulation induction. J Ultrasound Med 3: 341-346

Fleischer AC, Herbert CM, Sacks GA, Wentz AC, Entman SS, James AE jr (1986) Sonography of the endometrium during conception and nonception cycles of in vitro fertilization and embryo transfer. Fertil Steril 46: 442-447

Geisthövel F, Skubsch U, Zabel G, Schillinger H, Breckwoldt M (1983) Ultrasonographic and hormonal studies in physiologic and insufficient menstrual cycles. Fertil Steril 39: 277-283

Gleicher N, Friberg J, Fullan N, Giglia RV, Mayden K, Kesky T, Siegel I (1983) Egg retrieval for in vitro fertilization by sonographically controlled vaginal culdocentesis. Lancet II: 508-509

Glissant A, Mouzon J de, Frydman R (1985) Ultrasound study of the endometrium during in vitro fertilisation cycles. Fertil Steril 44: 786-790

Giorlandino C, Gleicher N, Nanni C, Vizzone A, Gentili P, Taramanni C (1987) The sonographic picture of endometrium in spontaneous and induced cycles. Fertil Steril 47: 508-511

Gonen Y, Caspers RF, Jacobson W, Blankier J (1989) Endometrial thickness and growth during ovarian stimulation: a possible predictor of implantation in in vitro fertilization. Fertil Steril 52: 446-450

Graf MJ, Reyniak JV, Battle-Mutter P, Laufer N (1988) Histologic evaluation of the luteal phase in women following aspiration for oocyte retrieval. Fertil Steril 49: 616-619

Granberg S, Wikland M (1987) Comparison between vaginal and transabdominal transducers for measuring ovarian volume. J Ultrasound Med 6: 649-653
Habermann PG, Hünlich T, Siebzehnrübl E, Trotnow S (1986) Die ultraschallkontrollierte Follikelpunktion in der "Zwei-Instrumente-Technik". Ultraschall 7: 179-182
Hackelöer BJ (1984) The role of ultrasound in female infertility management. J Ultrasound Med 10: 35-50
Hackelöer BJ (1986) What happened in follicular ultrasound (letter). Hum Reprod 1: 49
Hackelöer BJ, Hansmann M (1976) Ultraschalldiagnostik in der Frühschwangerschaft. Gynakologe 9: 108-122
Hackelöer BJ, Nitschke-Dabelstein S (1980) Ovarian imaging by ultrasound: an attempt to define a reference plane. J Clin Ultrasound 8: 497-500
Hackelöer BJ, Robinson HP (1978) Ultraschalldarstellung des wachsenden Follikels und Corpus luteum im normalen physiologischen Zyklus. Geburtshilfe Frauenheilkd 38: 163-168
Hackelöer BJ, Nitschke S, Daume E, Sturm G, Buchholz R (1977) Ultraschalldarstellung von Ovarveränderungen bei Gonadotropinstimulierung. Geburtshilfe Frauenheikd 37: 185-190
Hackelöer BJ, Fleming R, Robinson HP, Adam AH, Coutts JRT (1979) Correlation of ultrasonic and endocrinologic assessment of human follicular development. Am J Obstet Gynecol 135: 122-128
Hackelöer BJ, Dörfler R, Nitschke S, Buchholz R (1980) Ultraschalldarstellung des Follikelwachstums und Basaltemperaturmessung - Vergleich zweier Methoden zur Ovulationsbestimmung. Ultraschall 1: 133-139
Hammarberg K, Enk L, Nilsson L, Wikland M (1987) Oocyte retrieval under the guidance of a vaginal transducer: evaluation of patient acceptance. Hum Reprod 2: 487-490
Haning RV, Austin CW, Carlson IH, Kuzma DL, Shapiro SS, Zweibel WJ (1983) Plasma estradiol is superior to ultrasound and urinary estriol glucoronide as a predictor of ovarian hyperstimulation during induction of ovulation with menotropins. Fertil Steril 40: 31-36
Haning RV, Strawn EY, Nolten WE (1985) Pathophysiology of the ovarian hyperstimulation syndrome. Obstet Gynecol 66: 220-224
Hansmann M, Hackelöer BJ, Staudach A (1985) Ultraschalldiagnostik in Geburtshilfe und Gynäkologie. Springer, Berlin Heidelberg New York Tokyo
Hill GA, Herbert CM, Fleischer AC, Webster BW, Maxson WS, Wentz AC (1986) Enlargement of hydrosalpinges during ovarian stimulation protocols for in vitro fertilization and embryo replacement. Fertil Steril 45: 883-885
Hoffmann DI (1987) Evaluation of follicles. In: Sabbagha RE (ed) Diagnostic ultrasound. 2nd edn. Lippincott, Philadelphia, pp 454-460
Hoult IJ, Crespigny LC de, O'Herlihy C, Speirs AL, Lopata A, Kellow G, Johnston I, Robinson HP (1981) Ultrasound control of Clomiphene/human chorionic gonadotropin stimulated cycles for oocyte recovery and in vitro fertilization. Fertil Steril 36: 316-319
Hull ME, Moghissi KS, Magyar DM, Hayes MF, Zador I, Olson JM (1986) Correlation of serum estradiol levels and ultrasound monitoring to assess follicular maturation. Fertil Steril 46: 42-45
Imoedemhe DAG, Shaw RW, Kirkland A, Chan R (1987) Ultrasound measurement of endometrial thickness on different ovarian stimulation regimens during in-vitro fertilization. Hum Reprod 2: 545-547
Janssen-Caspers HAB, Kruitwagen RFPM, Wladimiroff JW, Jong FH de, Drogendijk AC (1986a) Diagnosis of luteinized unruptured follicle by ultrasound and steroid hormone assays in peritoneal fluid: a comparative study. Fertil Steril 46: 823-827
Janssen-Caspers HAB, Gent I van, Wladimiroff JW, Alberda AT, Zeilmaker GH (1986b) Ultrasonically guided percutaneous and laparoscopic oocyte aspiration: a comparative study. Europ J Obstet Gynecol Reprod Biol 21: 219-223
Kemeter P, Feichtinger W (1986) Vaginal oocyte retrieval using a vaginal sector scan probe with an automated puncture device. Hum Reprod 1: 21-24
Konickx PR (1983) The luteinized unruptured follicle syndrome (letter). Fertil Steril 39: 249-250
Kratochwil A (1969) Ein neues vaginales Ultraschall-Schnittbildverfahren. Geburtshilfe Frauenheilkd 29: 379-385
Kratochwil A, Urban G, Friedrich F (1972) Ultrasonic tomography of the ovaries. Ann Chir Gynaecol 61: 211-214
Kratochwil A, Jentsch K, Brezina KF (1973) Ultraschallanatomie des weiblichen Beckens und ihre klinische Bedeutung. Arch Gynecol 214: 273-275
Kun L, Bösze P (1973) Die Bedeutung der Ultraschalluntersuchung für die Diagnosestellung des polycystischen Ovars. Geburtshilfe Frauenheilkd 33: 452-454
Laatikainen T, Tulenheimo A, Andersson B, Kärkkäinen J (1983) Obesity, serum steroid levels, and pulsatile gonadotropin secretion in polycystic ovarian disease. Europ J Obstet Reprod Biol 15: 45-53
Leader A, Wiseman D, Taylor PJ (1985) The prediction of ovulation: a comparison of the basal body temperature graph, cervical mucus score, and real-time pelvic ultrasonography. Fertil Steril 43: 385-388
Leerentveld RA, van Gent I, van der Stoep M, Alberda BT, Wladimiroff JW (1984) Comparison of Graafian follicle dimensions as determined by static and real-time sector scanning. Fertil Steril 42: 929-931
Leeton J, Kerin J (1984) Embryo transfer. In: Wood C, Trounson A (ed) Clinical in vitro fertilization. Springer, Berlin Heidelberg New York, p 117
Lejeune B, Camus M, Deschacht J, Leroy F (1986) Differences in the luteal phases after failed od successful in vitro fertilization and embryo replacement. J In Vitro Fert Embryo Transfer 3: 358-365
Lenz S (1984) Ultrasonic-guided follicle puncture under local anesthesia. J In Vitro Fert Embryo Transfer 1: 239-243
Lenz S, Bang J (1986) Transvesical/transuterine collection of oocytes by ultrasound. J In Vitro Fert Embryo Transfer 3: 259-260
Lenz S, Lauritsen JG (1982) Ultrasonically guided percutaneous aspiration of human follicles under local anesthesia: a new method of collecting oocytes for in vitro fertilization. Fertil Steril 38: 673-677
Lenz S, Lauritsen JG, Kjellow M (1981) Collection of human ooctes for in vitro fertilization by ultrasonically guided follicular puncture. Lancet I: 1163-1164
Lenz S, Leeton J, Renou P (1987) Vaginal recovery of oocytes for in vitro fertilization using vaginal ultrasound. J In Vitro Fert Embryo Transfer 4: 51-55
Lewin A, Maragalioth EJ, Rabinowitz R, Schenker JG (1985) Comparative study of ultrasonically guided percutaneous aspiration with local anesthesia and laparoscopic aspiration of follicles in an in vitro fertilization program. Am J Obstet Gynecol 151: 621-625
Lewin A, Laufer N, Rabinowitz R, Margalioth EJ, Bar I, Schenker JG (1986a) Ultrasonically guided oocyte collection under local anesthesia: the first choice method for in

vitro fertilization - a comparative study with laparoscopy. Fertil Steril 46: 257-261
Lewin A, Laufer N, Rabinowitz R, Schenker JG (1986b) Ultrasonically guided oocyte recovery for in vitro fertilization: an improved method. J In Vitro Fert Embryo Transfer 3: 370-373
Lindemann HJ (1988) Hysteroscopy in relation to vaginosonography. 1st World Congress on vaginosonography in gynecology. June 9-12, 1988, Washington DC
Liukkonen S, Koskimies AI, Tenhunen A, Ylöstalo P (1984) Diagnosis of luteinized unruptured follicle (LUF) syndrome by ultrasound. Fertil Steril 41: 26-30
Lopata A, Johnston IWH, Leeton JF, Muchnicki D, Talbot JMcK, Wood C (1974) Collection of human oocytes at laparoscopy and laparotomy. Fertil Steril 25: 1030-1038
Mantzavinos T, Garcia JE, Jones HW (1983) Ultrasound measurement of ovarian follicles stimulated by human gonadotropins for oocyte recovery and in vitro fertilization. Fertil Steril 40: 461-465
Mao KR, Haines CJ, Tam PPL (1986) Successful oocyte retrieval and in vitro fertilization following ultrasound-directed aspiration of endometriomas. J In Vitro Fert Embryo Transfer 3: 333
McArdle C, Seibel M, Hann LE, Weinstein F, Taymor M (1983) The diagnosis of ovarian hyperstimulation (OHS): the impact of ultrasound. Fertil Steril 39: 464-467
Marinho AO, Sallam HN, Goessens LKV, Collins WP, Rodeck CH, Campbell S (1982) Real time pelvic ultrasonography during the periovulatory period of patients attending an artificial insemination clinic. Fertil Steril 37: 633-638
Martin J (1986) Follikulometrie mit Hilfe der Vaginosonographie. In: Popp LW (ed) Gynäkologische Endosonographie. Aktualisierter Kongreßband der 1. Arbeitstagung gynäkologische Endosonographie 1985 in Hamburg. Klemke, Quickborn, pp 143-146
Meldrum DR, Chetkowski RJ, Steingold KA, Randle D (1984) Vaginal ultrasound scanning of ovarian follicles. Fertil Steril 42: 803-805
Mettler L (1988) The role of pelviscopy, hysteroscopy and vaginosonography in IVF/ET and GIFT procedures. 1st World Congress on vaginosonography in gynecology. June 9-12, 1988, Washington DC
Nicolini U, Ferrazzi E, Bellotti M, Travaglini P, Elli R, Scaperrotta RG (1985) The contribution of sonographic evaluation of ovarian size in patients with polycystic ovarian disease. J Ultrasound Med 4: 347-351
Nitschke-Dabelstein S, Hackelöer BJ, Sturm G (1981) Ovulation and corpus luteum formation observed by ultrasonography. J Ultrasound Med 7: 33-39
Oelsner G, Pan SB, Barnea ER, Boyers SP, Tarlatzis BC, DeCherney AH (1986) The value of the cervical score in monitoring ovulation induction for in vitro fertilization: a prospective double-blind study. J In Vitro Fert Embryo Transfer 3: 366-369
O'Herlihy C, De Crespigny LC, Lopata A, Johnston I, Hoult I, Robinson H (1980a) Preovulatory follicular size: a comparison of ultrasound and laparoscopic measurements. Fertil Steril 34: 24-26
O'Herlihy C, De Crespigny LC, Robinson HP (1980b) Monitoring ovarian follicular development with real time ultrasound. Br J Obstet Gynaecol 87: 613-618
Orsini LF, Venturoli S, Lorusso R, Pluchinotta V, Paradisi R, Bovicelli L (1985) Ultrasonic findings in polycystic ovarian disease. Fertil Steril 43: 709-714
Parisi L, Tramonti M, Casciano S, Zurli A, Gazzarrini O (1982) The role of ultrasound in the study of polycystic ovarian disease. J Clin Ultrasound 10: 167-172
Parisi L, Tramonti M, Derchi LE, Casciano S, Zurli A, Rocchi P (1984) Polycystic ovarian disease: ultrasonic evaluation and correlations with clinical and hormonal data. J Clin Ultrasound 12: 21-26
Parsons J, Booker M, Goswamy J, Akkermans J, Riddle A, Sharma V, Wilson L, Withead M, Campbell S (1985) Oocyte retrieval for in vitro fertilization by ultrasonically guided needle aspiration via the urethra. Lancet I: 1076-1077
Parsons JH, Bolton VN, Wilson L, Campbell S (1987) Pregnancies following in vitro fertilization and ultrasound-directed surgical embryo transfer by perurethal and vaginal techniques. Fertil Steril 48: 691-693
Paulson JD, Speck G, Albarelli JN (1984) The use of ultrasonography in patients with unexplained infertility. Fertil Steril 42: 489-491
Picker RH, Smith DH, Tucker MH, Saunders DM (1983) Ultrasonic signs of imminent ovulation. J Clin Ultrasound 11: 1-2
Popp LW (1986) Möglichkeiten der vaginosonographisch gezielten Punktion in einem In-vitro Fertilisierungsprogramm. In: Popp LW (ed) Gynäkologische Endosonographie. Aktualisierter Kongreßband der 1. Arbeitstagung gynäkologische Endosonographie 1985, Hamburg. Klemke, Quickborn, pp 153-162
Popp LW, Müller-Holve W (1984) Die Indikationsstellung zur Vaginosonographie. In: Lutz H, Reichel L (eds) Ultraschalldiagnostik 83. Thieme, Stuttgart, p 74
Popp LW, Lemester S, Hinrichs S, Heesen D, Müller-Holve W, Martin K (1985) Intravaginale Ultraschalldiagnostik (Vaginosonographie) - erste Erfahrungen mit dem Panoramasektor. In: Judmaier G, Frommhold H, Kratochwil A (eds) Ultraschalldiagnostik 84. Thieme, Stuttgart, p 320
Queenan JT, O'Brian GD, Bains LM, Simpson J, Collins WP, Campbell S (1980) Ultrasound scanning of ovaries to detect ovulation in women. Fertil Steril 34: 99-105
Reinthaller A, Deutinger J, Müller-Tyl E, Riss P, Fischl F, Janisch H (1985) In vitro fertilization and embryo transfer program. 2nd Department of Obstetrics and Gynecology. J In Vitro Fert Embryo Transfer 2: 238-240
Riddle AF, Sharma V, Mason BA, Ford NT, Pampiglione JS, Parsons J, Campbell S (1987) Two years' experience of ultrasound-directed oocyte retrieval. Fertil Steril 48: 454-458
Ritchie WGM (1985) Ultrasound in the evaluation of normal and induced ovulation. Fertil Steril 43: 167-181
Roberts AB, Graham FM, Binkerd P (1985) Ultrasonic preovulatory follicular appearances as an indicator of the outcome of an in vitro fertilization cycle. Fertil Steril 43: 560-564
Robertson RD, Picker RH, Wilson PC, Saunders DM (1979) Assessment of ovulation by ultrasound and plasma estradiol determinations. Obstet Gynecol 56: 686-691
Rossavik IK, Gibbons WE (1986) Growth curve analyses of follicular growth in the in vitro fertilization program. Fertil Steril 45: 834-838
Schulman H, Fleischer A, Farmakides G, Bracero L, Rochelson B, Grunfeld L (1986) Development of uterine artery compliance in pregnancy as detected by Doppler ultrasound. Am J Obstet Gynecol 155: 1031-1036
Schulman J (1985a) Outpatient in vitro fertilization using vaginal oocyte retrieval and local anesthesia. N Engl J Med 312: 1639
Schulman JD (1985b) Laparoscopy for in vitro fertilization: end of an era (letter). Fertil Steril 44: 713

Schwimer SR, Lebovic J (1984) Vaginal pelvic ultrasonography. J Ultrasound Med 3: 381-386
Schwimer SR, Lebovic J (1985) Vaginal pelvic ultrasonography: accuracy in follicle and cyst size determination. J Ultrasound Med 4: 61-63
Schwimer SR, Rothman CM, Lebovic J, Oye DM (1984) The effect of ultrasound coupling gels on sperm motility in vitro. Fertil Steril 42: 946-947
Seifer DB, Collins RL, Paushter DM, George CR, Quigley MM (1988) Follicular aspiration: a comparison of an ultrasonic vaginal transducer with fixed needle guide and other retrieval methods. Fertil Steril 49: 462-467
Sharma V, Riddle A, Mason BA, Pampiglione J, Campbell S (1988) An analysis of factors influencing the establishment of a clinical pregnancy in an ultrasound-based ambulatory in vitro fertilization program. Fertil Steril 49: 469-478
Sheean LA, Goldfarb JJ, Kiwi R, Utian WH (1986) Arrest of embryo development by ultrasound coupling gels. Fertil Steril 45: 568-571
Smith B, Porter R, Ahuja K, Craft J (1984) Ultrasonic assessment of endometrial changes in stimulated cycles in an in vitro fertilization and embryo transfer program. J In Vitro Fert Embryo Transfer 1: 50-55
Speirs AL, Lopata A, Gronow MJ, Kellow GN, Johnston WIH (1983) Analysis of the benefits and risks of multiple embryo transfer. Fertil Steril 39: 468-471
Steptoe PC, Edwards RG (1970) Lararoscopic recovery of preovulatory human oocytes after priming of ovaries with gonadotropins. Lancet I: 683-689
Steptoe PC, Edwards RG (1978) Birth after the implantation of a human embryo. Lancet II: 336
Strickler RC, Christianson C, Crane JP, Curato A, Knight AB, Young V (1985) Ultrasound guidance for human embryo transfer. Fertil Steril 43: 54-61
Swanson M, Sauerbrei EE, Cooperberg PL (1981) Medical implications of ultrasonically detected polycystic ovaries. J Clin Ultrasound 9: 219-222
Tal J, Paz B, Samberg I, Lazarov N, Sharf M (1985) Ultrasonographic and clinical correlates of menotropin versus sequential clomiphene citrate: menotropin therapy for induction of ovulation. Fertil Steril 44: 342-349
Taylor PJ, Wiseman D, Mahadevan M, Leader A (1986) "Ultrasound rescue": a successful alternative form of oocyte recovery in patients with periovarian adhesions. Am J Obstet Gynecol 154: 240-244
Taylor KJW, Burns PN, Wells PNT, Conway DI, Hull MGR (1985) Ultrasound Doppler flow studies of the ovarian and uterine arteries. Br J Obstet Gynaecol 92: 240-246
Timor-Tritsch IE, Rottem S (1987) Vaginal ultrasonographic study of the Fallopian tube. Obstet Gynecol 70: 424-428
Timor-Tritsch IE, Rottem S, Thaler I (1988) Review of vaginal ultrasonography: a description with clinical application. Ultrasound Q 6: 1-34
Thickman D, Arger P, Tureck R, Blasco L, Mintz M, Coleman B (1986) Sonographic assessment of the endometrium in patients undergoing in vitro fertilization. J Ultrasound Med 5: 197-201
Trotnow S, Habermann PG, Hünlich T, Siebzehnrübl E (1985) A new set of instruments for sonographically controlled follicular puncture. Arch Gynecol 236: 211-217
Van Steirteghem AC, Smitz J, Camus M, Waesberghe L van, Deschacht J, Khan I, Staessen C, Wisanto A, Bourgain C, Devroey P (1988) The luteal phase after in-vitro fertilization and related precedures. Hum Reprod 3: 161-164
Vargyas JM, Marrs RP, Kletzky OA, Mishell DR (1982) Correlation of ultrasonic measurement of ovarian size and serum estradiol levels in ovulatory patients following clomiphene citrate for in vitro fertilization. Am J Obstet Gynecol 144: 569-573
Venturoli S, Paradisi R, Saviotti E, Porcu E, Fabbri R, Orsini LF, Bovicelli L, Flamigni C (1985) Ultrasound study of ovarian and uterine morphology in women with polycystic ovary syndrome before, during and after treatment with cyproterone acetate and ethinyloestradiol. Arch Gynecol 237: 1-10
Vermesh M, Kletzky OA, Davajan V, Israel R (1987) Monitoring techniques to predict and detect ovulation. Fertil Steril 47: 259-264
Werlin LB, Weckstein L, Weathersbee PS, Parenicka K, White D, Stone SC (1986) Ultrasound: a technique useful in determining the side of ovulation. Fertil Steril 46: 814-817
Wikland M, Nilsson L, Hansson R, Hamberger L, Janson PO (1983) Collection of human oocytes by the use of sonography. Fertil Steril 39: 603-608
Wikland M, Enk L, Hammarberg K, Nilsson L (1987) Use of a vaginal transducer for oocyte retrieval in an IVF/ET program. J Clin Ultrasound 15: 245-251
Wood C, Leeton JF, Talbot JMcK, Trounson AO (1981) Technique for collecting oocytes for in vitro fertilization. Br J Obstet Gynaecol 88: 756-759
Yee B, Barnes RB, Vargyas JM, Marrs RP (1987) Correlation of transabdominal and vaginal ultrasound measurements of follicle size and number with laparoscopic findings for in vitro fertilization. Fertil Steril 47: 828-832
Ying YK, Daly DC, Randolph JF, Soto-Albors CS, Maier DB, Schmidt CL, Riddick DH (1987) Ultrasonographic monitoring of follicular growth for luteal phase defects. Fertil Steril 48: 433-436
Ylöstalo P, Rönnberg L, Jouppila P (1979) Measurements of the ovarian follicle by ultrasound in ovulation induction. Fertil Steril 31: 651-655

Endosonographically Guided Punctures

1 General

Culdocentesis was an important diagnostic procedure before the advent of laparoscopy, but its importance has diminished with routine use. Before the advent of vaginosonographically guided follicular aspiration, endosonography played only a minor role as a means of directing invasive procedures in gynecologic patients, although some attempts were made to obtain cytologic or histologic confirmation of pelvic wall recurrence of cervical cancer using an endosonographically guided needle. The use of vaginosonography to demonstrate structures in the lesser pelvis has shown that a needle inserted vaginally must travel only a short distance to reach the ovaries. Because the ovaries are usually located very close to the vaginal wall in the region of the cul-de-sac, the vaginal approach offers an obvious solution to the problem of the "long-distance" aspiration.

The needle is directed through the vaginal vault under "vision." Several equipment manufacturers in cooperation with various groups of researchers have worked independently on the development of a vaginal transducer (Popp and Müller-Holve 1984; Schwimer and Lebovic 1984; Meldrum et al. 1984; Enk and Wikland 1985; Kemeter and Feichtinger 1986). These groups have reported on their results or described the development of new technical aids (Feichtinger and Kemeter 1986a, b; Cohen et al. 1986; Deutinger et al. 1986b, 1987a, b, c; Wikland et al. 1987; Lenz et al. 1987; Hammarberg et al. 1987). Vaginosonography and vaginosonographically guided follicular aspiration have today become indispensable tools for infertility treatment and especially for in vitro fertilization. These techniques benefit the patient by significantly reducing operative risk and treatment costs, and they benefit the therapist by simplifying the treatment and saving time.

2 Technical Aspects

An ultrasound probe for intracorporeal use must be sterilizable. While the probe for vaginosonography is covered with a sterile rubber sleeve before insertion, this cannot be done in vaginosonographically guided follicular aspiration because the needle, passed through a guide rigidly attached to the scanner, would have to pierce the sleeve as well. Therefore the scanner itself must be sterilizable, and the vagina must be aseptically prepared. Although the vagina cannot be made completely sterile, there is no evidence at present that contact between the aspiration needle and the vagina adversely affects the rate of oocyte division.

2.1 Scanner Types

In principle, an end-firing vaginal probe with a sector scan pattern can be used for ultrasound-guided punctures. Most transducers for this application are equipped with an attached needle guide that greatly facilitates needle insertion.

The main scanner types in current use are curved arrays, sector scanners with a 90°-110° field of view, and wide-angle "panoramic" scanners with a 240° forward field of view. The shaft of the panoramic scanner carries two needle guides that form an angle of approximately 30°. After the equipment has been calibrated, the monitor displays two on-screen lines for tracking the course of the needle. These biopsy lines can be stored and do not need to be recalibrated before each aspiration.

Linear scanners are not well suited for this application because they cannot display the course of a needle inserted vaginally, parallel to the scanner axis. This type of scanner can at best be used for a transperineal aspiration guided by a vaginal probe.

The panoramic scanner is optimum for vaginal follicular aspirations, because its wide-angle view gives the operator a clear, immediate impression of anatomic structures and relationships (Popp et al. 1985).

2.2 Freehand Needle Technique

In the freehand technique the transducer is held steady with one hand while the other hand inserts the needle at a point distant from the transducer and advances it into the sonographic field of view. Because varying resistances are encountered as the needle penetrates the tissue, and because the lesser pelvis contains vulnerable structures such as the ureters and blood vessels, this method should not be used routinely due to the exorbitant risk of injury. Because the target structures are often only a few millimeters in diameter, it is not unusual for the operator to miss the target on the initial attempt.

When this technique was first applied, it was found that the needle tended to push the ovary aside rather than pierce it (Popp 1986a). A thick thrusting or jabbing motion is needed to penetrate the cyst or follicular wall. Meanwhile, care must be taken that the depth of insertion does not go beyond the target; thus the success rate and complication rate of the freehand technique depend heavily on the experience of the operator.

For these reasons the freehand needle technique can no longer be recommended for endosonographically guided punctures.

Fig. 1. Vaginal scanner with needle guide. The needle is inserted through the fixed slots, which are set at a 30° angle to the scanner shaft ▼

2.3 Needle Guides

Needle guides provide a fixed attachment between scanner and needle. They are useful only for front-viewing sector scanners, the only type that can demonstrate the full length of the needle on the monitor screen. Most conventional scanner types are equipped with a guide groove or slot through which the needle can be inserted (Fig. 1)

A new type of needle guide is the automated puncture device designed by Kemeter and Feichtinger (1986). This device, mounted at a fixed angle, automatically controls the direction, depth, and force of needle insertion. It is rigidly attached to the scanner (Fig. 2). After the distance from the transducer to the center of the target structure (e.g., a follicle) has been measured, the spring on the puncture device is tightened, and the distance setting is made. The aspiration needle is attached to the device, and at a push of a button the needle is driven along the electronic biopsy line into the center of the follicle. The insertion is rapid enough that the inertia of the ovary keeps it from evading the needle tip. Suction should be initiated before the spring release is activated; otherwise the oocyte may be lost when the needle pierces the follicle wall.

The major advantage of this device is that even inexperienced operators can achieve a high rate of oocyte recovery. However, a serious disadvantage is that the device must be realigned each time a new follicle must be aspirated. This involves piercing the

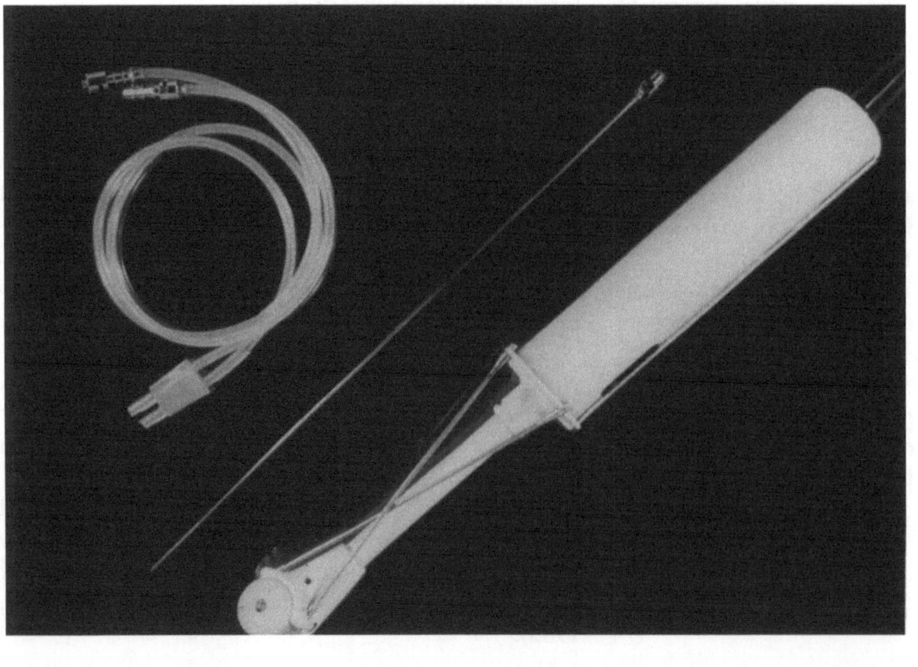

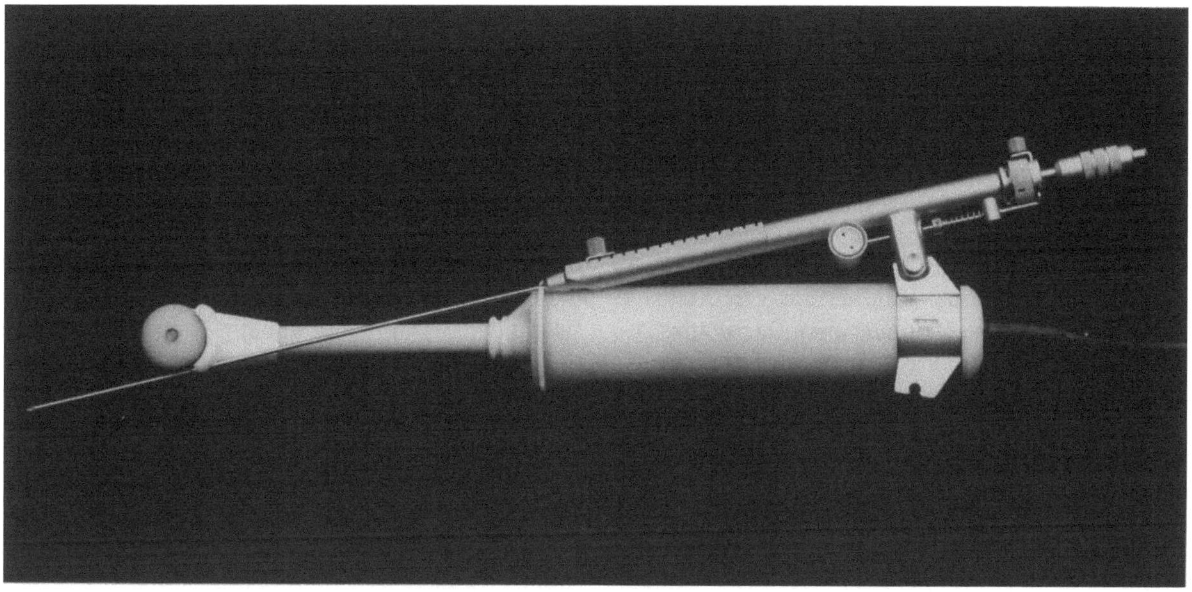

Fig. 2. Automated, spring-activated puncture device (Labotec, Göttingen, FRG) mounted on the vaginal scanner

vaginal mucosa at multiple sites, with an increased risk of injury and infection.

When a conventional needle guide is used, the follicle may disappear from the screen momentarily as the needle pierces the vaginal wall. Nevertheless, experienced examiners still prefer this method because once the needle has penetrated the vaginal wall, its position needs to be adjusted only slightly to aspirate multiple follicles; there is no need to pierce the vaginal mucosa more than once. Although the automated puncture device can be a great help to the novice in learning the technique of vaginosonographically guided aspiration, the needle guide represents a better long-term solution.

The first successful vaginal ultrasound visualization of structures of the lesser pelvis was accomplished in 1969 (Kratochwil 1969). Today, the performance of routine vaginosonographic folliculometry offers the best preparation for learning the technique of vaginal follicular aspiration (Deutinger et al. 1986a).

2.4 Preparation of the Patient

Basically the patient is prepared as for any minor surgical procedure. The vaginosonographically guided puncture is performed either under brief, superficial general anesthesia or under local anesthesia with mild sedation. In some cases we have not used any form of analgesia. For follicular aspiration the patient is placed in the lithotomy position. Immediately beforehand the patient empties her bladder, and the vagina is cleansed with physiologic saline. The use of coupling gel should be avoided in vaginal follicular aspiration until it can be proved that these gels are nonspermicidal (Schwimer et al. 1984).

In all current induction regimens for infertility treatment, the goal is to stimulate the growth of a maximum number of follicles and oocytes. The principal regimens for ovulation induction employ clomiphene, clomiphene/hCG, clomiphene/hMG, or clomiphene/hMG/hCG (Reinthaller et al. 1985a; Deutinger et al. 1985). A recent innovation is to initiate therapy by pituitary suppression with GnRH antagonists to abolish the effects of endogenous stimulation. Subsequent hormonal stimulation is initiated with FSH at the start of the cycle, followed by a combination of FSH and LH to mimic the physiologic cycle as closely as possible.

3 Indications

3.1 Follicular Aspiration

3.1.1 General

Today follicular aspiration is performed almost exclusively in stimulated cycles approximately 34–36 h after the administration of hCG. Oocytes for in vitro fertilization were initially retrieved by laparoscopy and later by a modified form of minilaparotomy (Lopata et al. 1974) and still later by laparoscopy (Steptoe and Edwards 1970; Feichtinger et al. 1981; Wood et al. 1981). All manipulations in these methods are performed under vision, and bleeding sites are easily recognized.

A disadvantage of laparoscopy in patients that have undergone several previous operations is the risk of injury to adherent bowel loops during insertion of the laparoscope. If adhesions are extensive or if there is massive endometriosis, the ovaries often cannot be satisfactorily visualized. Other disadvantages are the need for intubation anesthesia, the low head position, and the high intra-abdominal pressure and acidosis resulting from gas insufflation.

The techniques of ultrasound-guided transabdominal and transvesical follicular aspiration were developed in an effort to reduce these risks (Lenz et al. 1981). This type of aspiration can even be performed under local anesthesia. The transvesical route is preferred as less invasive because, unlike laparoscopy, it does not involve entry into the abdominal cavity. Injuries are unlikely to occur even in patients with extensive adhesions, as even bowel loops adherent to the anterior abdominal wall can easily evade the aspiration needle. Another advantage of this method is that the number and position of the follicles can be continuously monitored. Sector scanners are superior to linear scanners for transvesical aspiration in thin patients, whose prominent pelvic bones can make a linear scanner difficult to use.

In very obese patients it is often difficult to define the follicles with precision, and a "long-distance" needle insertion is required. The transvesical approach requires a full bladder, which can cause upward displacement of mobile ovaries, making it impossible to demonstrate the bladder and follicles on the same scan. Transvesical aspiration often cannot be performed if the ovaries are in the cul-de-sac.

Our oocyte recovery rate with the transvesical technique was 2.0 per aspiration - lower than with the laparoscopic method. The main reason for this is that we used the transvesical method whenever laparoscopy did not seem promising or was considered too risky because of extensive adhesions (Bernaschek and Deutinger 1988).

Potential complications of transvesical follicular aspiration include cystitis, microhematuria and macrohematuria. We observed no complications aside from three cases of mild lower abdominal discomfort lasting several days.

Besides transvesical ultrasound-guided follicular aspiration (Lenz et al. 1981; Lenz and Lauritsen 1982; Wikland et al. 1983; Lenz 1984; Bernaschek et al. 1984; Feichtinger and Kemeter 1984; Lenz and Bang 1986; Taylor et al. 1986; Mao et al. 1986; Lewin et al. 1986a, b; Janssen-Caspers et al. 1986), which usually employs a needle guide (Trotnow et al. 1985; Habermann et al. 1986), alternative ultrasound-guided techniques were developed. One of the earliest was follicular aspiration by culdocentesis guided by an abdominally placed transducer (Gleicher et al. 1983; Dellenbach et al. 1984, 1985; Schulman 1985a). However, this vaginal freehand technique did not yield satisfactory results (Feichtinger and Kemeter 1986b, 1987a). The needle cannot be directed to the follicle with perfect accuracy, and there is a significant risk of vascular injury (Drugan et al. 1988). Ultimately this method did not gain acceptance as a routine procedure.

Another technique, introduced before the development of vaginal scanners, was transurethral follicular aspiration (Parsons et al. 1985; Fateh et al. 1986; Riddle et al. 1987).

Aspirations performed under continuous ultrasound vision gained in importance when it was shown that they could provide an oocyte recovery rate comparable to (Lewin et al. 1985) and even exceeding (Feichtinger and Kemeter 1984) that of laparoscopic retrieval.

3.1.2 Vaginal Follicular Aspiration Guided by Vaginosonography

The development of vaginal transducers with an attached needle guide formed the decisive technical breakthrough for ultrasound-guided follicular aspiration (Figs. 3, 4) (Popp 1986b). The procedure begins by inserting the scanner into the vagina and orienting it so that a sagittal imaged is displayed on the monitor. The probe is then rotated 90° to obtain a horizontal section of the lesser pelvis, which often will demonstrate both ovaries simultaneously. The transducer has an attached, fixed needle guide that requires calibration only on initial use; the on-screen guides are stored and can be redisplayed at any time.

The 35-cm aspiration needle has a sharply beveled tip, a 1.4-mm outside diameter and 1.2-mm inside diameter. It is connected to a vacuum pump that develops a negative pressure of 180 mm Hg.

If an oocyte is not found in the initial aspirate, the follicle, which becomes conspicuous as it fills with blood, can be reaspirated. Usually it is unnec-

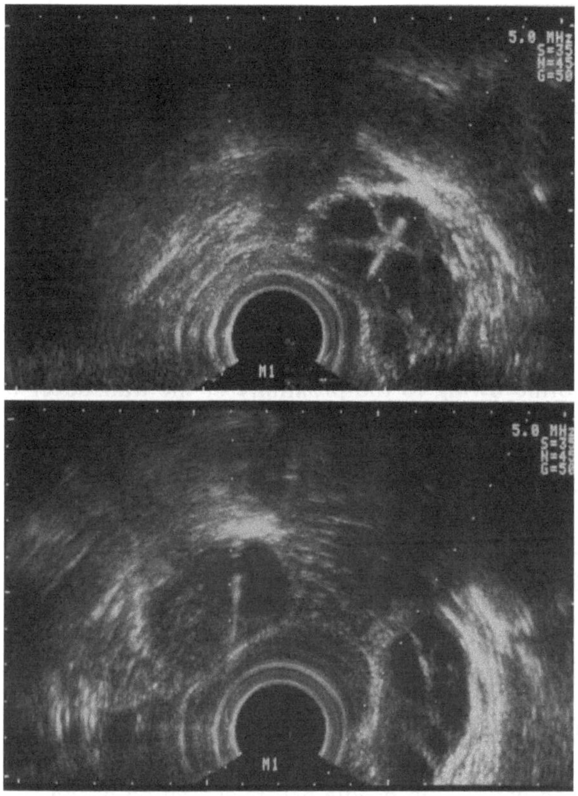

◀ **Fig. 3.** Vaginosonographic follicular aspiration. *Top:* aspiration of a follicle in the left ovary. *Bottom:* aspiration of a follicle in the right ovary

essary to flush the follicle with culture medium. Coagulation of the sometimes blood-tinged follicular fluid is avoided by collecting the aspirate in heparinized test tubes. The entire procedure rarely takes longer than 10 min.

This method of oocyte retrieval saves time, avoids bladder puncture, and is easy to perform even in obese women and patients with extensive postsurgical adhesions. Another advantage is that oocyte retrieval is still possible in case of premature follicular rupture by puncturing the cul-de-sac and aspirating the free fluid (Timor-Tritsch et al. 1988). The disadvantage of the vaginal route is the potential for contamination of the aspirate, since the vagina cannot be made absolutely sterile. However, we have found that the cleavage rate of oocytes retrieved vaginally is no less than that of oocytes collected by transvesical aspiration (Deutinger et al. 1987a).

It is known that retrieving a greater number of oocytes improves the chance of cultivating more transferrable embryos, which in turn improves the pregnancy rate. Today, vaginosonographically guided follicular aspiration provides a higher yield of oocytes, and thus of embryos, than any other technique. A comparison of laparoscopic and vaginal follicular aspirations in patients on identical in-

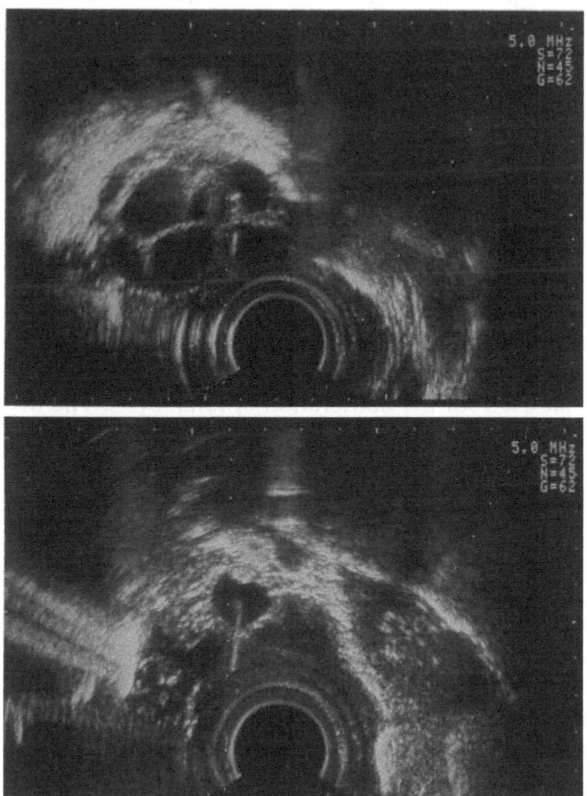

◀ **Fig. 4.** Vaginosonographic follicular aspiration in the right ovary. *Top:* numerous follicles are visible in the right ovary at the start of the aspiration. *Bottom:* the same ovary during aspiration of the last large follicle

Table 1. Comparison of laparoscopic and our first 21 cases of vaginal follicular aspiration

	Laparoscopic aspiration	Vaginal aspiration
Patients	21	21
Follicles	91	102
Retrieved oocytes	77	93
Oocytes per puncture	3.7	4.7
Oocytes per follicle	0.85	0.91
Embryos	40	62
Cleavage rate (%)	52.0	66.7**
Number of transfers	14	18
Embryos per transfer	2.9	3.4
Pregnancies	3	3
Pregnancies per transfer (%)	21.4	16.7*
Pregnancies per puncture (%)	14.1	14.1

Chi-square: * not significant, ** $p < 0.05$

duction regimens indicated comparable clinical parameters and hormone values, a comparable number of follicles, and a markedly higher yield of oocytes by vaginal aspiration than by laparoscopic retrieval (Deutinger et al. 1987a) (Table 1).

Janssen-Caspers et al. (1988) noted similar results in a comparison of laparoscopic and vaginal follicular aspirations.

Since October 1, 1985, all follicular aspirations at the Second University of Vienna Women's Hospital have been performed by the vaginal route. While the number of retrieved oocytes naturally depends on the type of induction regimen, the overall oocyte recovery rate for a variety of regimens has been 6.5 per puncture. Thus, 1387 oocytes were retrieved in a total of 221 punctures. The absence of a single complication in this series underscores the value of the method. Moreover, there was no case in which the aspiration failed to retrieve an oocyte.

Potential complications include the puncture of a major vessel in the lesser pelvis, ovarian abscess, pain, and fever. In a series of 100 vaginal follicular aspirations, Howe et al. (1988) noted three cases of subsequent infection in the lesser pelvis. The most significant source of infection is misdirection of the aspirating needle into a hydro- or pyosalpinx or an ovarian cyst. A hydrosalpinx is easily mistaken for a follicle by its sonographic appearance. Signs that distinguish hydrosalpinx are: mobility, lack of growth response to hormonal stimulation, absence of correlation with serum estradiol, and presence noted in a previous cycle. In addition, many of these lesions are septated, most have an elongated shape, and many can be identified as being separate from the ovary. The aspirate is usually turbid with a volume exceeding 10 cm^3. If the inadvertent puncture of a hydro- or pyosalpinx is noted, prophylactic antibiotics should be started right away.

Evers et al. (1988) reported on their first 181 vaginal follicular aspirations, with a 3% rate of failure to recover oocytes. These failures may relate to the low invasiveness of the procedure, which may have caused it to be applied in patients who did not respond optimally to induction therapy but who would not have been considered candidates for laparoscopic retrieval. At present no major complications are known to be associated with vaginal follicular aspiration.

Vaginal scanning has greatly simplified the conduct of in vitro fertilization. We use this method routinely for folliculometry and follicular aspirations at our center owing to its high patient acceptance and ease of use (Deutinger et al. 1987d). The vaginal method has replaced previous methods of follicular aspiration and oocyte collection and, we believe, has put an end to the era of laparoscopic oocyte retrieval for in vitro fertilization (Schulman 1985).

3.2 Aspiration of Cysts

Cysts can be aspirated under vaginosonographic guidance by the transabdominal route (Schwimer et al. 1985), but vaginal aspiration has gained wider acceptance because it is safer and easier to performe owing to the proximity of the transducer to the lesion (Fig. 5). When cystic masses are detected in the adnexal region, vaginal aspiration can spare even the pregnant patient the necessity of undergoing a more invasive procedure (Fig. 6) and in many cases can obviate the need for a second trimester laparotomy (Hess et al. 1988).

3.2.1 Cyst Wall Biopsy

If a cyst has been detected and malignant disease cannot be ruled out, the cyst wall should be biopsied with a thin needle before aspiration of the cystic fluid is attempted. The significance of a positive or negative finding is similar to that in a tumor biopsy (see chapter "Fine-Needle Aspiration Biopsy"), except that the cytologic examination of the aspirate can yield some additional information.

3.2.2 Sclerotherapy

In some cystic conditions such as recurrent cysts associated with chronic pelvic inflammatory disease, obliteration of the cyst can be attempted, following aspiration biopsy, by the injection of a sclerosing agent (ethoxysclerol). The procedure is similar to that used for the obliteration of renal cysts. The irritation of the cyst wall incites a reactive, aseptic inflammation resulting in adhesion of the wall surfaces.

3.2.3 Abscess Drainage

Abscesses like those following gynecologic operations in the region of the cul-de-sac are managed most easily by drainage via an ultrasound-directed needle (Smith and Annesley 1985). Vaginosonographic guidance is advantageous when drainage is performed by the vaginal route (Fig. 7).

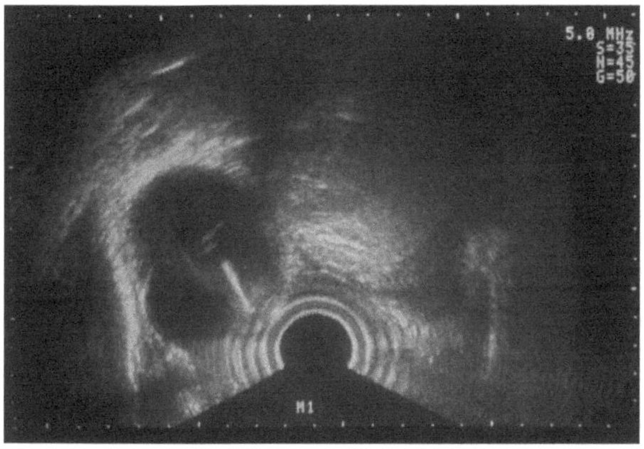

Fig. 5. Vaginosonographic aspiration of a lymphatic cyst located near the right pelvic wall 9 days after Wertheim's radical hysterectomy

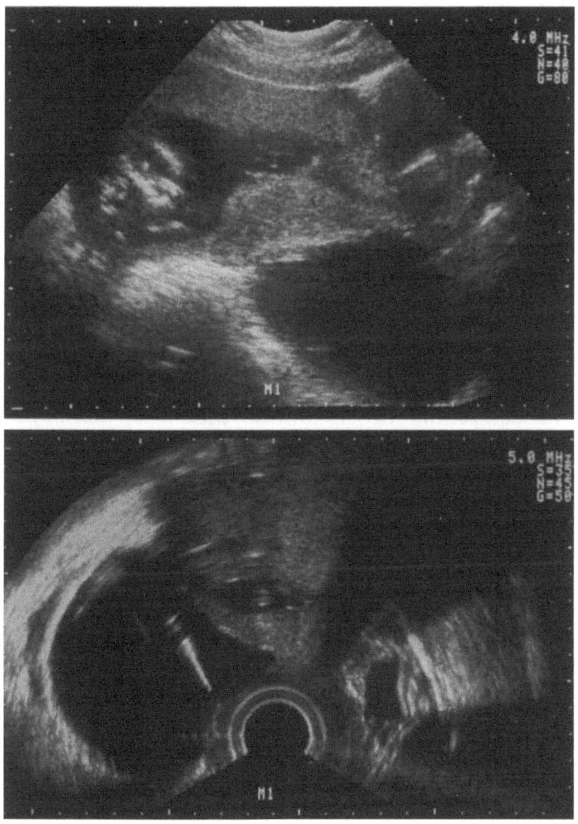

Fig. 6. *Top:* transabdominal sonogram (longitudinal scan) in pregnancy showing an approximately 10-cm cyst in the right adnexal region. *Bottom:* vaginal aspiration of the cyst during the pregnancy

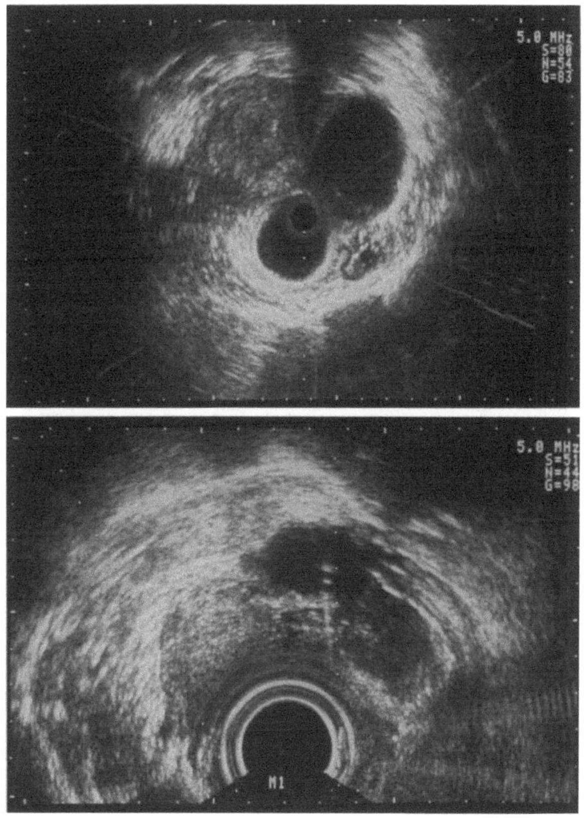

Fig. 7. *Top:* Rectal scan of an abscess in the left adnexal region. This abscess developed during radiotherapy for a stage III B cervical carcinoma. *Bottom:* vaginosonographic aspiration of the abscess

3.3 Puncture of Tubal Pregnancy

Ultrasound examination has significantly advanced the early diagnosis of ectopic pregnancy. This has made it possible to undertake surgical intervention at an earlier stage (Reinthaller et al. 1985b), in turn enabling the use of tube-conserving procedures that may allow the patient to have a subsequent intrauterine pregnancy. Without early diagnosis, endoscopic surgery for tubal pregnancy could scarcely have attained its current level of importance.

Early diagnosis by means of vaginosonography also allows for the early, "conservative" management of tubal pregnancy by a vaginal puncture. In this procedure the ectopic pregnancy is localized with ultrasound, and a needle is passed into the ectopic gestational sac under ultrasound guidance as in a vaginal follicular aspiration. Then prostaglandin or 10 mg of methotrexate is instilled into the sac (Feichtinger and Kemeter 1987). Regular serum β-hCG assays are recommended following the injection.

Experience with this procedure is still inadequate to draw conclusions about its efficacy or its effect on subsequent fertility. If it proves safe, it may revolutionize the treatment of tubal pregnancy, at least in selected patients. Treatment could be performed on an outpatient basis with no need for general anesthesia. We would propose the following approach to management: With a ruptured ectopic pregnancy, operative treatment is still the only option. If pregnancy is detected by a β-hCG assay but ultrasound fails to disclose an intrauterine gestational sac, does not show a definite ectopic sac, and the patient is asymptomatic, the examination should be repeated several days later. If a small ectopic pregnancy is demonstrated sonographically, vaginal puncture with instillation of methotrexate, prostaglandin, or glucose may be considered.

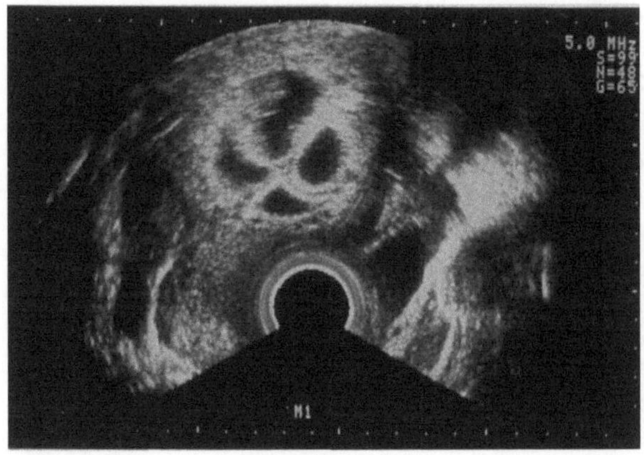

Fig. 8. Vaginal scan in the 9th week of a multiple pregnancy following in vitro fertilization. Four gestational sacs are visualized

3.4 Selective Embryocide

3.4.1 General

Hormonally induced superovulation leads to the maturation of multiple oocytes. Increasing the oocyte yield increases the likelihood of obtaining a greater number of embryos, and the transfer of multiple embryos (ideally no more than three) improves the chance that pregnancy will occur. If more than three embryos are transferred, there is a significant risk of multiple gestation. One of the most dreaded consequences of modern infertility therapy is the development of a multiple pregnancy with more than two embryos (Fig. 8). The obstetric outcome in these cases is often poor, with most pregnancies terminating without a live infant. The medical and social problems associated with multiple births are evident. The major reason for the high perinatal morbidity and mortality in these cases is premature labor caused by overdistention of the uterus (Berkowitz et al. 1988). Additionally there is a 20% maternal risk of preeclampsia and a 35% risk of severe postpartum hemorrhage. Patients must be counseled about the risk of continuing the pregnancy as well as the risk of selective embryocide (Hobbins 1988).

Only three options are available when a multiple pregnancy is diagnosed: (1) The patient elects to carry the pregnancy to term despite the high risk of a poor outcome. The delivery of a triplet pregnancy following in vitro fertilization was recently reported (Haake et al. 1988). (2) The patient elects pregnancy termination by curettage. This negates the effect of the often costly induction therapy and may impact on future reproductive performance. (3) The patient opts for selective embryocide in the hope of bearing at least one or two healthy infants. Breckwoldt et al. (1988) recently described a nontuplet pregnancy following in vitro fertilization that was successfully reduced to a triplet pregnancy by selective embryocide.

The prognosis of the embryos remaining after embryocide is uncertain due to lack of experience with large case numbers. Salat-Baroux et al. (1988) report on a series of 42 pregnancies reduced by selective embryocide, 30 of which resulted in the delivery of at least one infant. Evans et al. (1988) suggest reduction to no more than two embryos, considering the risk of a twin pregnancy to be acceptable. This also provides a safety margin in case one fetus is later found to have an abnormality, as there would still be hope for delivering one healthy baby.

3.4.2 Ultrasound-Guided Embryocide

Three different methods are available for ultrasound-guided embryocide: transabdominal sonography with transabdominal, transmural needle insertion (Kanhai et al. 1986) performed "freehand" or through a fixed needle guide; transcervical insertion guided by transabdominal sonography (Lopes et al. 1985); and transmural insertion guided by vaginosonography.

The most favorable time for undertaking embryocide is between the 8th and 12th weeks of pregnancy (Evans et al. 1988). At this stage the gestational sac is large enough to be punctured without much difficulty, and the resulting amniotic fluid loss is too small to jeopardize the remaining embryos by significantly altering the uterine volume (Salat-Baroux et al. 1988). It has been reported that embryocide is accomplished most rapidly and effectively by the intrathoracic injection of KCl or NaCl. Air injection (embolization) or puncture of the fetal circulation are still quite difficult in early pregnancy (Evans et al. 1988). The puncture may be performed in two sittings (Hansmann et al. 1985) or in one (Farquharson et al. 1988), the one-stage method being preferred because of its greater patient acceptance and comparable efficacy.

At present there is no way to prevent multiple pregnancy with absolute confidence. Nevertheless, the need for selective embryocide, which may be distressing or objectionable to many, can be kept to a minimum by following certain guidelines for in vitro fertilization in induced cycles:

- Reduce the hormone dose for ovulation induction.
- Do not aspirate all the follicles.
- Do not inseminate all the oocytes.
- Freeze and store the conceptus in the pronucleus or 4-cell stage.

Even when precautions are taken, of course, some incidence of multiple pregnancy is inevitable.

3.5 Fine-Needle Aspiration Biopsy

Indeterminate, circumscribed mass lesions in the lesser pelvis can be aspirated with a thin needle under sterile conditions using either local or general anesthesia to obtain material for cytologic study. Guidance of the procedure has been significantly improved by vaginosonography, which provides new information on tumor size and location in 50%-60% of gynecologic tumor patients even before the aspiration is performed (Vilaro et al. 1987). This in turn improves the result of the aspiration it-

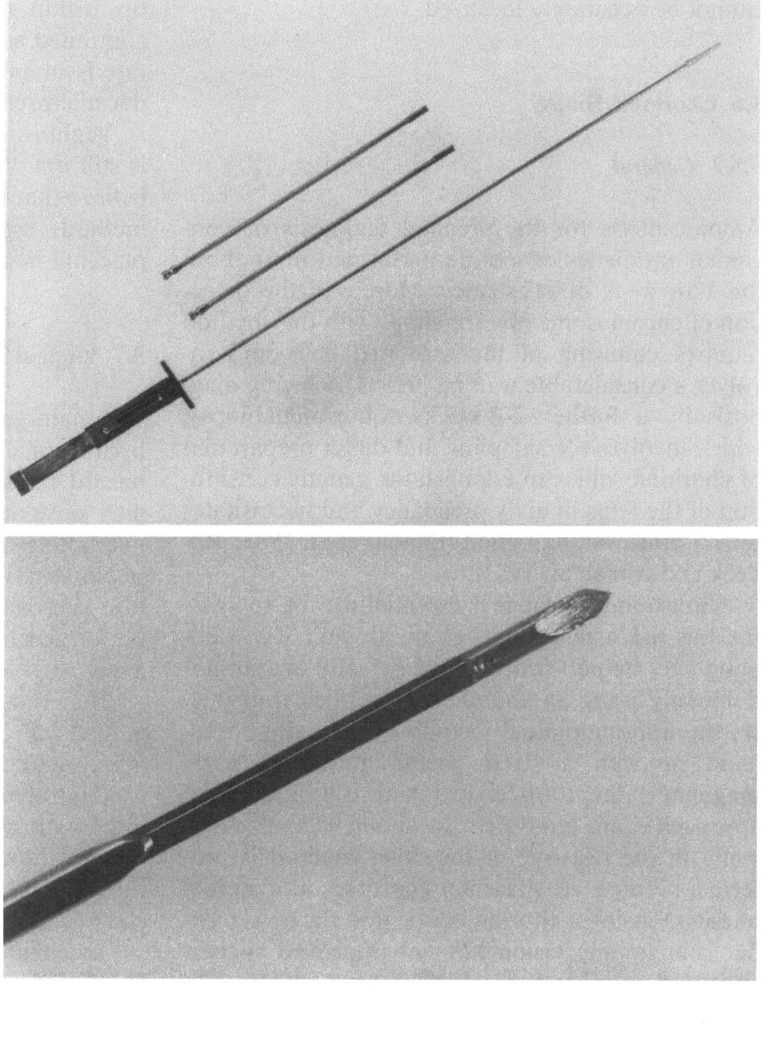

Fig. 9. Instrument set for vaginosonographically guided fine-needle biopsy. *Top:* biopsy needle with the guide slots of the vaginal scanner. The 30-cm needle is inserted into the slots. *Bottom:* tip of the biopsy needle

self. For the aspiration of a suspected pelvic wall recurrence following primary treatment for cervical carcinoma, a special needle has been constructed that fits into the guide slot of the vaginal scanner (Fig. 9). The procedure is done under continuous vaginosonographic control, analogous to a follicular aspiration. A suspected lymph node recurrence also can be biopsied with a vaginosonographically guided fine needle to retrieve material for histologic or cytologic evaluation.

There is a very low incidence of false-positive findings with fine-needle aspiration biopsy, yielding a specificity of 95% or more. Even 100% has been reported. However, the sensitivity is only 68%, i. e., a false-negative result may be expected in about one-third of cases (Nash et al. 1988). Hence a diagnosis cannot be made on the basis of negative findings, which warrant further diagnostic investigation or close clinical supervision.

Experiments with a transaxial 360° rectal scanner have not been rewarding, because the needle appears only as a bright spot, and its tip and shaft cannot be accurately localized.

3.6 Chorionic Biopsy

3.6.1 General

Amniocentesis for the prenatal diagnosis of congenital anomalies cannot be performed until about the 12th week of pregnancy. Moreover, the detection of chromosome abnormalities with this method requires culturing of the aspirated cells and involves a considerable waiting period, delaying diagnosis by a further 2-3 weeks. Chorionic biopsy, which involves the sampling and direct preparation of chorionic villi, can establish the genetic constitution of the fetus in early pregnancy and necessitates only a brief waiting period (Simoni et al. 1983; Rodeck and Morsman 1983).

Ultrasound guidance is essential for the successful, low-risk aspiration of chorionic villi. The sampling may be performed transcervically or transabdominally using an abdominally applied transducer; the transabdominal method may be done freehand or with a fixed needle guide. In early pregnancy, the transcervical and transabdominal approaches are equivalent in about 90% of cases, while in the rest one or the other method is preferred because of placental location or maternal obesity. Passing a chorionoscope into the cervix under sonographic vision has not improved success rates and is seldom practiced today. Ultrasound-guided aspiration is much simpler and avoids the abortion risk associated with cervical dilatation.

While the transcervical route was initially favored for chorionic biopsy, it was later shown that the transuterine (transmural) biopsy can be performed just as easily. This approach eliminates the risk of transcervical ascending infection (Brambati and Varotto 1985; Maxwell and Lilford 1985).

3.6.2 Chorionic Biopsy Using a Vaginal Scanner

Success with transabdominal ultrasound-guided chorionic biopsies led to initial trials with vaginosonographic guidance (Ghirardini et al. 1986). Following insertion of the vaginal scanner, the 0.9-mm needle is passed through the guide slot on the scanner and through the vaginal wall and myometrium to the placental margin. A 20-ml syringe containing 2-4 ml of medium is connected to the needle and, while continuous suction is applied, is moved gently inward and outward while keeping the needle tip within the chorion frondosum. Aspiration is continued as the needle is withdrawn, and the aspirate is immediately cleansed of residual blood under microscopic vision.

Vaginosonographically guided chorionic biopsy is still not widely practiced at the present time. We believe that it is a valuable adjunct to conventional methods, especially in the retroverted uterus with a placenta implanted on the fundus or posterior wall.

3.7 Vaginal Decompression of Hydrocephalus

The main goal of obstetric management in massive hydrocephalus is to enable a rapid delivery with no hazard to the mother. Severe fetal hydrocephalus may pose an absolute obstruction to delivery and necessitate an invasive procedure. In the past, cephalocentesis was done blindly, so that the puncture site could not be accurately localized, or was performed under transabdominal ultrasound guidance.

The first report on a vaginosonographically guided cephalocentesis was published in 1987 (Chayen and Rifkin 1987). An 18-gauge needle was used intrapartum to decompress hydrocephalus in a fetus with a biparietal diameter of 13 cm. At the time of the tap the cervical os was dilated to 2 cm, providing access for needle insertion, and 515 ml of clear cerebrospinal fluid was aspirated.

Encephalocentesis is an established procedure for decompressing massive fetal hydrocephalus to enable an uncomplicated vaginal delivery. With a fetus in the vertex presentation, the tap can be performed transabdominally or vaginally. The use of endosonography simplifies this procedure, and because the puncture is performed under visual control, cerebrospinal fluid can be safely withdrawn without risk of injury to maternal soft tissues. Of course, requirements for asepsis are the same as in a follicular aspiration.

References

Berkowitz RL, Lynch L, Chitkara U, Wilkins IA, Mehalek KE, Alvarez E (1988) Selctive reduction of multifetal pregnancies in the first trimester. N Engl J Med 318: 1043-1047

Bernaschek G, Deutinger J (1988) Endosonographie in Geburtshilfe und Gynäkologie. In: Gockel HP (ed) Jahrbuch der Radiologie 1988. Regensberg & Biermann, pp 105-119

Bernaschek G, Szalay S, Müller-Tyl E, Janisch H (1984) Ultraschallgeleitete Follikelpunktion zur Eizellgewinnung. In: Lutz H, Reichel L (eds) Ultraschalldiagnostik 83. Thieme, Stuttgart, p 68

Brambati B, Varotto F (1985) Infection and chorionic villus sampling. Lancet ii: 609

Breckwoldt M, Geisthövel F, Neulen J, Schillinger H (1988) Management of multiple conceptions after gonadotropin-releasing hormone analog/human menopausal gonadotropin/human chorionic gonadotropin therapy. Fertil Steril 49: 713-715

Chayen B, Rifkin MD (1987) Cephalocentesis. Guidance with a vaginal probe and vaginal needle placement. J Ultrasound Med 6: 221-223

Cohen J, Debache C, Pez JP, Junca AM, Cohen-Bacrie P (1986) Vaginal sonographically controlled ovarian puncture for oocyte retrieval for in vitro fertilization. J In Vitro Fertil Embryo Transfer 3: 309-313

Dellenbach P, Nisand I, Moreau L, Feger B, Plumere C, Gerlinger P, Brun B, Rumpler Y (1984) Vaginal sonographically controlled ovarian follicle puncture for egg retrieval. Lancet i: 1467

Dellenbach P, Nisand I, Moreau L, Feger B, Plumere C, Gerlinger P (1985) Vaginal sonographically controlled follicle puncture for oocyte retrieval. Fertil Steril 44: 656-662

Deutinger J, Reinthaller A, Müller-Tyl E, Riss P, Fischl F, Janisch H (1985) Vier Jahre erfolgreiche In vitro Fertilisierung an der II. Universitätsfrauenklinik Wien. Österr Ärzteztg 40: 52-58

Deutinger J, Reinthaller A, Riss P, Bernaschek G, Csaicsich P, Müller-Tyl E, Fischl F, Janisch H (1986a) Vaginosonography: an improvement for daily follicular measurement. Hum Reprod 1 [Suppl 2]: 7-8

Deutinger J, Reinthaller A, Bernaschek G, Riss P, Fischl F, Müller-Tyl E (1986b) Vaginal sonography and follicle puncture with a new vaginal scanner. In: 42nd Annual Meeting of the American Fertility Society, Toronto. Conjoint annual meeting - the American Fertility Society - the Canadian Fertility and Andrology Society, 27 Sept-2 Oct 1986, p 70

Deutinger J, Reinthaller A, Csaicsich P, Riss P, Fischl F, Bernaschek G, Müller-Tyl E (1987a) Follicular aspiration for in vitro fertilization: sonographically guided vaginal versus laparoscopic approach. Eur J Obstet Gynecol Reprod Biol 26: 127-133

Deutinger J, Riss P, Reinthaller A, Bernaschek G, Csaicsich P, Fischl F, Müller-Tyl E, Janisch H (1987b) Vergleich von vaginosonografischer, transabdomineller sonografischer und laparoskopischer Follikelpunktion zur Gewinnung von Eizellen im Rahmen eines In-vitro-Fertilisierungsprogrammes. Wien Med Wochenschr 5/6: 108-112

Deutinger J, Bernaschek G, Csaicsich P, Reinthaller A (1987c) Vaginal sonography and follicle puncture with a new vaginal scanner. In: 32nd Annual Convention of the American Institute of Ultrasound in Medicine, New Orleans, 6-9 Oct 1987, p 116

Deutinger J, Reinthaller A, Riss P, Bernaschek G, Csaicsich P, Fischl F, Müller-Tyl E (1987d) Comparison of the results of vaginal and abdominal follicle scans. Arch Gynecol Obstet 241: 171-176

Drugan A, Blumenfeld Z, Erlik Y, Timor-Tritsch IE, Brandes JM (1988) The use of vaginal sonography in infertility. In: Timor-Tritsch IE, Rottem S (eds) Vaginal sonography. Heinemann, London, pp 143-159

Enk L, Wikland M (1985) Preliminary report on a vaginal ultrasound transducer for oocyte collection. In: Fourth World Conference on In Vitro Fertilization, Melbourne, 1985. Brüel and Kjaer, Kopenhagen

Evans MI, Fletcher JC, Zador IE, Newton BW, Quigg MH, Struyk CD (1988) Selctive first-trimester termination in octuplet and quadruplet pregnancies: clinical and ethical issues. Obstet Gynecol 71: 289-296

Evers JLH, Larsen JF, Gnany GG, Sieck UV (1988) Complications and problems in vaginal sector scan-guided follicle aspiration. Fertil Steril 49: 278-282

Farquharson DF, Wittmann BK, Hansmann M, Yuen BH, Baldwin VJ, Lindahl S (1988) Management of quintuplet pregnancy by selectice embryocide. Am J Obstet Gynecol 158: 413-416

Fateh M, Ben-Rafael Z, Blasco L, Tureck RW, Meloni F, Mastroianni L (1986) Comparison of ultrasonographic transurethral and laparoscopic guided oocyte retrieval. Fertil Steril 46: 653-656

Feichtinger W, Kemeter P (1984) Laparoscopic or ultrasonically guided follicle aspiration for in vitro fertilization? J In Vitro Fertil Embryo Transfer 1: 244-249

Feichtinger W, Kemeter P (1986a) Vaginal sector scan sonography for needle guided vaginal follicle aspiration and other applications in gynecological routine and research. Fertil Steril 45: 722-725

Feichtinger W, Kemeter P (1986b) Erste Ergebnisse der vaginalen Follikelpunktion mit Hilfe eines vaginalen Ultraschallsektorscanners mit Punktionsnadelführung. In: Popp LW (ed) Gynäkologische Endosonographie. Aktualisierter Kongreßband der Ersten Arbeitstagung Gynäkologische Endosonographie 1985 in Hamburg. Klemke, Quickborn, pp 147-151

Feichtinger W, Kemeter P (1987a) Aspirationstechniken unter Ultraschallkontrolle. Verhandlungen der Deutschen Gesellschaft für Gynäkologie und Geburtshilfe. Arch Gynecol Obstet 242 (1-4): 141-146

Feichtinger W, Kemeter P (1987b) Conservative treatment of ectopic pregnancy by vaginal aspiration under sonographic control and methotrexate injection. Lancet ii: 381

Feichtinger W, Szalay S, Beck A, Kemeter P, Janisch H (1981) Die Gewinnung reifer menschlicher Eizellen mittels

Laparoskopie zum Zwecke der In-vitro-Fertilisierung. Geburtshilfe Frauenheilkd 41: 400-403
Ghirardini G, Popp LW, Camurri L, Stoeckenius M (1986) Vaginosonographic guided chorionic villi needle biopsy (vaginal chorionic villi sampling). Eur J Obstet Gynecol Reprod Biol 23: 315-319
Gleicher N, Friberg J, Fullan N, Giglia RV, Mayden K, Kesky T, Siegel I (1983) Egg retrieval for in vitro fertilization by sonographically controlled vaginal culdocentesis. Lancet ii: 508-509
Haake K-W, Alexander P, Baier D, Glander H-J, Weber W, Zimmermann G (1988) Zur Problematik einer Mehrlingsschwangerschaft nach In-vitro-Fertilisation und Embryotransfer. Geburtshilfe Frauenheilkd 48: 374-375
Habermann PG, Hünlich T, Siebzehnrübl E, Trotnow S (1986) Die ultraschallkontrollierte Follikelpunktion in der "Zwei-Instrumente-Technik". Ultraschall 7: 179-182
Hammarberg K, Enk L, Nilsson L, Wikland M (1987) Oocyte retrieval under the guidance of a vaginal transducer: evaluation of patient acceptance. Hum Reprod 2: 487-490
Hansmann M, Hackeloer B-J, Staudach A (1985) Ultrasound diagnosis in obstetrics and gynecology. Springer, Berlin Heidelberg New York, p 73
Hess LW, Peaceman A, O'Brian WF, Winkel CA, Cruikshank DP, Morrison JC (1988) Adnexal mass occurring with intrauterine pregnancy: report of fifty-four patients requiring laparotomy for definitive management. Am J Obstet Gynecol 158: 1029-1034
Hobbins J (1988) Selective reduction - a perinatal necessity? N Engl J Med 318: 1062-1063
Howe RS, Wheeler C, Mastroianni L Jr, Blasco L, Tureck R (1988) Pelvic infection after vaginal ultrasound-guided ovum retrieval. Fertil Steril 49: 726-728
Janssen-Caspers HAB, van Gent I, Wladimiroff JW, Alberda AT, Zeilmaker GH (1986) Ultrasonically guided percutaneous and laparoscopic oocyte aspiration: a comparative study. Eur J Obstet Gynecol Reprod Biol 21: 219-223
Janssen-Caspers HAB, Wladimiroff JW, van Gent I, Alberda AT, Leerentveld RA, Zeilmaker GH, Drogendijk AC (1988) Ultrasonically guided percutaneous and vaginal follicle aspiration; a comparative study. Hum Reprod 3: 337-339
Kanhai HH, van Russel EJC, Meerman RJ (1986) Selective termination in quintuplet pregnancy during first trimester. Lancet ii: 1447
Kemeter P, Feichtinger W (1986) Vaginal oocyte retrieval using a vaginal sector scan probe with an automated puncture device. Hum Reprod 1: 21-24
Kratochwil A (1969) Ein neues vaginales Ultraschall-Schnittbildverfahren. Geburtshilfe Frauenheilkd 29: 379-385
Lenz S (1984) Ultrasonic-guided follicle puncture under local anesthesia. J In Vitro Fertil Embryo Transfer 1: 239-243
Lenz S, Bang J (1986) Transvesical/transuterine collection of oocytes by ultrasound. J In Vitro Fertil Embryo Transfer 3: 259-260
Lenz S, Lauritsen JG (1982) Ultrasonically guided percutaneous aspiration of human follicles under local anesthesia: a new method of collecting oocytes for in vitro fertilization. Fertil Steril 38: 673-677
Lenz S, Lauritsen JG, Kjellow M (1981) Collection of human oocytes for in vitro fertilization by ultrasonically guided follicular puncture. Lancet i: 1163-1164
Lenz S, Leeton J, Renou P (1987) Vaginal recovery of oocytes for in vitro fertilization using vaginal ultrasound. J In Vitro Fertil Embryo Transfer 4: 51-55
Lewin A, Maragalioth EJ, Rabinowitz R, Schenker JG (1985) Comparative study of ultrasonically guided percutaneous aspiration with local anesthesia and laparoscopic aspiration of follicles in an in vitro fertilization program. Am J Obstet Gynecol 151: 621-625
Lewin A, Laufer N, Rabinowitz R, Schenker JG (1986a) Ultrasonically guided oocyte recovery for in vitro fertilization: an improved method. J In Vitro Fertil Embryo Transfer 3: 370-373
Lewin A, Laufer N, Rabinowitz R, Margalioth EJ, Bar I, Schenker JG (1986b) Ultrasonically guided oocyte collection under local anesthesia: the first choice method for in vitro fertilization - a comparative study with laparoscopy. Fertil Steril 46: 257-261
Lopata A, Johnston IWH, Leeton JF, Muchnicki D, Talbot JMcK, Wood C (1974) Collection of human oocytes at laparoscopy and laparotomy. Fertil Steril 25: 1030-1038
Lopes P, Talmat C, Thiery M, Defoort P, Dhont M (1985) Partial termination of a quintuplet pregnancy. Z Geburtshilfe Perinatol 189: 239-240
Mao KR, Haines CJ, Tam PPL (1986) Successful oocyte retrieval and in vitro fertilization following ultrasound-directed aspiration of endometriomas. J In Vitro Fertil Embryo Transfer 3: 333
Maxwell DJ, Lilford RJ (1985) An interesting ultrasonic observation following chorionic villus sampling. J Clin Ultrasound 13: 343-344
Meldrum DR, Chetkowski RJ, Steingold KA, Randle D (1984) Vaginal ultrasound scanning of ovarian follicles. Fertil Steril 42: 803-805
Nash JD, Burke TW, Woodward JE, Hall KL, Weiser EB, Heller PB (1988) Diagnosis of recurrent gynecologic malignancy with fine-needle aspiration cytology. Obstet Gynecol 71: 333-337
Parsons J, Booker M, Goswamy J, Akkermans J, Riddle A, Sharma V, Wilson L, Withead M, Campbell S (1985) Oocyte retrieval for in-vitro fertilization by ultrasonically guided needle aspiration via the urethra. Lancet i: 1076-1077
Popp LW (1986a) Endosonographische Methoden in Geburtshilfe und Gynäkologie. In: Popp LW (ed) Gynäkologische Endosonographie. Aktualisierter Kongreßband der Ersten Arbeitstagung Gynäkologische Endosonographie 1985 in Hamburg. Klemke, Quickborn, pp 35-50
Popp LW (1986b) Möglichkeiten der vaginosonographisch gezielten Punktion in einem In-vitro Fertilisierungsprogramm. In: Popp LW (ed) Gynäkologische Endosonographie. Aktualisierter Kongreßband der Ersten Arbeitstagung Gynäkologische Endosonographie 1985 in Hamburg. Klemke, Quickborn, pp 153-162
Popp LW, Müller-Holve W (1984) Die Indikationsstellung zur Vaginosonographie. In: Lutz H, Reichel L (eds) Ultraschalldiagnostik 83. Thieme, Stuttgart, pp 74-76
Popp LW, Lemester S, Hinrichs S, te Heesen D, Müller-Holve W, Martin K (1985) Intravaginale Ultraschalldiagnostik (Vaginosonographie) - erste Erfahrungen mit dem Panoramasektor. In: Judmaier G, Frommhold H, Kratochwil A (eds) Ultraschalldiagnostik 84. Thieme, Stuttgart, pp 320-322
Reinthaller A, Deutinger J, Müller-Tyl E, Riss P, Fischl F, Janisch H (1985a) In vitro fertilization and embryo transfer program, 2nd Department of Obstetrics and Gynecology. J In Vitro Fertil Embryo Transfer 2: 238-240
Reinthaller A, Deutinger J, Riss P (1985b) Wandel in Diagnose und Therapie der Extrauteringravidität. Wien Med Wochenschr 3: 61-64

Riddle AF, Sharma V, Mason BA, Ford NT, Pampiglione JS, Parsons J, Campbell S (1987) Two years' experience of ultrasound-directed oocyte retrieval. Fertil Steril 48: 454-458

Rodeck CH, Morsman JM (1983) First trimester chorion biopsy. Br Med Bull 39: 338-341

Salat-Baroux J, Aknin J, Antoine JM, Alamowitch (1988) The management of multiple pregnancies after induction for superovulation. Hum Reprod 3: 399-401

Schulman J (1985a) Outpatient in vitro fertilization using vaginal oocyte retrieval and local anesthesia. N Engl J Med 312: 1639

Schulman J (1985b) Laparoscopy for in vitro fertilization: end of an era (letter). Fertil Steril 44: 713

Schwimer S, Lebovic J (1984) Vaginal pelvic ultrasonography. J Ultrasound Med 3: 381-386

Schwimer S, Rothman C, Lebovic J, Oye D (1984) The effect of ultrasound coupling gels on sperm motility in vitro. Fertil Steril 42: 946-947

Schwimer S, Marik J, Lebovic J (1985) Percutaneous ovarian cyst aspiration using continuous vaginal ultrasonography monitoring. J Ultrasound Med 4: 259-261

Simoni G, Brambati B, Rossella DF, Terzoli GL, Ferrari M, Farccaro M (1983) Efficient direct chromosome analyses and enzyme determinations from chorionic villi samples in the first trimester. Hum Genet 63: 349-357

Smith DF, Annesley W (1985) Ultrasound guided manual lysis of abscess loculations. J Clin Ultrasound 13: 350-353

Steptoe PC, Edwards RG (1970) Laparoscopic recovery of preovulatory human oocytes after priming of ovaries with gonadotropins. Lancet i: 683-689

Taylor PJ, Wiseman D, Mahadevan M, Leader A (1986) "Ultrasound rescue": a successful alternative form of oocyte recovery in patients with periovarian adhesions. Am J Obstet Gynecol 154: 240-244

Timor-Tritsch IE, Rottem S, Thaler I (1988) Review of vaginal ultrasonography: a description with clinical application. Ultrasound Q 6: 1-34

Trotnow S, Habermann PG, Hünlich T, Siebzehnrübl E (1985) A new set of instruments for sonographically controlled follicular puncture. Arch Gynecol 236: 211-217

Vilaro MM, Rifkin MD, Pennell RG, Baltarowich OH, Needleman L, Kurtz AB, Goldberg BB (1987) Vaginal ultrasound. A technique for evaluation of nonfollicular pelvic masses. J Ultrasound Med 6: 697-701

Wikland M, Nilsson L, Hansson R, Hamberger L, Janson PO (1983) Collection of human oocytes by the use of sonography. Fertil Steril 39: 603-608

Wikland M, Enk L, Hammarberg K, Nilsson L (1987) Use of a vaginal transducer for oocyte retrieval in an IVF/ET program. J Clin Ultrasound 15: 245-251

Wood C, Leeton JF, Talbot JMcK, Trounson AO (1981) Technique for collecting oocytes for in vitro fertilization. Br J Obstet Gynaecol 88: 756-759

Vaginal Doppler Techniques

1 Basic Principles of Doppler Ultrasound

The Doppler effect, named for the Austrian physicist Christian Johann Doppler, is based on the fact that all wave phenomena, including sound and light, undergo a change of frequency and wavelength when the source of the wave and the observer are moving relative to each other. In acoustics, the number of sound vibrations that reach the listener per unit time change according to whether the sound source is moving toward or away from the listener. The original basis on which Doppler formulated his law – the color change observed in distant, receding stars (Doppler 1843) – applies with equal validity to acoustic waves (Buys Ballot 1845).

Today the Doppler effect is used in medicine to provide information on moving structures such as blood cells, which in turn can furnish indirect information on the status of the cardiovascular system. Ultrasound waves emitted from a stationary source are reflected at the interface between the red blood cells and the blood plasma. The motion of the cells produces a change of frequency in the reflected wave, whose frequency shift is given by the formula

$$f(0) - f(r) = f(d)$$

where $f(0)$ is the transmitted sound frequency, $f(r)$ is the sound frequency picked up by the receiver, and $f(d)$ is the frequency shift. The receiver compares the frequencies of the reflected signals with those of the transmitted signals and determines the frequency shift. When the sound velocity, wavelength, and incidence angle of the Doppler beam are known, the frequency shift can be used to determine the velocity of the moving blood cells. The magnitude of the frequency shift depends on the ultrasound frequency and the blood flow velocity.

The Doppler frequency shift is related to the velocity of blood cell movement by the formula

$$f(d) = \frac{2\, f(0)\ v \cos \alpha}{c}$$

where $f(d)$ is the frequency shift, $f(0)$ is the transmitted frequency, v is the velocity of the reflector, α is the angle between the incident sound wave and the direction of motion of the reflector (the insonation angle), and c is the velocity of sound in the medium (average 1540 m/s).

It is significant that the frequency shift depends on the cosine of the insonation angle. The best results are obtained with an angle α of 45°, whose cosine equals 1. If the incident sound wave is perpendicular to the study vessel, the cosine equals zero and no signal is obtained.

2 Vaginal Probes

2.1 Continuous Wave Doppler

In a continuous-wave Doppler system, the transducer is designed so that one crystal transmits sound continuously while a second crystal functions continuously as a receiver. The main advantage of continuous-wave devices is that they can measure extremely high flow velocities. The frequency shifts can be made audible and used in the diagnosis of vascular stenosis. The main disadvantage of these devices is that they receive signals from all blood vessels in the path of the Doppler beam, regardless of their distance from the transducer. Due to their lack of depth selectivity, continuous-wave devices have only limited applications in obstetrics, and the superimposition of frequency shifts from maternal and fetal vessels at different depths can lead to confusing results.

Continuous-wave Doppler instruments are far simpler in their design and operation than devices which function in the pulsed mode.

2.2 Pulsed Doppler

The problems noted above can be overcome by employing a pulsed Doppler sound source. Pulsed systems have the capacity of "range gating," which allows Doppler signals to be recorded selectively from a sample volume located a specified distance from the transducer. This eliminates the reception of unwanted signals originating from structures other than the intended target. To be useful in obstetrics, pulsed Doppler ultrasound must be combined with B-mode imaging of the selected vessel (Bernaschek 1980; Diwan 1983; Jeanty 1984). Simultaneous B-mode imaging of the region of interest makes it possible to receive and record Doppler signals from a precisely defined, range-gated tissue volume, which can be positioned within a single blood vessel.

Pulsed Doppler transducers are technically more complex and far more expensive than continuous-wave devices. Pulsed devices emit very short sound pulses about 1 μs in duration. The time delay between sound transmission and echo reception defines the range from which signals are to be received. Thus, if the delay between successive pulses is about 100 times the pulse length, or 100 μs, this would correspond to a distance of 7.5 cm between the gate and transmitter. The width of the gate is determined by the time span during which the receiver is able to receive a returning signal. Thus, the distance of the gate from the transducer can be calculated from the time delay between pulse transmission and signal reception, since the sound velocity in tissues (1540 m/s) is known to an accuracy of ±6%. The relation between velocity and wavelength is given by the formula

$$c = f\lambda$$

where c is the sound velocity in the medium (cm/s), f is the frequency (vibrations per second, by definition >2 MHz for ultrasound), and λ is the wavelength in cm.

The higher the selected frequency (8-10 MHz), the shorter the wavelength and the greater the absorption rate. Higher frequencies provide better resolution for the examination of superficial blood vessels. The frequency range of 2-4 MHz is advantageous for deeper vessels, as it gives the best trade-off between resolution (accuracy of measurement) and sound absorption (loss of information). By combining real-time ultrasound imaging with pulsed Doppler, it is possible to position the sample gate precisely over the area of interest and analyze flow in specific vessels.

In theory, it is possible to make the Doppler gate extremely narrow (2-3 mm) and position it within the lumen of an artery to eliminate unwanted signals from the moving vessel walls. In practice, however, a high-pass filter is needed to eliminate wall signals in all but the largest vessels. Filters that eliminate frequencies of 125 and 250 Hz have proven most valuable in studies of fetal blood flow. Given the proximity of the arteries to their accompanying veins, there is always a possibility that venous "noise" will be superimposed on the arterial signals. The latest generation of pulsed Doppler systems helps avoid this problem through bidirectional signal processing. This indicates whether the direction of the flow is toward or away from the transmitter, making it possible to differentiate between arterial and venous flow.

High flow velocities can complicate the analysis of Doppler waveforms. Higher velocities lead to an aliasing or fold-over effect in which the peaks of the Doppler waves are "folded over" into another channel. This phenomenon is observed in in vitro models when the flow velocity equals or exceeds 50 cm/s (Graf et al. 1985; Kaesemann and Trenkel 1986). It can be corrected by restoring the wave peaks electronically or by increasing the pulse repetition frequency (PRF), which can be controlled by the operator to some degree.

2.3 Doppler Waveforms

The flow velocity waveforms recorded with the Doppler probe are usually displayed concurrently with the sectional image (Fig. 1). The rate at which the waveforms are displayed, and thus the number of recorded cardiac cycles, is controlled by a selector switch. A 2.5-s display is adequante for visual assessment. A waveform is accepted for analysis only if it displays constant amplitudes over several cardiac cycles and presents a uniform signal density.

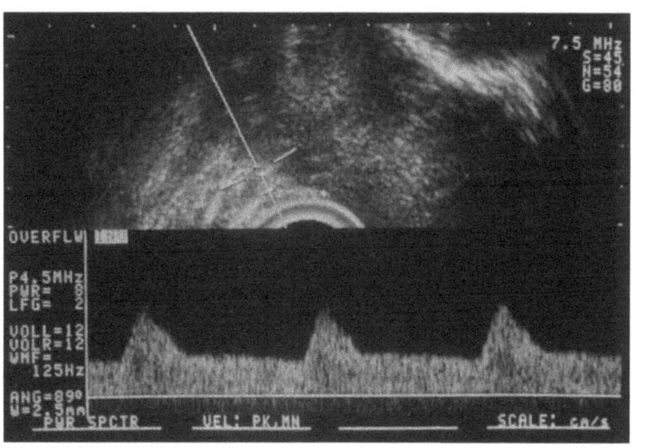

Fig. 1. Original image from a vaginal pulsed Doppler study of the right uterine artery. The flow velocity waveform of the gated vessel is displayed below the real-time image

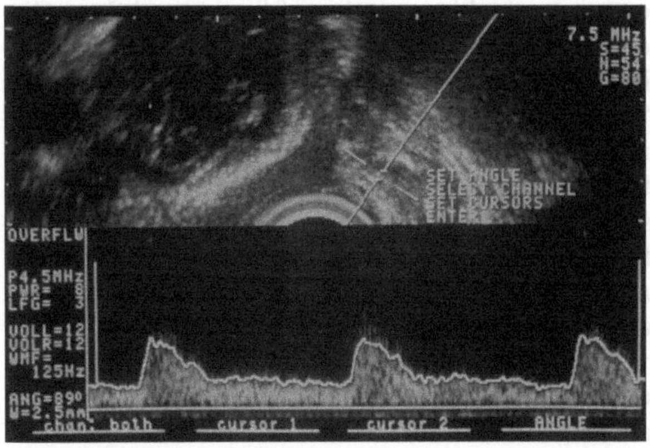

Fig. 2. Vaginal pulsed Doppler study of the left uterine artery. Integrated software automatically outlines the flow velocity waveform envelope

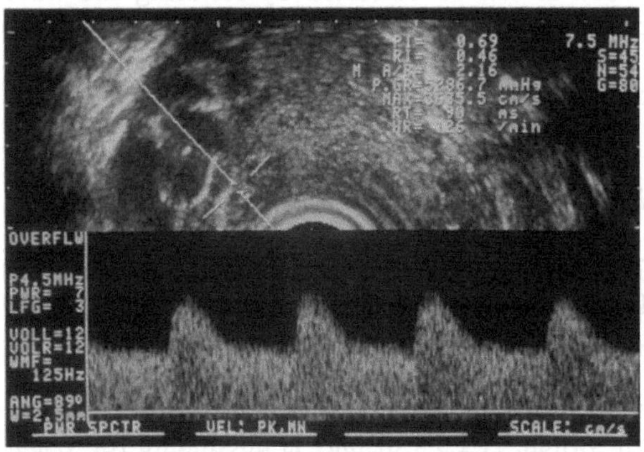

Fig. 3. Vaginal pulsed Doppler study of the right uterine artery. The flow velocity waveform is evaluated by integrated software. The results of the computation are displayed in the upper part of the image

Flow-related frequency changes are analyzed by an on-line computer program (Bernaschek and Deutinger 1988). This software allows for automatic outlining or manual tracing (with a light pen) of the Doppler waveform envelope (Fig. 2). When the insonation angle is entered, the program can determine the height of the maximum frequency shift in systole (A), the minimum frequency shift in diastole (B), the mean blood flow velocity (area integral), and the heart rate. All these data are stored by the system and are displayed on screen when the study is completed (Fig. 3). A permanent record can be made on video tape. Only good-quality waveforms that portray at least four successive cardiac cycles are evaluated. Doppler waveform analysis centers on the relation between the systolic and diastolic blood flow velocities in the study vessel, with clinical and experimental evidence showing that diastolic flow changes have the greater significance (Giles et al. 1982).

These quantities can be used to calculate parameters that are independent of the insonation angle, most notably the pulsatility index (PI), the resistance index (RI), and the A/B ratio. The PI is given by the formula

$$\frac{A-B}{\text{mean}}$$

in which the difference between the peak systolic flow and least diastolic flow is divided by the mean flow velocity (Gosling and King 1975). The RI is defined by the formula $(A-B)/A$. Although the A/B ratio, in common with the RI, expresses only two points on the waveform (Fitzgerald and Drumm 1977), it is the major parameter used today for Doppler waveform analysis (Schulman 1987 a).

In recent years Doppler studies of fetal blood flow have been used increasingly in high-risk pregnancies for the detection of impending fetal compromise. Abnormal waveforms in the umbilical vein and aorta (Eik-Nes et al. 1980), umbilical artery (Gill and Kossoff 1979), and internal carotid artery (Wladimiroff et al. 1986) have been shown to correlate with fetal growth retardation.

Initially, Doppler flow studies focused on the quantitative determination of blood volumes. But blood volume can be measured only if the cross section of the study vessel is known (Gill and Kossoff 1979; Eik-Nes et al. 1980). This introduces an additional parameter, the calculated cross-sectional area of the vessel, since volume flow represents the product of the blood flow velocity and the vessel cross-sectional area. Although the cross section can be estimated by averaging several measurements of the vascular diameter, the result is at best an approximation, if only because of pulsatile fluctuations in the vessel size. It must also be considered that an error of only 1 mm in measuring the diameter of a small vessel becomes greatly magnified, since the linear dimension is raised to the third power to calculate volume. As a result, interest in the measurement of volume flow has been confined to the largest fetal blood vessels such as the aorta or umbilical vein. In vitro studies have shown significant variations in calculated volume flow as a function of blood flow velocity and the type of filter used (Kaesemann and Trenkel 1986). The clinical measurement of volume flow will not be feasible until computer-controlled equipment is available that can monitor and make allowance for dynamic changes in vessel diameters during the cardiac cycle. One such application is currently under development (J. W. Wladimiroff 1987, personal communication). At present, however, the changes in vessel diameter during the cardiac cycle make this method too uncertain for routine use. Even a small error in measurement can lead to a 30% error in the calculated flow (Fendel et al. 1983).

The absolute blood flow velocity in a vessel can be measured only if the course of the vessel is precisely known over a significant portion of its length (Gill and Kossoff 1977). Thus, absolute flow measurements on the fetal side of the placenta are feasible only for the aorta, since long segments of the umbilical vessels are seldom visualized on longitudinal scans. For absolute flow measurements it is necessary to know precisely the angle between the Doppler beam and the selected vessel, for a discrepancy of only a few degrees can significantly alter the result (Fendel et al. 1983; Eik-Nes et al. 1984; Watson et al. 1987). However, the insonation angle can be determined only if the course of the vessel is precisely known, and the angle can change unpredictably as a result of fetal movements. Given the numerous sources of error, absolute flow measurements are usually dispensed with in favor of a description of the flow velocity waveform during a complete cardiac cycle (Stuart et al. 1980). The waveforms are analyzed according to mathematical rules, and angle-dependent sources of error are eliminated.

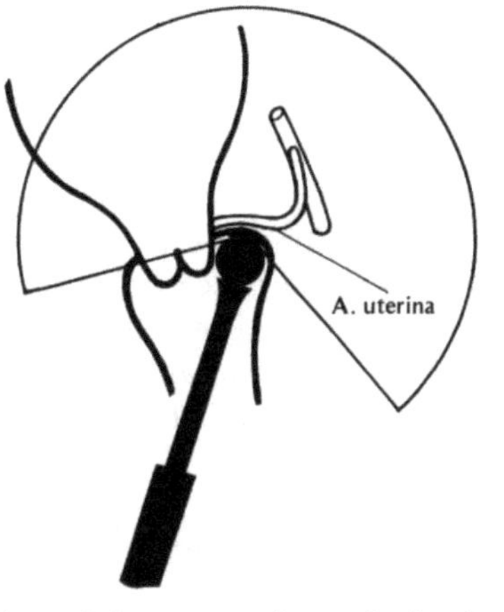

Fig. 4. Position of the scanner for vaginal pulsed Doppler sampling of the left uterine artery

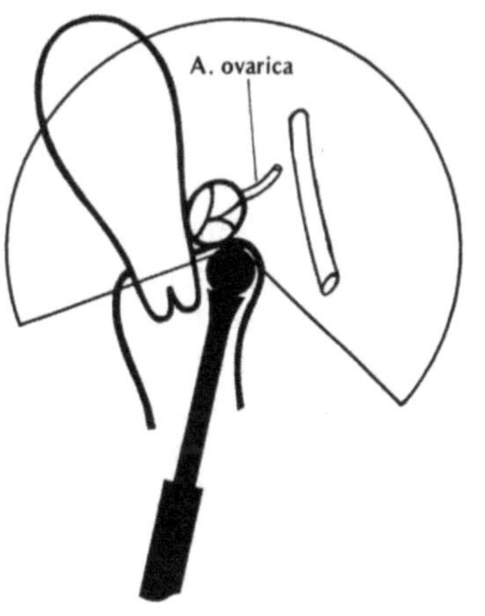

Fig. 5. Position of the scanner for vaginal pulsed Doppler sampling of the left ovarian artery

3 Vaginal Pulsed Doppler Techniques

3.1 Examination Procedure

For vaginal pulsed Doppler studies we use a vaginal probe with a 240° forward scan angle. This gives coverage of the entire lesser pelvis and makes orientation simple (Deutinger and Bernaschek 1986, 1987). The imaging frequency is 7.5 MHz, the Doppler frequency 4.5 MHz. The Doppler beam can be moved from the center of the image to any point within the 240° field of view. For most studies the optimum pulse repetition frequency is 5.2 kHz with a focusing depth of 7–8 cm and a 125-Hz filter. The power output according to the manufacturer is less than 100 mW/cm^2 (Rudelstorfer et al. 1988 a).

The patient is placed in the lithotomy position, which allows the greatest probe mobility. First the probe is positioned so that a sagittal scan is obtained. When the examiner has become oriented in the lesser pelvis and has identified essential landmarks (cervix, gestational sac, ovary, pelvic wall), he positions the probe in the lateral vaginal fornix and rotates it 90° onto a coronal plane to locate the uterine artery in the parametrium (Fig. 4). With some practice the examiner can consistently demonstrate a pulsating, elongated echo 2–4 mm in diameter. The uterine artery can be recognized by its typical course, a variable portion of which is delineated, but its major distinguishing feature is its conspicuous pulsation (Rudelstorfer et al. 1988 b).

The ovarian artery and iliac artery also are visualized from the lateral vaginal fornix (Fig. 5). The area between the adnexa and pelvic wall is demonstrated on a cross-sectional scan, whereupon the ovarian and iliac arteries are located by rotating the probe slightly (Deutinger et al. 1987 d). With the Doppler beam trained on the target vessel, the sample gate is opened until signals are received from the entire vessel lumen.

Very little pressure should be applied to the probe during the examination. When the study vessel has been located on the real-time image, the Doppler gate is positioned over the vessel, and the frequency shift signals are received and recorded. It takes 5-15 min to record waveforms from both uterine arteries or both ovarian arteries depending on the gestational age or follicle size.

3.2 Advantages of Vaginal Doppler

Initial results of Doppler flow studies in obstetric patients were reported in 1977 (Gill 1977). To date, however, there have been few reports on Doppler studies of blood vessels in the female pelvis (Ishihara et al. 1984; Taylor et al. 1985; DuBose et al. 1985). This may be largely because such studies are difficult to perform with conventional abdominally applied transducers (linear-array scanner with a fixed-angle pulsed Doppler beam) in patients with prominent anterior iliac spines. In obese patients, too, it can be difficult to position the beam at an optimum angle with respect to the study vessel.

"Duplex" scanners that combine real-time imaging and pulsed Doppler have now been developed for vaginal use. These scanners offer several advantages for blood flow studies in the lesser pelvis (Deutinger et al. 1987 a; Rudelstorfer et al. 1987). A major advantage is the proximity of the vaginal probe to the structures of interest. The mobility of the probe enables the transducer to be positioned in the lateral vaginal fornix, where it is within about 2 cm of the target vessels. This enables the use of higher ultrasound frequencies (5 or 7.5 MHz), resulting in improved resolution.

In 1985 initial attempts were made to record flow velocity waveforms from arterial vessels of the lesser pelvis by the vaginal route. The recorded signals were compared with signals obtained intraoperatively by direct application of the transducer to the vessels in order to investigate the significance of different waveform patterns (Taylor et al. 1985). There are isolated reports of transabdominal Doppler studies of the uterine artery in nonpregnant patients (DuBose et al. 1985; Bowie et al. 1987). In the past, vaginal Doppler studies were performed exclusively with continuous-wave transducers (Schulman et al. 1986). But these measurements are susceptible to interference from other vessels in the beam path such as the venous plexus, and the accuracy of the results is often unsatisfactory, although the waveform "signatures" of major pelvic vessels such as the external iliac artery, internal iliac artery, uterine artery, and ovarian artery can usually be distinguished from one another even without B-mode imaging (Taylor et al. 1985).

Vaginal imaging of the vessel followed by pulsed Doppler interrogation serves to identify the vessel uniquely, while the pulsed-mode Doppler eliminates extraneous signals (Deutinger et al. 1987 b). In clinical use this method has been found to have very high patient acceptance in terms of comfort and duration when adequate preliminary counseling is given.

One problem with the vaginal method in pregnant patients near term is that the descent of the presenting fetal part can make it difficult or impossible to place the probe in the lateral vaginal fornix. This in turn can make it difficult to sample the uterine artery or the fetal internal carotid artery.

4 Clinical Applications

4.1 The Uterine Artery

4.1.1 Normal Pregnancy

Doppler ultrasound shows great promise as a technique for enhancing our ability to assess fetal well-being. The simple, noninvasive method of Doppler ultrasound has been used for more than 10 years to measure blood flow velocities in fetal vessels (Gill and Kossoff 1977, 1979; McCallum et al. 1978; Eik-Nes et al. 1980). Although it is customary in obstetrics always to speak of the "fetomaternal unit," Doppler ultrasound has been used for some time to study flow patterns on the fetal side of the placenta while largely disregarding the no less important maternal vascular system. So far the problem of evaluating the maternal component, i. e., the blood flow in the pregnant uterus, by Doppler ultrasound has not been satisfactorily resolved. While the fetal aorta and umbilical vessels have been studied with considerable success, Doppler investigations on the uterine side have been limited to the arcuate arteries, even though the diagnostic value of these studies is questionable. Clearly, the tendency to sample the arcuate vessels is due to the fact that the study of other maternal vessels, especially the uterine artery, has been extremely difficult and is only now becoming feasible through application of the vaginal method.

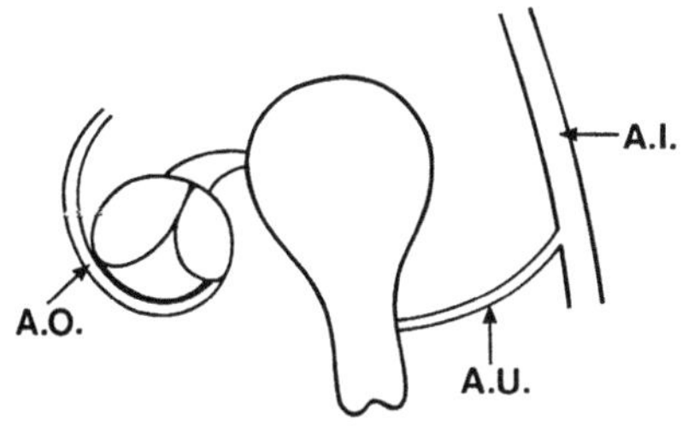

Fig. 6. Schematic representation of the blood supply of the female genital organs. *A. I.*, internal iliac artery; *A. U.*, uterine artery; *A. O.*, ovarian artery

The uterus receives most of its blood supply from the two uterine arteries; small anastomoses with the ovarian vessels are hemodynamically insignificant (Fig. 6). After entering the myometrium, the uterine arteries divide into 10-15 arcuate arteries on each side. These vessels encircle the uterus and anastomose with one another on the anterior and posterior midline. Radial arteries arise from the arcuates and pass more deeply into the myometrium. The radial arteries divide to form the basal arteries, which supply blood to the basal endometrium and the spiral arteries. Blood is carried by the spiral arteries to the placenta, where it perfuses the decidua and intervillous space (Figs. 7 and 8).

To determine the normal limits for uterine artery waveforms, we performed vaginal pulsed Doppler flow studies in 88 women with uncomplicated singleton pregnancies between the 7th and 40th weeks of gestation. All patients accepted for evaluation in the study had their dates confirmed by ultrasound in early pregnancy and delivered an infant of normal birthweight by the Hohenauer chart. Patients with premature deliveries were excluded. All examinations were performed in patients who had not smoked for at least 3 h. To evaluate total uterine perfusion in the various stages of pregnancy, we averaged the findings in the right and left uterine arteries.

In very early pregnancy we found a high pulsatility index with high systolic blood flow velocities and minimal end-diastolic flow velocities (Fig. 9). As the pregnancy progresses there is an increase in blood flow velocities and especially in the diastolic flow velocity, leading to a significant decrease in the A/B ratio. The greatest changes in our studies are recorded at the start of the second trimester (Fig. 10). After the end of the second trimester there is no longer an appreciable increase in diastolic flow velocity (Figs. 11, 12) (Deutinger et al. 1988 c).

Comparing the relative contributions of the right and left uterine arteries to the uterine blood supply,

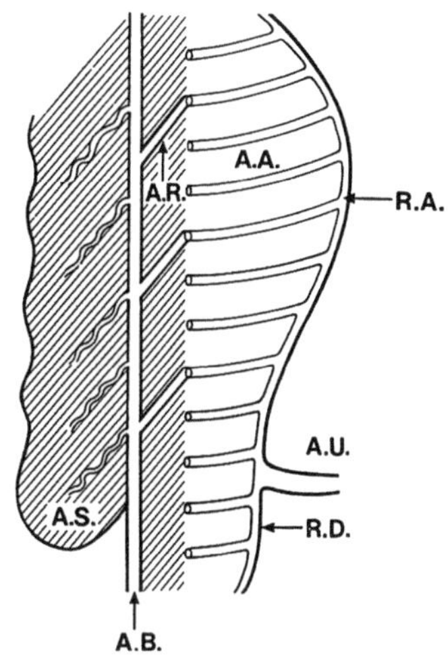

Fig. 7. Schematic longitudinal section of the uterine blood supply. The uterine artery *(A. U.)* devides on both sides into an ascending branch *(R. A.)* and a descending branch *(R. D.)*. These give rise to 10-15 arcuate arteries *(A. A.)*, which course within the uterine wall, parallel to its surface. The radial arteries *(A. R.)* arise from the arcuates and supply the basal arteries *(A. B.)* and the spiral arteries *(A. S.)*

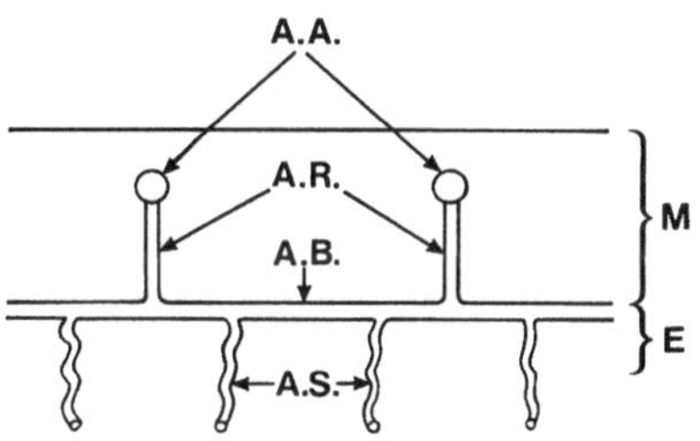

Fig. 8. Schematic cross-section of the arterial vessels in the uterine wall. *M*, myometrium; *E*, endometrium; *A. A.*, arcuate artery; *A. R.*, radial artery; *A. B.*, basal artery; *A. S.*, spiral artery

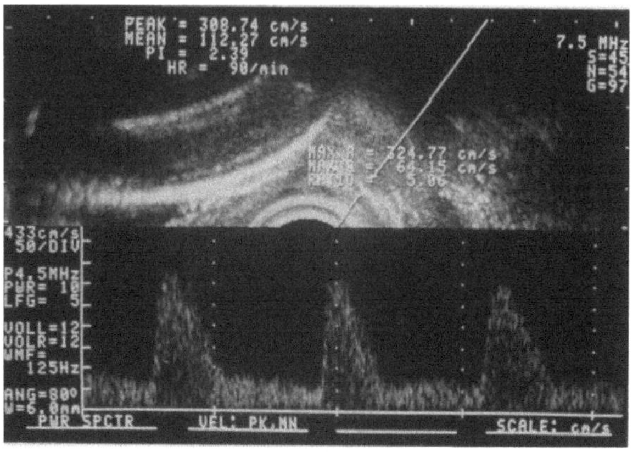

Fig. 9. Vaginal pulsed Doppler study of a left uterine artery in the first trimester. There is a very large difference between the maximum systolic and end-diastolic flow velocities

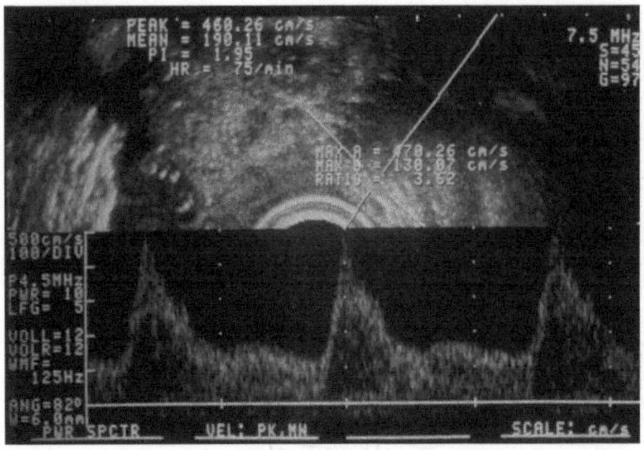

Fig. 10. Vaginal pulsed Doppler study of a left uterine artery in the second trimester. The diastolic flow velocity is markedly higher than in the first trimester. The slight diastolic notch is caused by the "windkessel" action of the aorta

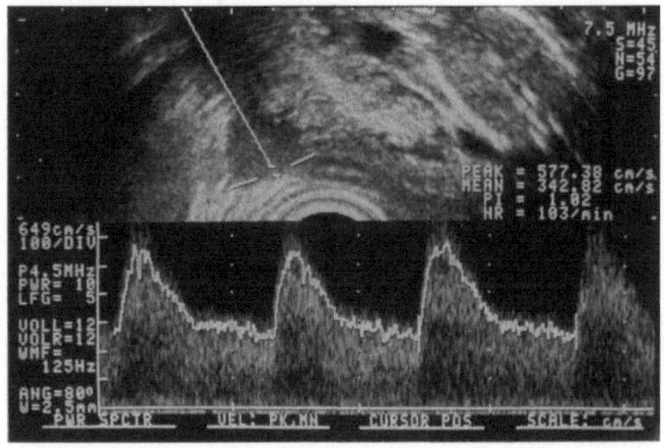

Fig. 11. Vaginal pulsed Doppler study of a right uterine artery in the third trimester. There is no significant change in diastolic blood flow velocity compared with the second trimester

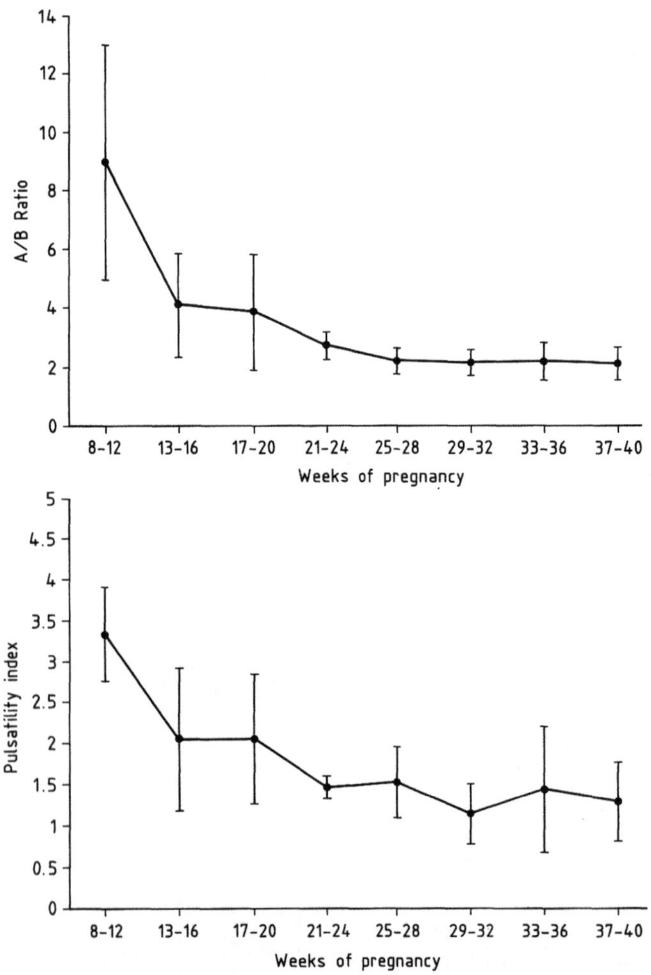

Fig. 12. *Top:* the A/B ratio show a significant decline in the first 25 weeks of pregnancy and remains fairly constant after 26 weeks. *Bottom:* the pulsatility index shows a less pronounced decline, but as with the A/B ratio there is no significant change after 26 weeks

we note some marked disparities in early pregnancy that could account for the high scatter of individual values measured in the first and second trimesters. However, when we consider the average values for both vessels, no discrepancies are observed. The difference between the A/B ratios of the right and left uterine arteries decreases in the second trimester and especially in the third trimester (Schulman et al. 1987; Deutinger et al. 1988 c, e; Rudelstorfer et al. 1988 c). By the third trimester the difference is only 0.4 ± 0.3 (Fig. 13). Thus, abnormal perfusion of the uterus should be suspected in the third trimester if the discrepancy between the left and right uterine artery A/B ratios is greater than 1.

The change of blood flow velocity in the uterine artery with continued gestation can be explained by the physiologic trophoblastic invasion of the spiral arteries and the associated dilatation of those vessels (Hustin and Schaaps 1987). This leads to a tremendous increase in the total cross section of the vascular bed and transforms the uterine circulation into a very low-resistance system in which diastolic flow velocities are correspondingly raised.

Methods of assessing uterine blood flow remain a subject of controversy. There is much evidence to show that changes in the flow pattern of the arcuate arteries provide an earlier signal of impending fetal compromise than any other known parameter for fetal surveillance (Campbell et al. 1983, 1984, 1986; Trudinger et al. 1985 b). Disadvantages of transabdominal Doppler arcuate flow studies are poor reproducibility, uncertainty regarding the exact anatomic source of the signals, and the fact that each uterine artery divides into 10–15 arcuates. The shape of the flow velocity waveform also depends

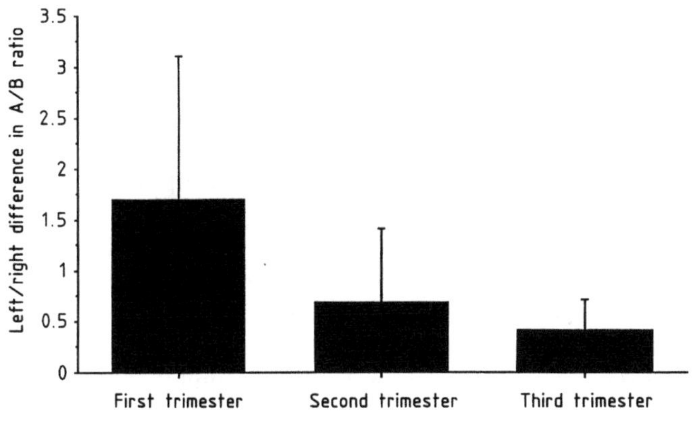

Fig. 13. The difference between the A/B ratios in the right and left uterine arteries decreases over the course of the normal pregnancy

on whether the vessels are sampled at the inside or outside of the uterine wall (Arabin 1987; Arabin et al. 1987 a, b; Schulman 1987 a, b). In patients who develop preeclampsia, the pathogenetic mechanism is based on morphologic and functional changes at the level of the radial arteries (Brosens 1964), so altered flow patterns are found predominantly in the preplacental bed. This change in flow patterns provides a means for classifying hypertension in pregnancy (Ducey et al. 1987). The subplacental bed is most difficult to visualize with ultrasound in cases where the placenta is implanted anteriorly (Marx et al. 1985). Trudinger et al. (1985 a) performed their studies in arcuate vessels located close to the placenta. However, physiologic trophoblast invasion occasionally affects the distal portions of the radial arteries as well, so that sampling close to the placenta can yield increased diastolic flow velocities and can simulate normal perfusion.

Reports on transabdominal Doppler studies of the uterine arteries illustrate the difficulties of this technique. In many cases it is not possible to visualize the uterine artery on the real-time scan. Thus, Taylor et al. (1985), using a sample volume placed just lateral to the cervix, were able to record signals from the ascending and descending branches of the uterine artery but not from the main arterial trunk. The vaginal approach appears to eliminate this source of error by allowing the transducer to be placed close to the study vessel and in this respect is superior to the transabdominal method.

Unlike Doppler sampling of an arcuate artery, the vaginal pulsed method permits a comprehensive assessment of uterine blood flow, since the uterine arteries represent the sum of flow conditions in all the arcuates. This is especially advantageous in patients with extensive placental infarcts and in patients who smoke. The sampling of a single arcuate artery supplying an infarcted area of the placenta could easily lead to erroneous results (Deutinger et al. 1988 d).

The vaginal Doppler assessment of uterine blood flow is an important part of the overall evaluation of fetomaternal blood flow (Fleischer et al. 1986; Schulman 1987 b; Schulman et al. 1987). We expect that this new method will enhance our understanding of the physiology of placentation and open up new possibilities in the diagnosis and management of high-risk pregnancies. It is hoped that the normal reference values established in our study will help vaginal Doppler to gain an established role in the evaluation of high-risk pregnancies (preeclampsia, placental insufficiency, diabetes mellitus, hypertension). Doppler studies of uteroplacental vessels during delivery are of interest for recording physiologic changes in fetal blood flow but have not yet acquired importance as a routine method of fetal surveillance (Fendel et al. 1985, 1987; Fleischer et al. 1987).

4.1.2 Complicated Pregnancy

Doppler flow studies have shown great promise in the very early detection of increased fetal risk (Erskine and Ritchie 1985; Trudinger et al. 1987; Wladimiroff et al. 1987). So far the problem of evaluating the maternal component of the fetomaternal unit (i. e., blood flow in the pregnant uterus) has not been satisfactorily resolved using the Doppler technique (Neilson 1987; Deutinger et al. 1988 a). The deficiencies of the continuous-wave method were noted in Sect. 2.1 of this chapter. Pulsed Doppler sampling of an arcuate artery is widely practiced (Campbell et al. 1983), but the arcuate arteries often cannot be visualized with ultrasound, and this can lead to inconsistent results in the same patient. There appear to be two main reasons for this: First, the shape of the arcuate waveform depends on the location of the sampled vessel, i. e., different waveform patterns may be recorded in vessels located closer to the placenta or closer to the outer uterine wall. Second, the blood supply to different portions of the placenta can vary. Thus, experimental studies have shown significant qualitative and quantitative differences in the blood flow to the individual placental cotyledons (Howard et al. 1987).

In one study, vaginal pulsed Doppler recordings from the uterine artery were compared with trans-

Table 1. Ultrasound biometry below the 10th percentile (Hohenauer) and actual birthweight

	n
Below 10th percentile (estimated)	30
Normal birthweight at delivery	6
Below 10th percentile (birthweight)	19
Chromosome abnormality	2
Unknown pregnancy outcome	3

Table 2. Clinical data (mean ±SD) on 25 patients with suspected fetal growth retardation

Doppler (week of gestation)	34.9± 4.5
Delivery (week of gestation)	38.4± 3.0
Fetal weight (g)	2354 ±384
Interval from Doppler to delivery (weeks)	3.4± 3.9

abdominal Doppler studies of an arcuate artery in growth-retarded fetuses, and it was determined how frequently abnormal flow velocities in the uterine vessels were associated with compromise of the fetal circulation (Deutinger et al. 1988 d).

The Doppler studies were performed in 30 patients between the 26th and 41st weeks of gestation with suspected fetal growth retardation (Table 1). The criterion for acceptance in the study was a fetus below the 10th percentile of our normal population according to routine ultrasound fetal abdominal biometry (Deutinger et al. 1988 b). Five patients were excluded from the study: three women examined as outpatients were lost to follow-up during the subsequent course of their pregnancy, and in two cases the fetal growth retardation was associated with a chromosome abnormality verified by chorion biopsy. The clinical data for the remaining 25 cases are shown in Table 2.

After routine initial sampling of the umbilical artery (Fig. 14), signals were recorded from an arcuate artery using an abdominally applied duplex scanner (Combison 320, Kretztechnik, Zipf, Austria) following the method of Trudinger et al. (1985 a). The sample gate was opened to maximum size, and the transducer was placed on the left or right lower quadrant of the maternal abdomen and moved slowly until a characteristic arcuate artery wave-

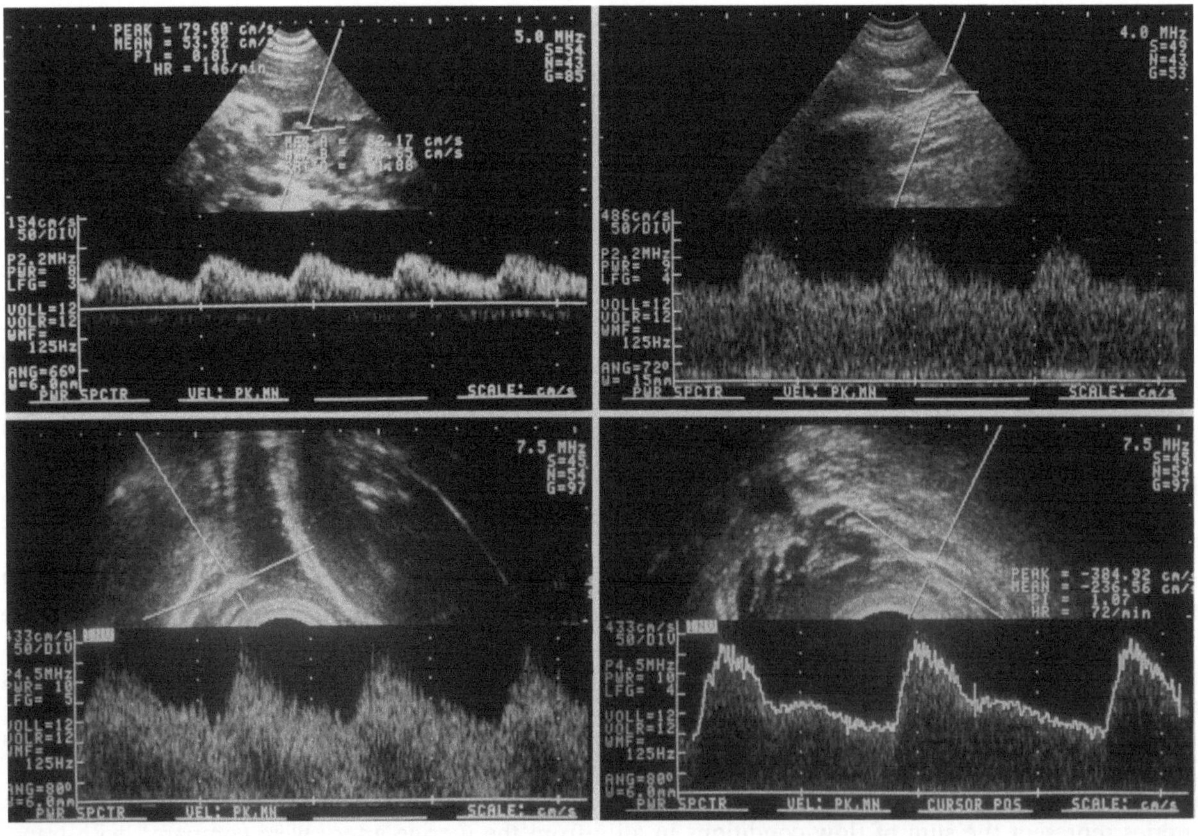

Fig. 14. Normal flow velocity waveforms in the third trimester of pregnancy. Umbilical artery (*upper left*, transabdominal), arcuate artery (*upper right*, transabdominal), left uterine artery (*lower right*, vaginal), right uterine artery (*lower left*, vaginal) ▼

form was obtained. On completion of the transabdominal examination, we recorded uterine artery waveforms by the vaginal approach.

As there is no significant change in the summation diastolic flow velocities in the uterine arteries after 25 weeks' gestation (Schulman 1987 b), we considered an A/B ratio greater than 2.8 to be abnormal (Rudelstorfer et al. 1988 a). An A/B ratio greater than 2 is presumed to be abnormal for the arcuate artery (Campbell et al. 1983), and greater than 3 for the umbilical artery (Trudinger et al. 1985 a).

Of the 25 cases that were evaluated, the uterine artery waveform was abnormal in 18 cases, and the arcuate artery waveform was abnormal in 20 cases (Table 3).

Six infants from this series had a normal birthweight by the Hohenauer chart following appropriate therapy. There was one case where the uterine artery waveform was abnormal but the birthweight, at 2750 g, was within normal limits. In this case, however, the waveform was recorded in only the 31st week of pregnancy, while the infant was delivered in week 37. In three cases the A/B ratio in the arcuate artery was increased, and in one case the umbilical artery A/B ratio was increased. In the two cases with chromosome abnormalities, both the umbilical artery and uterine artery waveforms appeared normal. The importance of pulsed vaginal recordings of the uterine arteries is underscored by the four cases (Table 4), which showed a marked discrepancy between the flow patterns recorded in the arcuate and uterine arteries.

Table 3. Number of abnormal flow velocity waveforms recorded in 25 patients with suspected fetal growth retardation

Vessel	*n*
Uterine artery	18
Arcuate artery	20
Uterine + arcuate artery	16
Umbilical artery	21
Uterine + umbilica artery	17
Arcuate + umbilical artery	13
Arcuate + umbilical + uterine artery	12

Case 1: In this case the umbilical artery waveform and uterine artery waveform were both abnormal, while the arcuate artery waveform was normal. This patient delivered an infant weighing 2430 g in the 39th week (Fig. 15).

In three other cases this situation was reversed: the arcuate artery waveform was abnormal while the umbilical and uterine artery waveforms were normal (Fig. 16).

Cases 2-4 are typified by one patient who was a heavy smoker. Imaging revealed scattered hyperechoic areas in the placenta. Flow velocity waveforms were recorded in the 37th week of pregnancy, and in the 40th week the patient delivered a healthy infant weighing 3250 g. Although the baby had a normal birthweight and showed no other signs of growth retardation, the arcuate artery waveform was abnormal.

In our studies we found a good correlation between an abnormal A/B ratio in the umbilical and uterine arteries. In some cases, however, there was a marked discrepancy between findings in the arcuate artery and the uterine arteries. In transabdominal examinations it is customary to sample any arcuate vessel that happens to be covered by the gate. But as our four cases illustrate, the flow velocity in a randomly selected arcuate artery is not necessarily representative of the perfusion of the uterus as a whole. For example, in a case where calcium deposits occupy a portion of the placenta, the Doppler beam may sample an arcuate artery supplying the calcified portion while in fact the overall blood flow is entirely adequate. It has also been shown experimentally that local vasoconstrictions in the placenta can elicit the compensatory dilatation of other vessels (Howard et al. 1987). For these reasons, waveforms recorded in one arcuate artery cannot be considered representative of flow conditions in the uterus as a whole. Transabdominal Doppler invariably samples only one side branch of the uterine artery (an arcuate artery), which frequently leads to inconsistent results (Schulman 1987 b; Deutinger et al. 1988 b).

Table 4. Contrast of flow velocity waveforms in the arcuate artery and uterine artery

Case	Doppler in week	Uterine artery	Arcuate artery	Umbilical artery	Delivery in week	Fetal weight (g)
1	36	abnormal	normal	abnormal	39	2430
2	37	normal	abnormal	normal	40	3250
3	34	normal	abnormal	normal	41	3050
4	35	normal	abnormal	normal	40	2900

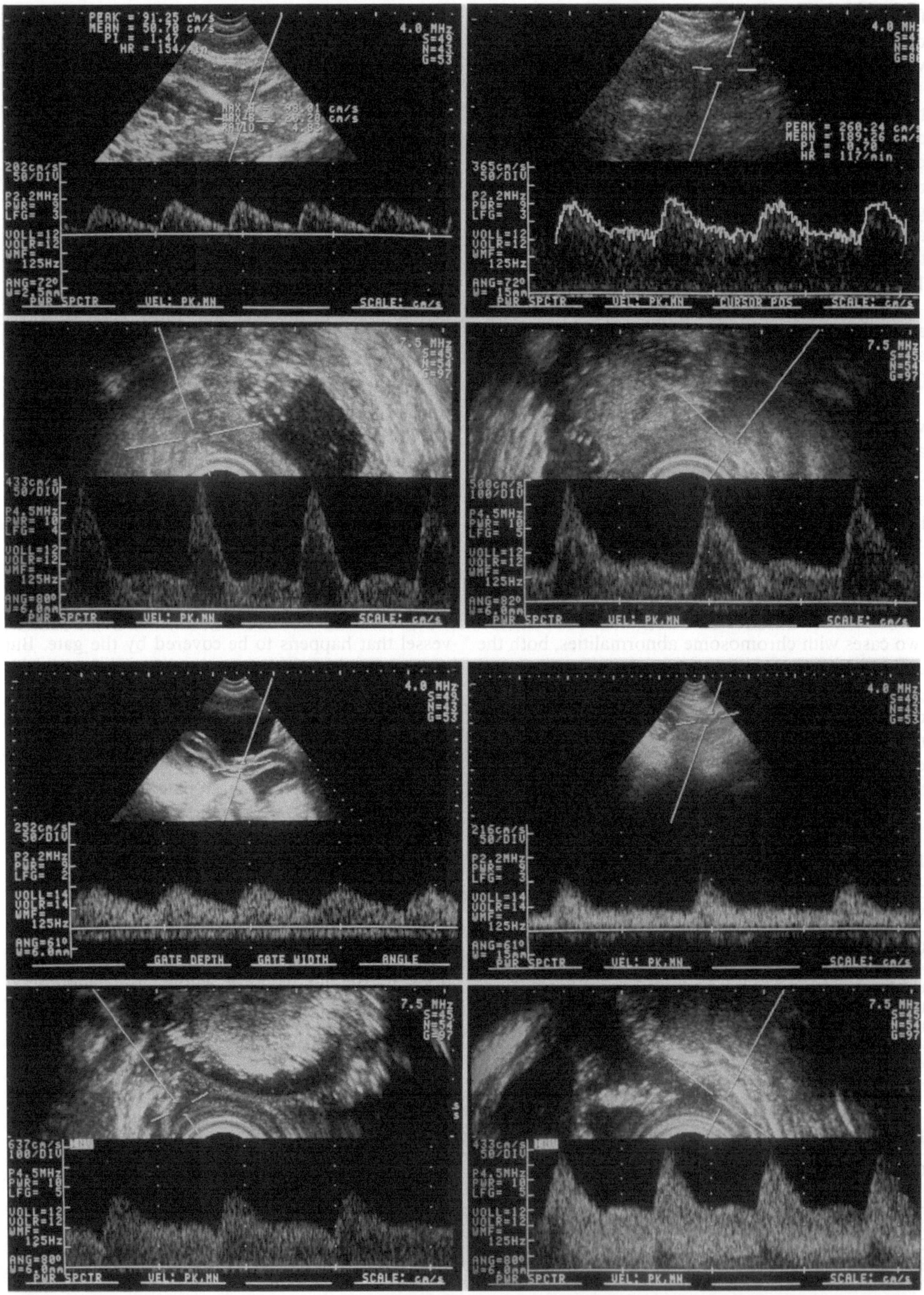
PEAK = 91.25 cm/s
MEAN = 50.70 cm/s
PI = 1.47
HR = 154/min
4.0 MHz
PEAK = 260.24 cm/s
MEAN = 189.26 cm/s
PI = 0.70
HR = 117/min
7.5 MHz
PWR SPCTR
VEL: PK,MN
CURSOR POS
SCALE: cm/s
GATE DEPTH
GATE WIDTH
ANGLE

◀ **Fig. 15.** Doppler study in the 36th week of this pregnancy showed an abnormal umbilical artery waveform *(upper left)*, a normal arcuate artery waveform *(upper right)*, and abnormal waveforms in the uterine arteries *(below)*

By comparison, vaginal Doppler studies of the uterine arteries offer several advantages. The recordings are made with a pulsed beam under vision, allowing for better reproducibility. The vaginal method evaluates the summation flow in both uterine arteries, allowing assessment of the total uterine perfusion. Transabdominal studies are often complicated by obesity, oligohydramnios, or pronounced hydramnios - none of which pose an obstacle to the vaginal method. Furthermore, large blood flow discrepancies between the right and left uterine arteries provide an additional criterion for suspicion of impending fetal compromise (Deutinger et al. 1988 f).

Many authors have come to recognize the value of Doppler fetal blood flow studies in the assessment of fetal well-being. Changes in the normal waveform patterns of the umbilical veins and aorta (Eik-Nes et al. 1980), umbilical arteries (Gill and Kossoff 1979), and internal carotid artery (Wladimiroff et al. 1986) have all been found to correlate with fetal growth retardation.

Hackett et al. (1987) found an abnormal A/B ratio in uterine vessels and in the umbilical arteries of growth-retarded fetuses. In our study as well, we found a good correlation between waveform findings in these vessels and fetal growth retardation. The reduced diastolic flow velocity in these vessels can be explained by the decrease in placental vascular perfusion that is associated with abnormal blood flow velocities in the umbilical arteries (McCowan et al. 1987).

The observable discrepancy of blood flow in the two uterine arteries is most pronounced in early pregnancy, is independent of placental location, and underscores the complexity of the physiologic processes of early gestation. This discrepancy almost disappears in the course of the normal pregnancy, so its persistence is an additional criterion for fetal jeopardy.

In summary it may be said that, with some experience, vaginal pulsed Doppler ultrasound is a technically simple, reproducible method of assessing uterine arterial flow that can be performed even in early pregnancy without difficulty. The vaginal examination is well tolerated after appropriate counseling (Deutinger et al. 1987 c) and provides a new adjunctive study for evaluating the fetomaternal circulation during pregnancy. Besides the assessment of fetal well-being (Trudinger and Cook 1985; Deutinger et al. 1988 b), vaginal pulsed Doppler provides an additional parameter for the early diagnosis of fetal malformations. Many of these anomalies are associated with growth retardation, but their normal fetal and uterine blood flow waveforms serve to distinguish them from idiopathic growth retardation. In the two cases in our series where growth retardation coexisted with chromosome anomalies, the pulsed Doppler study revealed normal flow velocity waveforms.

◀ **Fig. 16.** Doppler study in the 37th week of this pregnancy showed a normal umbilical artery waveform *(upper left)*, an abnormal arcuate artery waveform *(upper right)*, and normal waveforms in the uterine arteries *(below)*

4.2 Fetal Vessels

4.2.1 Early Pregnancy

Fetal signals can be recorded with a vaginal Doppler probe in the earliest weeks of pregnancy. The heart rate can be measured as early as the 6th week (Hogaki et al. 1987), making it possible to confirm fetal life at a very early stage even in the retroverted uterus. Studies on fetal heart rate in very early pregnancy have shed light on physiologic heart rate patterns (Hogaki et al. 1987). However, pulsed Doppler examination of the conceptus is not generally advised at such an early stage (Campbell 1987) because of the significantly higher energy output compared with ultrasound imaging.

4.2.2 The Fetal Internal Carotid Artery

It has been demonstrated that the fetal cerebral vessels, unlike the peripheral vessels, show a decline in diastolic flow velocity during the normal course of pregnancy (Wladimiroff et al. 1986). In fetal growth retardation, on the other hand, the increased peripheral vascular resistance in the fetal body is accompanied by a compensatory lowering of resistance in the cerebral vessels with a resultant acceleration of diastolic flow. It appears that states of fetal deprivation initiate compensatory mechanisms which ensure the maintenance of an adequate cerebral oxygen supply. This effect underlies the well-known phenomenon of asymmetrical growth retardation in placental insufficiency (Campbell 1974). Doppler waveforms recorded from the cerebral interpreted as a "brain-sparing ef-

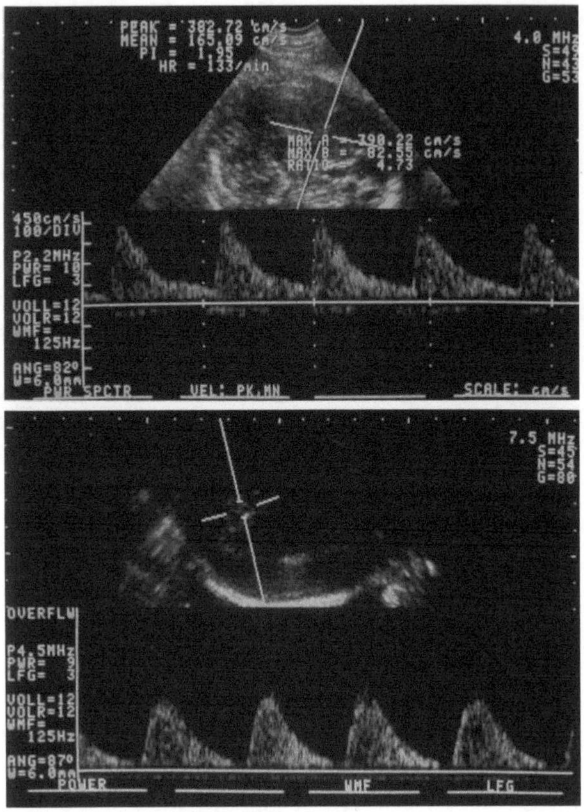

Fig. 17. *Top:* transabdominal pulsed Doppler study of a fetal internal carotid artery. *Bottom:* vaginal Doppler study of a fetal internal carotid artery

fect" (Marsal et al. 1984; Wladimiroff et al. 1986). Such findings have previously been reported in the common carotid artery and internal carotid artery, but sampling of the internal carotid artery is preferable for two reasons: First, the internal carotid artery is more representative of the hemodynamics of cerebral blood flow than the common carotid artery because the latter also gives rise to the external carotid. Second, the common carotid artery is more difficult to visualize due to the curved position of this area.

When the fetus is in a vertex presentation, recordings can be made from the internal carotid artery using an vaginal probe. Changes in the internal carotid artery waveform are of particular interest near term (Fig. 17), although descent of the fetal head can hamper or prevent correct placement of the transducer at this time.

4.3 Uterine Artery Flow in Cervical Carcinoma

Invasion of the parametria by inoperable cervical carcinoma is demonstrated best by rectosonography (Bernaschek and Janisch 1983). A vaginal probe is of limited use for this application because of the forward scan angle, which cannot insonate the whole parametrial surface at right angles, and because vaginal infiltration may obstruct probe insertion into the lateral fornix. In the presence of parametrial invasion, we have noted a discrepancy of blood flow velocities between the uterine arteries in favor of the less heavily affected side, i.e., the diastolic flow velocity decreases, and the A/B ratio increases, with the degree of parametrial invasion (Fig. 18). At first this may seem surprising, since areas affected by carcinoma tend to have a very rich blood supply. But because the parametrium ensheaths the uterine artery, infiltration of the parametrium reduces the elasticity of the vessel wall, leading to a relative decrease in the diastolic flow

Fig. 18. Vaginal pulsed Doppler study of the uterine arteries in a patient with left parametrial invasion by cervical carcinoma. The waveform in the right uterine artery *(left)* shows a higher diastolic blood flow velocity than the waveform in the left artery ▼

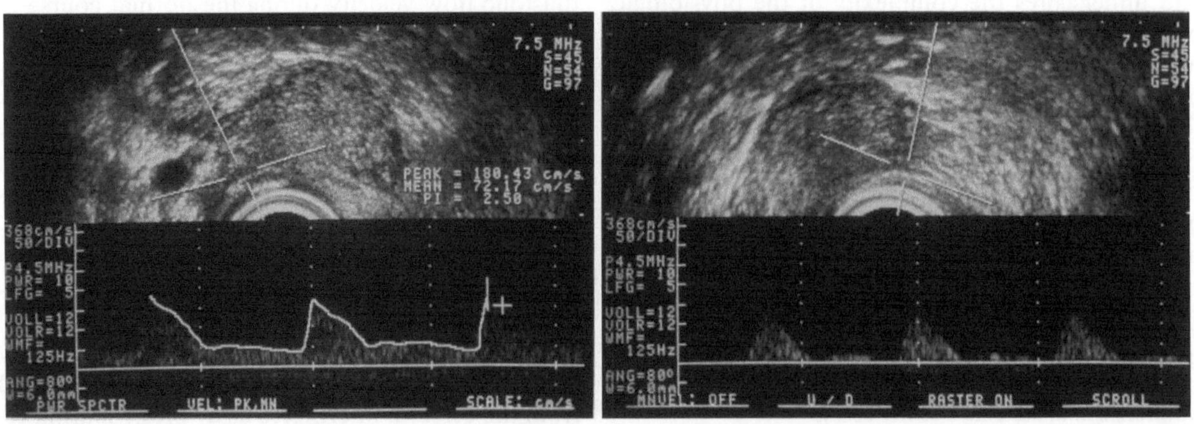

velocity (which is the physical parameter of vascular elasticity). Follow-up measurements after a course of radiotherapy demonstrate a relative increase in the diastolic flow velocity with an associated decrease in the A/B ratio.

4.4 Infertility

4.4.1 Failure of Implantation After Embryo Transfer

The arcuate arteries can only occasionally be visualized with ultrasound in nonpregnant women (DuBose et al. 1985). It is technically easier to demonstrate the uterine artery, although the problems and disadvantages of transabdominal scanning (full bladder, obesity, etc.) can be very troublesome in this application.

Another application of Doppler ultrasound in the uterine artery is the measurement of blood flow velocities in infertility patients. Successful implantation after embryo transfer relies upon synchronicity of the endometrium, i. e., transformation of the endometrium from the proliferative to the secretory phase. The rise of estrogens and progesterone toward the middle of the cycle is associated with a change in the perfusion rate of the ovary (Deutinger et al. 1989a) and uterus (Goswamy and Steptoe 1987). One effect of the rising progesterone level is an increase in the uterine blood flow, which is manifested in Doppler studies by an increased diastolic blood flow velocity.

When implantation fails to occur after in vitro fertilization and embryo transfer, abnormal uterine artery flow patterns can be demonstrated in more than 50% of cases in both the follicular and luteal phases (Goswamy and Steptoe 1987). These results were obtained transabdominally in an ascending branch of the uterine artery in a small patient population and should be regarded as preliminary. Vaginal Doppler will provide a means of studying physiologic and pathologic flow patterns in the ovarian and uterine arteries of infertility patients and will aid in testing the efficacy of hormonal replacement therapy and various induction regimens.

4.4.2 The Ovarian and Iliac Arteries in Stimulated Cycles

The ovarian artery and internal iliac artery provide most of the blood supply to the female internal genital organs. During the follicular phase of the cycle, proliferative processes are associated with an increase of blood flow to the affected organs. In an effort to determine whether blood flow velocities change in response to hormonal stimulation, we used the new vaginal scanner to record waveforms from the internal iliac artery and ovarian artery during the follicular phase in patients receiving induction therapy.

Table 5. Number of patients and measurements of the ovarian artery

Number of patients	10	10	7	6	3
Day of cycle	8	9	10	11	12
Number of measurements	19	18	13	12	6

Table 6. Number of follicles larger than 14 mm vs the A/B ratio in the corresponding ovarian artery

Follicles	2	3	4	5
A/B ratio	3.85 (±1.40)	3.5 (±1.54)	2.81 (±1.04)	2.71 (±1.33)

Table 7. Serum estradiol levels vs the A/B ratio in the ovarian arteries (averaged for both vessels)

Estradiol (pg/ml)	600	1100	1600
A/B ratio	3.62 (±1.57)	2.95 (±1.42)	2.36 (±1.14)

Fourteen patients with tubal infertility undergoing perparation for in vitro fertilzation were examined. All were on a clomiphene/hMG/hCG induction regimen (Deutinger et al. 1985). A total of 68 measurements were performed in ten patients; in four cases a signal could be recorded from only one ovarian artery (Table 5). The results from ten patients were evaluated. Two patients each failed to present for examination on 2 days, and in two other patients induction had to be discontinued due to an insufficient rise of serum estradiol.

During the follicular phase of the stimulated cycle, several follicles undergo mutaration, causing the ovarian volume to increase. This alters not only the blood distribution within the ovary but also the perfusion rate of the ovary as a whole (Edwards 1980). These physiologic processes lead to a change in the physical parameters of the blood vessels, with a corresponding change of flow velocities in the vessels supplying the ovary (Fig. 19).

As the number of follicles larger than 14 mm increased, we noted a significant decrease in the A/B ratio in the ovarian artery ($p < 0.05$). Thus, the A/B

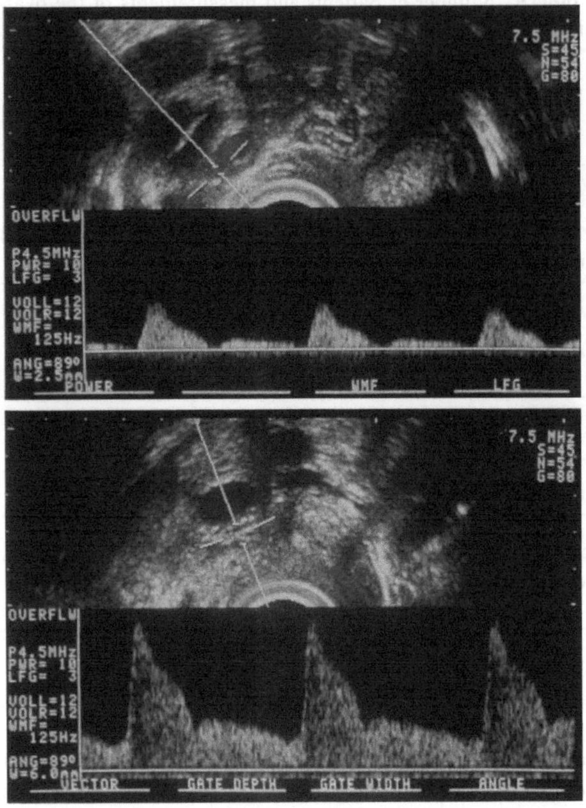

Fig. 19. Vaginal pulsed Doppler study of the ovarian vessels in a stimulated cycle. *Top:* right ovarian artery on cycle day 8, showing little or no diastolic blood flow. *Bottom:* right ovarian artery before follicular aspiration, showing a marked increase in diastolic blood flow velocity

ratio was 3.85 in the presence of two follicles and only 2.71 in the presence of five follicles (Table 6).

The A/B ratio also correlated with serum estradiol levels, decreasing from 3.62 at a level of 600 pg/ml to 2.36 at 1600 pg/ml (Table 7).

The A/B ratio in the internal iliac artery ranged from 5 to 13.

This relatively high value reflects the "hemodynamic proximity" of the abdominal aorta, whose pulse wave propagates almost unchanged into the iliac artery. As in the aorta, retrograde flow is observed in early diastole. No correlation was found between the A/B ratio in the iliac artery and follicular growth or serum estradiol levels. Due to the low diastolic blood flow velocity in ovarian arteries in early cycle days, the pulsatility index seems to be preferable for evaluation. We currently use this index for quantifying this signals (Deutinger et al. 1989b).

5 Clinical Significance of Vaginal Pulsed Doppler Blood Flow Studies

Vaginal pulsed Doppler is the first noninvasive procedure that can furnish information on physiologic changes in uterine blood flow during pregnancy and on ovarian blood flow during the menstrual cycle. The advantages of this method are briefly summarized below:

Recordings are made with a pulsed beam under vision and are not distorted by extraneous signals from other vessels. The proximity of the transducer to the study vessel allows the use of higher frequencies, resulting in better resolution of the real-time image. Since the average flow in both uterine arteries is evaluated, the total uterine perfusion can be assessed in pregnant patients.

Normally there is only one direction of blood flow, forward, throughout the cardiac cycle in fetal vessels and in the uterine arteries during pregnancy. This means that these vessels offer little resistance to blood flow in diastole. As gestation progresses the diastolic flow velocity increases further, signifying a decrease in diastolic resistance.

By contrast, in a pregnancy complicated by fetal growth retardation, hypertension, or preeclampsia, vascular resistance increases, and there is a resultant decrease in the Doppler-recorded diastolic flow velocity. In very severe cases a complete cessation of blood flow may be observed in diastole. This abnormal waveform may be recorded in the uterine artery, the umbilical artery, or in both vessels. Abnormal flow velocity waveforms show a definite correlation with fetal outcome and the severity of maternal proteinuria and hypertension (Griffin et al. 1983). A reduction of diastolic flow velocity in the uterine vessels can be recorded well before change is noted in any other parameters used for the surveillance of high-risk pregnancies (Campbell et al. 1983).

During development of the placenta, trophoblastic cells begin to migrate into the spiral arteries. A second wave of trophoblastic invasion occurs between the 14th and 20th weeks of gestation, extending to the intramyometrial parts of the vessels and even reaching the distal segments of the radial arteries.

Cytotrophoblastic cells replace the endothelium and also destroy tissue in the media, resulting in further dilatation of the spiral arteries (Brosens 1964). In patients who later develop preeclampsia, the second wave of cytotrophoblastic invasion fails to occur, so resistance at the spiral arterial level remains high (Robertson 1976; Robertson et al. 1984). This high vascular resistance leads to diminished perfusion of the placenta, especially in diastole, and may culminate in fetal malnutrition and hypoxia. These abnormal processes of placentation can be demonstrated in preeclampsia as well as in idiopathic fetal growth retardation (Khong et al. 1986).

The constricted spiral arteries lead to a decrease in fetoplacental perfusion and cause several compensatory mechanisms to become active. If Doppler indicates a reduced diastolic flow velocity, this does not necessarily mean that the fetus is suffering acute hypoxia at the time of the study. By reducing its metabolic demands, mobilizing reservers, and redistributing flow, the fetal body can cope with the oxygen and nutrient deficit caused by the diminished flow. Only as the reserves become exhausted and fetal demands increase with continued gestation do we observe the picture of intrauterine growth retardation with a negative oxytocin loading test.

The fall of diastolic flow velocity in the umbilical arteries may be caused by a second effect that occurs somewhat later, as indicated by the finding of Giles et al. (1985) that the rise of diastolic resistance in the umbilical artery correlates with the degree of obliteration of the chorionic villi.

The main importance of the Doppler examination in pregnancy - especially in the uterine vessels, where the earliest changes are detected - is its ability to provice early identification of potential high-risk pregnancies (Campbell et al. 1983; Nicolaides et al. 1987; Pearce 1987; Rightmire and Campbell 1987). It is the first noninvasive method capable of measuring the preplacental resistance. Not every pregnancy with an abnormal Doppler waveform is destined for an adverse fetal outcome. Conversely, a decrease in diastolic blood flow resistance has not been recorded in every case of fetal growth retardation or severe hypertension. This may be due to the fact that most Doppler studies have been performed in arcuate arteries, whose flow patterns vary according to the site at which the sampling is done. There is no question, however, that every pregnancy with an abnormal Doppler finding should be given close and intensive surveillance. We feel that the time proposed by Campbell et al. (1983) for a screening examination, the 18th week of pregnancy, is too early. It is better to screen the pregnancy no earlier than about the 25th week. By that time cytotrophoblastic invasion is complete, and in normal pregnancies there should be no further changes in flow velocity waveforms until term. Also, studies performed too early in pregnancy have no therapeutic implications. Both uterine arteries should be sampled, for it is common to find a marked flow discrepancy between the left and right arteries in early pregnancy. Flow conditions in the uterus as a whole are assessed by averaging the recordings from both arteries. Large discrepancies are no longer observed in the second trimester of the normal pregnancy, and the persistence of such a discrepancy indicates potential fetal jeopardy and might be combined with alteration of fetal hemodynamics (Deutinger et al. 1990).

At this stage of pregnancy the presenting fetal part is still well above the lesser pelvis and does not pose an obstacle to the vaginal examination.

Our current knowledge of the value of Doppler ultrasound in pregnancy is still insufficient to draw direct clinical conclusions based on the results of a single examination. Rather, an abnormal finding an Doppler screening should prompt the attending physician to supervise the pregnancy more closely. Hard statistical data based on prospective, randomized studies are needed before we can definitively assess the value of this method (Reuwer et al. 1986; Herrmann et al. 1986; Neilson 1987).

One intriguing aspect of Doppler waveform analysis is the possibility of improving an originally diminished diastolic blood flow velocity as a treatment for fetal growth retardation.

References

Arabin B (1987) Die Bedeutung von Perzentilenwerten uteroplazentarer und fetaler Blutflußparameter für die klinische Praxis. Ultraschall Klin Prax Suppl 1: 51

Arabin B, Bergmann PL, Saling E (1987a) Pathophysiologische und klinische Aspekte der Blutflußmessung in uteroplazentaren Gefäßen, in der Nabelschnurarterie, in der fetalen Aorta und in der fetalen Arteria carotis communis. Geburtshilfe Frauenheilkd 47: 587-593

Arabin B, Bergmann PL, Saling E (1987b) Qualitative Analyse von Blutflußspektren uteroplazentarer Gefäße, der Nabelschnurarterie, der fetalen Aorta und der fetalen Arteria carotis communis in normaler Schwangerschaft. Ultraschall Klin Prax 2: 114-119

Bernaschek G, Deutinger J (1988) Endosonographie in Geburtshilfe und Gynäkologie. In: Gockel HP (ed) Jahrbuch der Radiologie 1988. Regensberg & Biermann, pp 105-119

Bernaschek G, Janisch H (1983) Eine Methode zur Objektivierung des Parametrienbefundes beim Zervixkarzinom. Geburtshilfe Frauenheilkd 43: 498-500

Bernaschek G, Dadek C, Kratochwil A (1980) Echographische Darstellung der großen fetalen Gefäße. Ultraschall 1: 101-105

Bowie JD (1987) Sonographie of the uterus. In: Sabbagha RE (ed) Diagnostic ultrasound, 2nd edn. Lippingcott, Philadelphia, pp 474-496

Brosens K (1964) A study of the spiral arteries of the decidua basalis in normotensive and hypertensive pregnancies. J Obstet Gynecol Brit Commonw 71: 222-230

Buys Bailot CD (1845) Akustische Versuche auf der Niederländischen Eisenbahn nebst gelegentlichen Bemerkungen zur Theorie des Hrn. Prof. Doppler. Pogg Ann 66: 321-351

Campbell S (1974) The assessment of fetal development by diagnostic ultrasound. In: Milunsky A (ed) The pregnancy at risk. Clinics in perinatology: vol 1/2. Saunders, Philadelphia, p 507

Campbell S (1987) Personal communication

Campbell S, Diaz-Recasens J, Griffin DR, Cohen-Overbeek TE, Pearce JM, Willson K, Teague MJ (1983) New Dopler technique for assessing uteroplacental blood flow. Lancet I: 675-677

Campbell S, Hernandez CJ, Cohen-Overbeck TA, Pearce JMF (1984) Assessment of fetoplacental and uteroplacental blood flow using duplex pulsed Doppler ultrasound in complicated pregnancies. J Perniat Med 12: 262-265

Campbell S, Pearce MF, Hackett G, Cohen-Overbeek T, Hernandez C (1986) Qualitative assessment of uteroplacental blood flow: early screening test for high risk pregnancies. Obstet Gynecol 68: 649-653

Deutinger J, Bernaschek G (1986) Die vaginosonographische Pelvimetrie als neue Methode zur sonographischen Bestimmung der inneren Beckenmaße. Geburtshilfe Frauenheilkd 46: 345-347

Deutinger J, Bernaschek G (1987) Vaginosonographical determination of the true conjugate and the transverse diameter of the pelvic inlet. Arch Gynecol Obstet 240: 241-246

Deutinger J, Reinthaller A, Müller-Tyl E, Riss P, Fischl, Janisch H (1985) Vier Jahre erfolgreiche In vitro Fertilisierung an der 2. Universitätsfrauenklinik Wien. Österr Ärzteztg 40: 52-58

Deutinger J, Rudelstorfer R, Bernaschek G (1987 a) Transvaginale Flowmessungen an der Arteria uterina während der Schwangerschaft. XIII. Kongreß der Deutschen Gesellschaft für Perinatale Medizin, Berlin, Dec 1-4

Deutinger J, Rudelstorfer R, Bernaschek G (1987 b) Transvaginale gepulste Dopplermessungen in der Arteria Uterina bei pathologischen Schwangerschaften. Ultraschall Klin Prax 2 Suppl 1: 13

Deutinger J, Reinthaller A, Riss P, Bernaschek G, Csaicsich P, Fischl F, Müller-Tyl E (1987 c) Comparison of the results of vaginal and abdominal follicle scans. Arch Gynecol Obstet 241: 171-176

Deutinger J, Rudelstorfer R, Bernaschek G (1987 d) Transvaginale gepulste Dopplerultraschalluntersuchungen an den Ovarialgefäßen. Ultraschall Klin Prax Suppl 1: 45-46

Deutinger J, Rudelstorfer R, Bernaschek G (1988 a) Doppler-Strömungsmessungen am schwangeren Uterus. Ultraschall Klin Prax 3: 19-24

Deutinger J, Rudelstorfer R, Bernaschek G (1988 b) Doppler Strömungsmessungen in der Arteria arcuata und in der Arteria uterina bei fetaler Wachstumsretardation. Geburtshilfe Frauenheilkd 48: 863-868

Deutinger J, Rudelstorfer R, Bernaschek G (1988 c) Pulsed Doppler measurement and visual vessel recognition of the uterine artery during pregnancy. Am J Obstet Gynecol 159: 1072-1076

Deutinger J, Rudelstorfer R, Bernaschek G (1988 d) Gepulste Strömungsmessungen in Arteria uterina und in Arteria arcuata. Ber Gynäkol Geburtsh 125: 653-654

Deutinger J, Rudelstorfer R, Bernaschek G (1988 e) Transvaginale gepulste Strömungsmessungen in den Uterinarterien: Normwerte und Vergleich mit der Messung in einer Arteria arcuata. Ultraschall Klin Prax 3 [Suppl 1]: 35

Deutinger J, Rudelstorfer R, Bernaschek G (1988 f) Endovaginal pulsed Doppler velocimetry of the main uterine arteries during pregnancy. First World Congress on Vaginosonography in Gynecology, June 9-12, Washington D.C.

Deutinger J, Rudelstorfer R, Bernaschek G (1989 a) Transvaginale gepulste Dopplermessungen von Strömungsgeschwindigkeiten in Beckengefäßen nach Zyklusstimulation. Geburtshilfe Frauenheilkd 49: 33-36

Deutinger J, Reinthaller A, Bernaschek G (1989 b) Transvaginal pulsed Doppler measurement of blood flow velocity in the ovarian arteries during cycle stimulation and after follicle puncture. Fertil Steril 51: 466-470

Deutinger J, Rudelstorfer R, Bernaschek G (1990) Vaginosonographic Doppler velocimetry in both uterine arteries of high-risk pregnancies: relationship to fetal hemodynamics. Am J Obstet Gynecol (submitted)

Diwan RV, Brennan JN, Selim MA, McGrew TL, Rashad FA, Rustia MU, Bellon EM (1983) Sonographic diagnosis of arteriovenous malformation of the uterus and pelvis. J Clin Ultrasound 11: 295-298

Doppler C (1843) Über das farbige Licht der Dopplersterne und einiger anderer Gestirne des Himmels. Abhandl Königl Böhm Gesellsch 2: 465-482

DuBose TJ, Hill LW, Hennigan HW Jr, Nichols DH, Mezaraups GG, Porter L, Marley L, Butschek CM, Karnaze GC, Walser E, Reyes K, Cunyust JA (1985) Sonography of arcuate uterine blood vessels. J Ultrasound Med 4: 229-233

Ducey J, Schulman H, Farmakides G, Rochelson B, Bracero L, Fleischer A, Guzman E, Winter D, Penny B (1987) A classification of hypertension in pregnancy based on Doppler velocimetry. Am J Obstet Gynecol 157: 680-685

Edwards RG (1980) Conception in the human female. Academic, London

Eik-Nes SH, Brubakk AO, Ulstein MK (1980) Measurement of human fetal blood flow. Br Med J 1: 283-284

Eik-Nes SH, Marsal K, Kristoffersen K (1984) Methodology and basic problems related to blood flow studies in the human fetus. J Ultrasound Med 10: 329-337

Erskine RL, Ritchie JWK (1985) Umbilical blood flow characteristics in normal and growth retarded fetuses. Br J Obstet Gynaecol 92: 605-610

Fendel H, Fendel M, Warnking R (1983) Fehlermöglichkeiten der gepulsten Dopplermethode zur Blutflußmessung am Feten. Z Geburtshilfe Perinatol 187: 83-87

Fendel H, Pauen A, Jung H (1985) Veränderung des uterinen Blutstromes unter der Geburt. Arch Gynecol 253: 197-198

Fendel H, Fettweis P, Billet P, Werdin R, Sohn C, Giani G, Freiberg C (1987) Doppleruntersuchungen des arteriellen utero-feto-plazentaren Blutflusses vor und während der Geburt. Z Geburtshilfe Perinatol 191: 121-129

Fitzgerald DE, Drumm JE (197) Non-invasive measurement of the fetal circulation using ultrasound: a new method. Br Med J 2: 1450-1451

Fleischer A, Schulman H, Farmakides G, Bracero L, Grunfeld L, Rochelson B, Koenigsberg M (1986) Uterine artery Doppler velocity in pregnant women with hypertension. Am J Obstet Gynecol 154: 806-813

Fleischer A, Anyaegbunam AA, Schulman H, Farmakides G, Randolph G (1987) Uterine and umbilical artery velocimetry during normal labor. Am J Obstet Gynecol 157: 40-43

Giles WB, Trudinger BJ, Cook CM (1982) Fetal umbilical artery velocity waveforms. Ultrasound Med Biol 1: 98

Giles WB, Trudinger BJ, Baird PJ (1985) Fetal umbilical artery velocity waveforms and placental resistance: pathological correlation. Br J Obstet Gynaecol 92: 31-38

Gill RW, Kossoff G (1977) Quantitative blood flow measurement in deep-lying vessels. In: Chef R (ed) Real time ultrasound in perinatal medicine. Karger, Basel, pp 139-141

Gill RW, Kossoff G (1979) Pulsed Doppler combined with B-Mode imaging for blood flow measurement. Contrib Gynecol Obstet 6: 139-141

Gosling RG, King DH (1975) Ultrasound angiology. In: Maraous AW, Adamson L (eds) Arteries and veins. Churchill Livingstone, Edinburgh, pp 61-98

Goswamy RK, Steptoe PC (1987) Abnormal uterine blood flow patterns - a possible cause for failed implantation in IVF patients? Hum Reprod 2 Suppl 1: 56

Graf HP, Klosa W, Kurz CS, Schillinger H (1985) Der Einfluß des "aliasing"-Effektes und des "Filters" auf die quantitative Bestimmung der mittleren Flußgeschwindigkeit und des Flußvolumens pro Zeiteinheit bei der gepulsten Ultraschall-Doppler-Methode. Ultraschall 6: 237-244

Griffin D, Cohen-Overbeek T, Campbell S (1983) Fetal and uteroplacental blood flow. Clin Obstet Gynecol 10: 565-602

Hackett GA, Nicolaides KH, Campbell S (1987) Doppler ultrasound assessment of fetal and uteroplacental circulations in severe second trimester oligohydramnios. Br J Obstet Gynaecol 94: 1074-1077

Herrmann F, Neuerburg-Heusler D, Nissen P (1986) Statistische Grundbegriffe für die Validierung diagnostischer Verfahren. Ein Beitrag zur Qualitätssicherung am Beispiel der Doppler-Sonographie. Ultraschall Med 7: 59-69

Hogaki M, Ohkawa T, Takizawa M, Takeuchi Y (1987) Transvaginal detection of fetal signals by Doppler ultrasound. J Perinat Med 15 Suppl 1: 167

Howard RB, Hosakawa T, Maguire MH (1987) Hypoxia-induced fetoplacental vasoconstriction in perfused placental cotyledons. Am J Obstet Gynecol 157: 1261-1266

Hustin J, Schaaps J-P (1987) Echographic and anatomic studies of the maternotrophoblastic border during the first trimester of pregnancy. Am J Obstet Gynecol 157: 162-168

Ishihara K, Tsuzaki T, Maeda K (1984) Blood flow measurement of maternal pelvic artery and umbilical cord artery with the ultrasonic pulsed Doppler flow metry. Organisation Gestosis, Aachen

Jeanty P, Romero R, Hobbins JC (1984) Vascular anatomy of the fetus. J Ultrasound Med 3: 113-122

Kaesemann H, Trenkel K (1986) Kombinierte B-Bild-Doppler-Sonographie-Messungen am In-vitro-Modell. Z Geburtshilfe Perinatol 190: 43-48

Khong TY, Wolf F de, Robertson WB, Brosens I (1986) Inadequate maternal vascular response to placentation in pregnancies complicated by pre-eclampsia and by small-for-gestational age infants. Br J Obstet Gynaecol 93: 1049-1059

McCallum WD, Williams CS, Nagel S, Daigle RE (1978) Fetal blood flow velocity waveforms. Am J Obstet Gynecol 132: 425-429

McCowan LM, Mullen BM, Ritchie K (1987) Umbilical artery flow velocity waveforms and the placental vascular bed. Am J Obstet Gynecol 157: 900-902

Marsal K, Lingman G, Giles WB (1984) Evaluation of the carotid, arotic and umbilical blood velocity waveforms in the human fetus. Proceedings of the Society for the Study of Fetal Physiology. XIth Annual Conference, Oxford, p 33

Marx M, Casola G, Scheible W, Deutsch A (1985) The subplacental complex: Further sonographic observations. J Ultrasound Med 4: 459-461

Neilson JP (1987) Doppler ultrasound. Commentary. Br J Obstet Gynaecol 94: 929-934

Nicolaides KH, Clewell WH, Rodeck CH (1987) Measurement of human fetoplacental blood volume in erythroblastosis fetalis. Am J Obstet Gynecol 157: 50-53

Pearce JM (1987) Uteroplacental and fetal blood flow. Obstet Gynaecol 1/1: 157-184

Reuwer PJHM, Nuyen WC, Beijer HJM, Heethar RM, Haspels AA, Bruinse HW (1986) Feto-placental circulatory competence. Eur J Obstet Gynecol 21: 15-26

Rightmire DA, Campbell S (1987) Fetal and maternal Doppler Blood flow parameters in postterm pregnancies. Obstet Gynecol 69: 891-894

Robertson WB (1976) Uteroplacental vasculature. J Clin Pathol 29: 9-17

Robertson WB, Brosens I, Pijnenborg R, Wolf F de (1984) The making of the placental bed. Eur J Obstet Gynecol 18: 255-266

Rudelstorfer R, Deutinger J, Bernaschek G (1987) Vaginosonographische Darstellung der Arteria uterina zur Doppler-Blutflußmessung während der Schwangerschaft. Ultraschall Klin Prax Suppl 1: 12-13

Rudelstorfer R, Deutinger J, Katz E, Bernaschek G (1988 a) Transvaginale gepulste Doppler Ultraschalluntersuchungen der Arteria uterina in der Schwangerschaft: Vorteil der gepulsten Methode. Z Geburtshilfe Perinatol 192: 14-18

Rudelstorfer R, Deutinger J, Bernaschek G (1988 b) Gepulste Doppler-Strömungsmessungen mittels vaginaler Ultraschallsonde - Integration in die fetomaternelle Hämodynamik. Ultraschall 9: 25-29

Rudelstorfer R, Deutinger J, Bernaschek G (1988 c) Vaginosonographische Strömungsmessungen in der Arteria uterina: pathologische Seitendifferenz und ihre Auswirkung auf die fetale Hämodynamik sowie fetal outcome. Ultraschall Klin Prax 3 [Suppl 1]: 35

Schulman H (1987 a) The clinical implications of Doppler ultrasound analysis of the uterine and umbilical arteries. Am J Obstet Gynecol 156: 889-893

Schulman H (1987 b) Doppler velocimetry of fetal and maternal blood vessels. In: Sabbagha RE (ed) Diagnostic ultrasound, 2nd edn. Lipincott, Philadelphia, pp 156-164

Schulman H, Fleischer A, Farmakides G, Bracero L, Rochelson B, Grunfeld L (1986) Development of uterine artery compliance in pregnancy as detected by Doppler ultrasound. Am J Obstet Gynecol 155: 1031-1036

Schulman H, Ducey J, Farmakides G, Guzman E, Winter D, Penny B, Lee C (1987) Uterine artery Doppler velocimetry: The significance of divergent systolic/diastolic ratios. Am J Obstet Gynecol 157: 1539-1542

Stuart B, Drumm J, Fitzgerald DE, Duignan NM (1980) Fetal blood velocity waveforms in normal pregnancy. Br J Obstet Gynaecol 87: 780-785

Taylor KJW, Burns PN, Wells PNT, Conway DI, Hull MGR (1985) Ultrasound Doppler flow studies of the ovarian and uterine arteries. Br J Obstet Gynaecol 92: 240-246

Trudinger BJ, Cook CM (1985) Umbilical and uterine artery flow velocity waveforms in pregnancy associated with major fetal abnormality. Br J Obstet Gynaecol 92: 666-670

Trudinger BJ, Giles WB, Cook CM, Bombardieri J, Collins L (1985 a) Fetal umbilical artery flow velocity waveforms and placental resistance: clinical significance. Br J Obstet Gynaecol 92: 23-30

Trudinger BJ, Giles WB, Cook CM (1985 b) Uteroplacental

blood flow velocity-time waveforms in normal and complicated pregnancy. Br J Obstet Gynaecol 92: 39-45

Trudinger BJ, Cook CM, Giles WB, Conelly A, Thompson RS (1987) Umbilical artery flow velocity waveforms in high-risk pregnancy. Randomised controlled trial. Lancet II: 188-190

Watson PT, Young WP, Hegge FN (1987) Doppler measurements of maternal and fetal blood flow. Clin Obstet Gynecol 30: 948-955

Wladimiroff JW, Tonge HM, Stewart PA (1986) Doppler ultrasound assessment of cerebral blood flow in the human fetus. Br J Obstet Gynaecol 63: 471-475

Wladimiroff JW, Wijngaard JAGW vd, Degani S, Noordam MI, Eyck J v, Tongue HM (1987) Cerebral and umbilical arterial blood flow velocity waveforms in normal and growth-retarded pregnancies. Obstet Gynecol 69: 705-709

Subject Index

MIX
Papier aus verantwortungsvollen Quellen
Paper from responsible sources
FSC® C105338

If you have any concerns about our products,
you can contact us on
ProductSafety@springernature.com

In case Publisher is established outside the EU,
the EU authorized representative is:
Springer Nature Customer Service Center GmbH
Europaplatz 3, 69115 Heidelberg, Germany

Printed by Libri Plureos GmbH
in Hamburg, Germany